CW00727579

Nanolithography of Biointerfaces

Burlington House, London, UK

3–5 July 2019

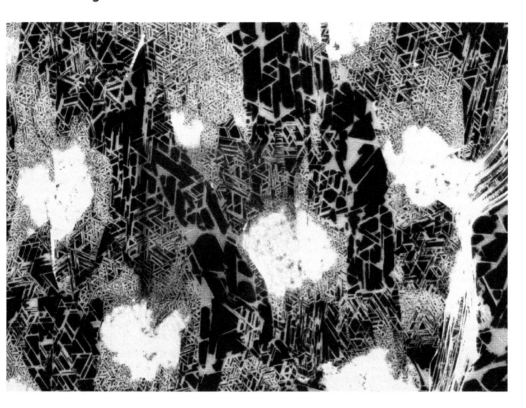

FARADAY DISCUSSIONS
Volume 219, 2019

The **Faraday Division** of the Royal Society of Chemistry, previously the Faraday Society, was founded in 1903 to promote the study of sciences lying between chemistry, physics and biology.

Editorial Staff

Executive Editor
Richard Kelly

Deputy Editor
Maria Southall

Editorial Production Manager
Claire Darby

Publishing Editors
Lucy Balshaw, Michael Spencelayh

Editorial Assistant
Harriet Kent

Publishing Assistants
Allison Holloway, Julie-Ann Roszkowski

Publisher
Jamie Humphrey

Faraday Discussions (Print ISSN 1359-6640, Electronic ISSN 1364-5498) is published 8 times a year by the Royal Society of Chemistry, Thomas Graham House, Science Park, Milton Road, Cambridge, UK CB4 0WF.

Volume 219 ISBN 13: 978-1-78801-676-6

2019 annual subscription price: print+electronic £1220 US $2148; electronic only £1162, US $2046.
Customers in Canada will be subject to a surcharge to cover GST. Customers in the EU subscribing to the electronic version only will be charged VAT.

All orders, with cheques made payable to the Royal Society of Chemistry, should be sent to the Royal Society of Chemistry Order Department, Royal Society of Chemistry, Thomas Graham House, Science Park, Milton Road, Cambridge, CB4 0WF, UK
Tel +44 (0)1223 432398; E-mail orders@rsc.org

If you take an institutional subscription to any Royal Society of Chemistry journal you are entitled to free, site-wide web access to that journal. You can arrange access via Internet Protocol (IP) address at www.rsc.org/ip

Customers should make payments by cheque in sterling payable on a UK clearing bank or in US dollars payable on a US clearing bank.

Whilst this material has been produced with all due care, the Royal Society of Chemistry cannot be held responsible or liable for its accuracy and completeness, nor for any consequences arising from any errors or the use of the information contained in this publication. The publication of advertisements does not constitute any endorsement by the Royal Society of Chemistry or Authors of any products advertised. The views and opinions advanced by contributors do not necessarily reflect those of the Royal Society of Chemistry which shall not be liable for any resulting loss or damage arising as a result of reliance upon this material. The Royal Society of Chemistry is a charity, registered in England and Wales, Number 207890, and a company incorporated in England by Royal Charter (Registered No. RC000524), registered office: Burlington House, Piccadilly, London W1J 0BA, UK, Telephone: +44 (0) 207 4378 6556.

Printed in the UK

Faraday Discussions

Faraday Discussions are unique international discussion meetings that focus on rapidly developing areas of chemistry and its interfaces with other scientific disciplines.

Scientific Committee volume 219

Chair
Adam Braunschweig, City University of New York, United States

Deputy Chair
Ryan Chiechi, University of Groningen, Netherlands
Lee Cronin, University of Glasgow, United Kingdom
Laura Hartmann, Heinrich-Heine-Universität, Germany
Stephan Schmidt, Heinrich-Heine-Universität, Germany
Sébastien Vidal, Université de Lyon, France

Faraday Standing Committee on Conferences

Chair
John M. Seddon, Imperial College London, UK

Secretary
Susan Weatherby, Royal Society of Chemistry, UK
Graeme M. Day, University of Southampton, UK
David Lennon, University of Glasgow, UK

Angelos Michaelides, University College London, UK
Susan Perkin, University of Oxford
Ivana Radosavljević Evans, Durham University, UK
Paul Raithby, University of Bath, UK
Pat Unwin, University of Warwick, UK
Claire Vallance, University of Oxford, UK

Advisory Board

Vic Arcus, The University of Waikato, New Zealand
Dirk Guldi, University of Erlangen-Nuremberg, Germany
Luiz Liz-Marzán, CIC biomaGUNE, Spain
Guy Lloyd-Jones, University of Edinburgh, UK
Marina Kuimova, Imperial College London, UK

Andrew Mount, University of Edinburgh, UK
Frank Neese, Max Planck Institute for Chemical Energy Conversion, Germany
Michel Orrit, Leiden University, The Netherlands
Zhong-Qun Tian, Xiamen University, China
Siva Umapathy, Indian Institute of Science, Bangalore, India
Mike Ward, University of Sheffield, UK
Bert Weckhuysen, Utrecht University, The Netherlands
Julia Weinstein, University of Sheffield, UK
Sihai Yang, University of Manchester, UK

Information for Authors

This journal is © the Royal Society of Chemistry 2018 Apart from fair dealing for the purposes of research or private study for non-commercial purposes, or criticism or review, as permitted under the Copyright, Designs and Patents Act 1988 and the Copyright and Related Rights Regulation 2003, this publication may only be reproduced, stored or transmitted, in any form or by any means, with the prior permission in writing of the Publishers or in the case of reprographic reproduction in accordance with the terms of licences issued by the Copyright Licensing Agency in the UK. US copyright law is applicable to users in the USA.

⊖ The paper used in this publication meets the requirements of ANSI/NISO Z39.48-1992 (Permanence of Paper).

Registered charity number: 207890

MIX
Paper from responsible sources
FSC
www.fsc.org   FSC® C013604

ROYAL SOCIETY OF CHEMISTRY   CELEBRATING IYPT 2019

Nanolithography of Biointerfaces

Faraday Discussions
www.rsc.org/faraday_d

A General Discussion on Nanolithography of Biointerfaces was held in London, UK on the 3rd, 4th, and 5th of July 2019.

RSC Publishing is a not-for-profit publisher and a division of the Royal Society of Chemistry. Any surplus made is used to support charitable activities aimed at advancing the chemical sciences. Full details are available from www.rsc.org

CONTENTS

ISSN 1359-6640; ISBN 978-1-78801-676-6

Cover
See Claridge *et al.*, *Faraday Discuss.*, 2019, **219**, 229–243.

Microcontact printing of amphiphiles can generate 1 nm wide functional stripes; here, striking morphological differences are evident between striped and standing phases.

Image reproduced by permission of Shelley Claridge from *Faraday Discuss.*, 2019, **219**, 229.

SPONSORS

THE US ARMY

SWISSLITHO AG

This journal is © The Royal Society of Chemistry 2019

CONCLUDING REMARKS

ADDITIONAL INFORMATION

This journal is © The Royal Society of Chemistry 2019

PAPER

Multivalent glycan arrays

Marco Mende,[a] Vittorio Bordoni, [ID][a] Alexandra Tsouka,[ab]
Felix F. Loeffler, [ID][a] Martina Delbianco [ID][a] and Peter H. Seeberger [ID] *[ab]

Received 11th June 2019, Accepted 26th June 2019

DOI: 10.1039/c9fd00080a

Glycan microarrays have become a powerful technology to study biological processes, such as cell–cell interaction, inflammation, and infections. Yet, several challenges, especially in multivalent display, remain. In this introductory lecture we discuss the state-of-the-art glycan microarray technology, with emphasis on novel approaches to access collections of pure glycans and their immobilization on surfaces. Future directions to mimic the natural glycan presentation on an array format, as well as *in situ* generation of combinatorial glycan collections, are discussed.

1. Introduction

Glycans decorate the surface of many cells, forming a thick layer (glycocalyx) that mediates a variety of important cellular processes.[1] This 100 nm–1 mm thick glycan layer comprises highly diverse structures, including glycoproteins, glyco-lipids, and glycopolymers. Several complex biological processes, such as protein folding, cell–cell interaction, cell adhesion, and signaling, are the result of the interactions of glycans with themselves (carbohydrate–carbohydrate interactions, CCIs) or with glycan binding proteins (carbohydrate–protein interactions, CPIs).[2–4] In addition, pathogens use these glycans as receptors for the attachment to host cells and subsequent invasion.[5,6] At the same time, pathogenic glycans are recognized by the immune system, which initiate the immune response.[7,8] Pathological events, such as tumor metastasis, inflammation, and infections, are all mediated by glycan–protein interactions.

A better understanding of these CPIs is of fundamental importance. Yet, in comparison to polynucleotides and proteins, the study of glycans and CPIs has been slower for multiple reasons: the complexity of carbohydrate synthesis and their difficult isolation from natural sources has hampered a detailed analysis of such compounds. The limited access to collections of pure materials precluded high-throughput screening formats. Even though glycan arrays have become extremely popular and primary analytical tools for the study of CPIs,[9–11] they are

[a]Department of Biomolecular Systems, Max Planck Institute of Colloids and Interfaces, Am Mühlenberg 1, 14476 Potsdam, Germany. E-mail: peter.seeberger@mpikg.mpg.de

[b]Department of Chemistry and Biochemistry, Freie Universität Berlin, Arnimallee 22, 14195 Berlin, Germany

limited due to glycan availability. Additionally, CPIs are very weak (typically in the micromolar range) and glycan binding proteins can often interact with many substrates with low specificity. Nature's strategy to enhance binding strength and specificity is multivalency, where multiple carbohydrate units bind to one protein to gain stronger affinity than the sum of the single contributions.[12–14] Chemists have aimed to reproduce nature, developing several synthetic multivalent systems that mimic natural supramolecular interactions.[15] Nevertheless, recreating the binding thermodynamics of natural interfaces in a microarray format, is extremely challenging.[16] Different approaches aimed to mimic the natural glycan presentation on an array surface. The most common approach involves the direct printing of glycans, controlling the density by varying the concentration or by surface functionalization.[17–19] Alternatively, prearranged multivalent systems, based on natural or unnatural scaffolds, can be immobilized on surfaces, aiming at more defined glycan presentation.[9,20]

A challenging approach is the direct synthesis of glycans or multivalent glycan systems on the array. However, in comparison to other biomolecules, chemical carbohydrate synthesis on surfaces is far more difficult, due to the demanding reaction parameters. Only the synthesis of disaccharides has been achieved to date.[21] Enzymatic synthesis on surfaces is more common, *e.g.*, for the synthesis of *N*-glycans or the discovery of glycosyltransferases.[22–25] Yet, whether glycans can also be synthesized *in situ* in a molecularly defined and multivalent fashion, remains to be shown.

We review the state of the art of glycan microarrays, from access to glycan collections, to surface immobilization, and analysis. We will focus on current approaches to mimic natural interfaces and new directions in surface functionalization. Moreover, we will describe how simple glycans and more complex multivalent scaffolds are printed or grown from surfaces to elucidate important cellular processes.

2. Access to glycan collections

The first step towards the production of a glycan microarray is the identification of suitable glycans. Two approaches are available and currently used to access glycans (see Fig. 1): isolation from natural sources and/or synthesis (enzymatic or chemical). Natural glycans can be readily obtained from animal tissues, plant material, or from cultured pathogens.[26] Large collections in terms of size and diversity could be accessed, when completely uncharacterized binders need to be identified.[10] Nevertheless, the isolation procedures and characterization of the final carbohydrates could be extremely challenging, often resulting in mixtures of compounds. Heterogeneous samples, often containing minor impurities, could culminate in non-reproducible results. Moreover, extracted glycans generally require an extra functionalization step for immobilization on surfaces. Compound collections obtained from chemical synthesis are generally smaller, more focused, and less diverse. Generating a set of related glycans, with the possibility of including non-natural glycans, is of great interest for the elucidation of structure–activity relationships. Chemically obtained compounds are highly pure, reducing the possibility of false results. A reactive linker can be easily installed during synthesis, facilitating subsequent immobilization. Using these two approaches, many glycans were prepared and printed on arrays.[9,10] The

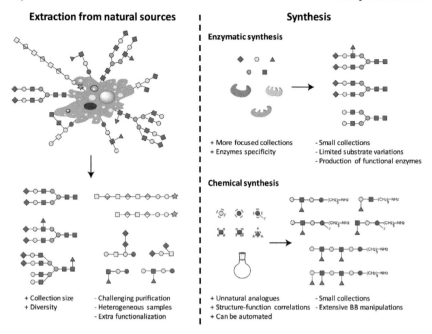

Fig. 1 Different approaches to access glycan collections.

microarray with the currently largest diversity is represented by the mammalian array (version 5.3) of the Consortium for Functional Glycomics (CFG), which includes more than 600 synthetic and isolated compounds.[26] A microbial glycan microarray is also available, including more than 300 carbohydrates. However, covering the huge diversity of microbial glycans, often containing rare sugars, remains a major challenge.[27] Efforts to access such glycans in a well-defined manner are still needed.

2.1. Automated glycan assembly

Collections of well-defined glycans are fundamental tools to elucidate glycan interactions. With the aim to explore the natural and unnatural diversity of glycans, systematic strategies for the chemical, enzymatic, and chemo-enzymatic synthesis of glycans were developed. Nevertheless, the challenging installation of the glycosidic linkage, that requires regio- and stereo-control, poses a bottleneck. Enzymatic synthesis relies on the specificity of the enzymes to form the desired glycosidic linkage, but to date, it is limited by the availability of suitable glycosyltransferases.[28] Such enzymes are extremely efficient with natural substrates, but often tolerate only limited substrate variations, hindering access to chemically modified glycans and unnatural structures. In addition, the *in vitro* production of functional enzymes is sometimes troublesome.[29] Despite several challenges, enzymatic synthesis remains a powerful option, when poorly reactive monosaccharides such as sialic acid or particularly challenging linkages such as β-mannosides need to be installed.[30] Efforts to standardize this process resulted in two fully auto-mated systems.[31,32]

Chemical synthesis offers the unique opportunity to access well-defined natural and unnatural structures. Collections of complex synthetic glycans, including heparin sulfate glycans, GPI-anchors, and high-mannose oligosaccharides, were used to create custom arrays to characterize lectin and antibody specificity and to study the human response to infections and allergies.[33] The biggest drawback of this approach is the enormous synthetic effort required. Automated Glycan Assembly (AGA) speeds up the process, allowing for quick and reliable access to glycans.[34,35] The sequential addition of sugar building blocks (BBs) on a solid support replaces the purification steps with simple washing cycles. The coupling cycle, consisting of glycosylation, capping, and deprotection, has been optimized to achieve nearly quantitative conversion in around 1.5 h.[36,37] Moreover, the glycan is attached to the solid support through a linker that, upon UV irradiation, liberates the target glycan already equipped with an amino-linker for subsequent surface functionalization.[38] Collections of natural and unnatural glycans found applications in vaccine development,[39,40] materials science,[36,41,42] and structural studies[37] (see Fig. 2). Well-defined linear β(1,3) and branched β(1,3) β(1,6) glucans permitted to conclude that most individuals form antibodies that bind to both linear (protective) and branched (non-protective) epitope.[43] Synthetic keratan sulfate (KS) analogues, with different sulfation patterns, helped to identify the specific interaction between the disulfated KS tetrasaccharide and the adeno-associated virus AAVrh10 gene-therapy vector (see Fig. 2).[44] Frameshifts of the *S. pneumoniae* serotype 8 (ST8) capsular polysaccharides were used to identify the glycotopes recognized by antibodies against ST8. The insights were essential for the preparation of a semisynthetic *Streptococcus pneumoniae* serotype 8 glycoconjugate vaccine candidate.[40] AGA was exploited to determine the binding epitopes of many plant cell-wall glycan-directed mAbs.[45,46] A total of 88 synthetic

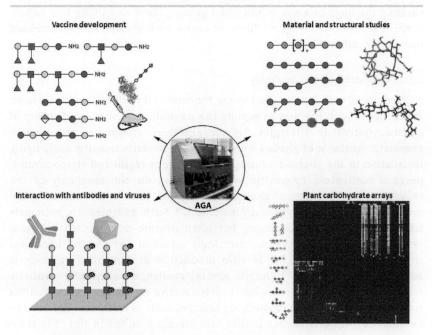

Fig. 2   Applications of glycan collections synthesized with AGA.

oligosaccharides, including arabinogalactan-, rhamnogalacturonan-, xylan-, and xyloglucan were printed on a microarray aiming to comprehensively map the epitopes of plant cell-wall glycan-directed antibodies (see Fig. 2).[47]

3. Printing on surfaces

Once the glycan collection has been produced, the glycans are printed onto the array surface. High accuracy and reproducibility are essential for a reliable microarray. Two technologies are mainly used to deposit bioactive molecules, such as carbohydrates, on a reactive surface. These rely on contact and non-contact printing (see Fig. 3).[48]

Contact technologies (see Fig. 3A) are naturally more precise, mainly relying on pin printing and microstamping of arrays. A pin printing setup consists of a robotically controlled print head, equipped with one to dozens of differently shaped solid pins. The pins soak a certain volume of a spotting solution (dissolved biomolecule or building blocks) from wells of a microtiter plate upon dipping. Nanoliters of the solution can then be deposited as a droplet on the reactive surface by bringing the pins in contact with the surface. The transfer process relies on favorable surface energies between the spotting solution, the surface, and the pin. An alternative to pin printing is microcontact printing, where crosslinked polydimethylsiloxane (PDMS) microstamps with micro-features are used.[49] Spray-on or robotic feature–feature ink transfer is applied to coat the stamp with the spotting solution. The substance is then transferred to the surface upon contact between the stamp and the surface. This technique is mainly used to array one compound on a surface, while pin printing allows for the deposition of different molecules at the same time. In both cases, extensive washing steps and refilling after iterative cycles are necessary.

Non-contact printing technologies (see Fig. 3B) rely on the ejection of spotting solution from a reservoir through an orifice as a droplet or stream onto the microarray surface. Common inkjet printing technology uses a solution of a dissolved biomolecule or building block, serving as the "ink". The solutions are ejected from a cartridge by a print head nozzle at a distance of 1–5 mm from the surface. The ejection process can be triggered by mainly three different methods: piezo actuation, valve-jet, or thermal inkjet. All three methods are based on a reversible and rapid change of pressure within the cartridge to release small droplets of the spotting solution. Non-contact printing approaches are highly flexible, since they allow for a fast switching between various cartridges and frequent refilling is avoided. Furthermore, because the method is contact-free,

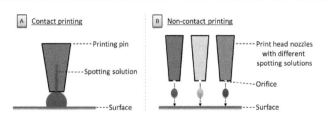

Fig. 3   Schematic illustration of contact (A) and non-contact printing (B).

there is no risk of surface disruption. Possible clogging of the nozzle and process-related contaminations are the main drawback of this technique.

The combinatorial laser-induced transfer method (cLIFT) helps to circumvent contamination and clogging issues.[50] Currently the method is restricted to the chemical synthesis of peptide and peptoid[51] arrays, but may be expanded to glycan array synthesis. Novel micro- and nanoprinting technologies exploit cantilevers from atomic force microscopy to pattern surfaces, such as the well-known dip-pen nanolithography.[52] Meanwhile, the technique evolved to sophisticated microfluidic and lithographic setups, enabling photochemical patterning of surfaces with different monosaccharides in high resolution.[53] Another recent scanning probe approach shows the layer-by-layer printing and synthesis of peptides with a resolution of 50 μm.[54] After deposition of the compounds onto the surface, immobilization can be achieved in many different approaches.

3.1. Non-covalent immobilization

The first immobilization of glycans onto a surface was reported in 2002, following a non-covalent adsorption approach.[55] Non-covalent attachment of either modified or unmodified glycans to a surface is mediated by electrostatic interactions, hydrogen bonds or van der Waals forces. Today the selective covalent attachment of sugar molecules to a microarray is preferred, because it results in more stable and well-defined binding sites, enabling more precise biomolecular interactions.

Site-nonspecific immobilization. The easiest way to immobilize a glycan on a surface is the non-covalent, site-nonspecific approach (see Fig. 4A). Since no extra-functionalization of the sugar is required, this method is only suitable for longer glycans that maintain a large contact area with the surface. The binding site of the molecule to the surface is random, which makes screening of biomolecular interaction less precise. Moreover, there is a constant risk of losing the compounds during the washing steps.

With this approach, unmodified polysaccharides were spotted onto a nitrocellulose-coated glass slide.[55,56] Charged polysaccharides like heparin are particularly suitable for this approach, since the negatively charged sulfate groups can be efficiently attached to positively charged poly-L-lysine coated glass slides *via* electrostatic interactions.[57,58]

Site-specific immobilization. Reproducibility can be enhanced by specific binding to a surface at a distinct position of the glycan (see Fig. 4B). The chemical modification of the glycan is generally carried out at the reducing end. Glycoconjugates, such as glycolipids, can be easily immobilized on a surface, resembling the natural presentation of glycans.

Nitrocellulose or PVDF (polyvinylidene difluoride) membranes were used to immobilize lipid-conjugated glycans (neoglycolipids) *via* hydrophobic interactions (van der Waals forces). The neoglycolipids were prepared by reductive amination of the sugar compound and an amino-conjugated lipid.[59–68] A similar attachment strategy used fluorous tagged glycans for immobilization on a Teflon/epoxy coated glass slide.[69–71] The perfluorinated alkyl chain allows for easy purification and permits strong binding to the surface that survives extensive washing steps. Another fluorous approach was carried out on aluminum oxide coated glass slides, which were covalently functionalized with a phosphonate, tagged with

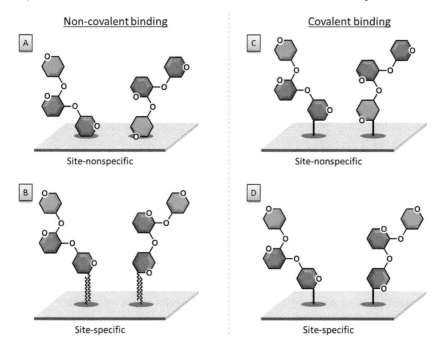

Fig. 4 Different immobilization strategies for glycan microarray production. (A) Non-covalent, site-nonspecific glycan binding; (B) non-covalent site-specific glycan binding; (C) covalent site-nonspecific glycan binding; (D) covalent site-specific glycan binding.

a perfluoroalkyl chain.[72,73] On-spot analysis *via* mass spectrometry was also possible. Importantly, when hydrophobic surfaces are used, a blocking step before biomolecular screenings is required.

The strong biotin–streptavidin interaction ($K_d \sim 10^{-15}$ M) was exploited to manufacture glycan arrays. Streptavidin-coated surfaces in combination with biotinylated glycans were utilized.[74-77] Similarly, DNA hybridization was employed to prepare glycan microarrays. The glycans were functionalized with an oligonucleotide that was hybridized with the complementary oligonucleotide attached to a surface.[78]

3.2. Covalent immobilization

The covalent attachment of a glycan to a surface is usually preferred, because it minimizes the risk of compound leaching during the washing steps. Glass slides coated with a silane or thin polymer film are employed, which are functionalized with various functional groups for the coupling reaction.

Site-nonspecific immobilization. The simplest and fastest way to couple unmodified glycans to a surface is the covalent site-nonspecific approach (see Fig. 4C). However, the random binding of the sugar can be problematic for the validity of the biomolecular binding screenings. Photochemical reactions, where the functionalized glass slide bears a photo-activatable group, are commonly used.

Photo-labile groups such as aryl(trifluoromethyl)diazirine[79] or 4-azido-2,3,5,6-tetrafluorophenyl[80] are common functionalities that, upon UV irradiation, turn into reactive carbene or nitrene species. These reactive compounds are able to react easily with the spotted unprotected sugar compounds *via* simple insertion reactions to form stable covalent bonds. Another approach makes use of phthalimide-modified surfaces.[81,82] Upon UV irradiation, the carbonyl groups of phthalimide can readily undergo a photochemical hydrogen abstraction reaction with the desired sugar which ends in stable covalent bonds between the compound and the surface. The reaction between boronic acid functionalized surfaces and diols of the sugars was also exploited to produce carbohydrate microarrays.[83]

Site-specific immobilization. The covalent site-specific attachment of chemically modified glycans is now the method of choice for carbohydrate microarray production (see Fig. 4D), with many different available reactions. These coupling reactions have to be highly selective, easy to manipulate and mild. Selective surface attachment renders the binding studies with biomolecules more reliable when compared to site-nonspecific approaches. The nature of the linker between the sugar and the surface plays a crucial role, influencing protein binding. Hydrophilic oligo or poly(ethylene glycol)-based linkers often show better results compared to the hydrophobic analogues. Additionally, the linker affects the nonspecific adsorptions of the proteins and its length is important for the accessibility of the attached glycan.[84,85] The most challenging part of the covalent site-specific method is the functionalization of the sugar, which often requires multiple steps and well-wrought synthetic strategies.

A very powerful strategy exploits the thiol–maleimide chemistry. The reaction is very fast under mild conditions and highly selective. Glycans are either functionalized with maleimide groups and coupled to thiol-coated surfaces,[84–86] or thio-sugars are attached to maleimide-coated surfaces (see Fig. 5A).[87–94] With this approach, even challenging glycans like glycosylphosphatidylinositols (GPIs) could be easily printed onto microarrays.[95,96] The formation of disulfide bonds, either between thiosulfonate-conjugated glycans and thiol-functionalized surfaces or between thiol-conjugated glycans and pyridyl disulfide-modified surfaces, was also successful.[97,98] Nevertheless, the possible oxidation of the thiol due to exposure to air can create problems.

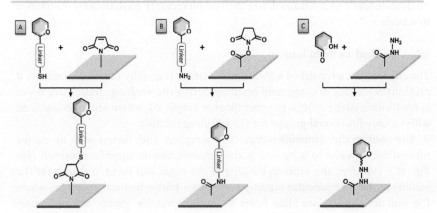

Fig. 5 Important coupling reactions for site-specific covalent bond formation. (A) Thiol-functionalized sugar and maleimide surface; (B) amine-functionalized sugar and *N*-hydroxysuccinimide ester surface; (C) free reducing end glycan and hydrazide surface.

The most widely used immobilization strategy employs amino-functionalized glycans and *N*-hydroxysuccinimide (NHS) ester-coated surfaces (see Fig. 5B). The reaction between these two compounds under slightly basic conditions (pH ~ 8.5) leads to the formation of a very stable amide bond with very good selectivity.[11,99–102] The NHS ester-coated glass slides are commercially available and the synthesis of amine-functionalized glycans follows standard protocols.[103] Amine-modified glycans can be readily accessed also from automated strategies.[35] Alternatively, the amino-modified glycans can be attached to cyanuric chloride-functionalized surfaces *via* nucleophilic aromatic substitution.[104,105]

Unprotected linker-free glycans can be immobilized in a site-specific way, using hydrazide- (see Fig. 5C) or oxyamine-modified surfaces.[106,107] These functional groups are highly nucleophilic and able to react easily with the reducing ends of the glycans to form stable adducts. Similarly, an aldehyde-functionalized surface and oxyamine-modified sugars were used to prepare glycosaminoglycan microarrays.[108] Another glycosaminoglycan microarray was produced on an amine-coated surface, using a deaminated heparin, bearing an aldehyde functionality.[109]

Epoxide-coated surfaces in combination with hydrazide-functionalized sugars offer a valuable alternative to form stable covalent bonds (see Fig. 6C).[110–115] Moreover, epoxide-coated surfaces are very versatile and can be used in combination with many different nucleophiles such as amine- or thiol-conjugated carbohydrates (see Fig. 6A and B).[116,117] Cycloadditions with dienophile-conjugated carbohydrates also show good selectivity and can be carried out under mild conditions. Diels–Alder reaction between a cyclopentadiene-linked

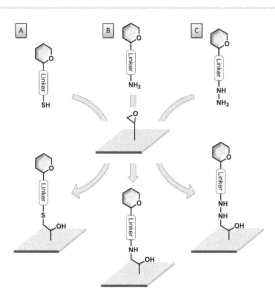

Fig. 6 Versatility of epoxide-coated surfaces. (A) Thiol-functionalized sugar; (B) amine-functionalized sugar; (C) hydrazide-functionalized sugar.

sugar and a benzoquinone-coated surface,[118] as well as the very fast tetrazine–norbornene inverse electron demand Diels–Alder reaction were applied.[119] Nevertheless, the lack of long-term stability of some of these compounds limits their applications.

The copper-catalyzed azide–alkyne click reaction (CuAAC) was applied in glycan microarray production, because of its high selectivity and compatibility with a broad range of functional groups. Glycans functionalized with azide groups are coupled to alkyne-functionalized surfaces (or inverted functionalization).[120–124] In addition, azide-modified glycans were used to prepare microarrays through chemoselective Staudinger ligation.[125] Photochemical attachment of a 4-azido-2,3,5,6-tetrafluorophenyl-conjugated sugar to a polymer monolayer offered a very mild alternative.[80,126] After spotting the compound onto the surface, irradiation with UV light converts the azide functionality to a reactive nitrene species, which is able to react with the polymer monolayer to from a stable covalent bond.

4. Multivalent presentation

Carbohydrate–protein interactions are very weak. However, usually multiple simultaneous interactions between several carbohydrate ligands and one receptor occur, which increases the binding strength. This concept is called multivalency. For a multivalent interaction to take place, the spatial distribution and orientation of the sugar groups are crucial.[127–129] To translate this to carbohydrate microarrays, the glycan presentation and, especially the density and orientation, need to be considered in detail.

In conventional array platforms, single monovalent glycans are randomly attached to the array surface *via* a linker, yielding a certain – but uncontrolled – multivalent display, which may be sufficient to elicit a high-avidity binding event. These systems usually rely on two-dimensional arrangements of monovalent glycans, with very little control over spatial organization. However, carbohydrate–protein interactions vary quite significantly and high glycan density may either enhance it *via* multivalency or suppress it *via* steric hindrance. Therefore, several studies were conducted to identify the optimal presentation of carbohydrates by varying the glycan concentration during printing and the flexibility of the attachment point.[130,131] To date, full control on spatial organization is still a big challenge and clustering effects can cause unreproducible results.

To improve control over glycan presentation, multivalent glycoconjugates with various valencies and spatial arrangements have been designed and immobilized on arrays. Scaffolds, based on natural glycoproteins, neoglycoproteins/neoglycopeptides, glycodendrimers, multivalent display on DNA, glycoclusters, and glycopolymers, have been used (Fig. 7).[13,132–144]

Natural glycoproteins, such as the heavily glycosylated mucins, were used as multivalent glycan systems for the production of arrays. This microarray retained the three-dimensional presentation of mucin oligosaccharides, without modifications of the protein backbone and permitted the discovery of biologically important motifs for bacterial–host interactions.[145] A similar approach uses natural proteins, such as bovine serum albumin (BSA) or human serum albumin (HSA), for the production of neoglycoproteins/neoglycopeptides (proteins or peptides with glycans covalently attached *via* non-native linkage), which display multiple copies of each glycan. Presynthesized glycoconjugates can be

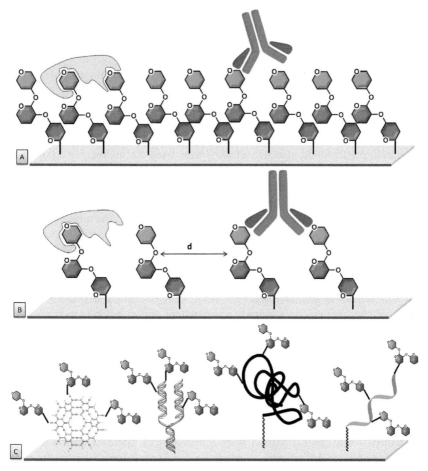

Fig. 7 Schematic glycan presentation on microarrays. (A) High density arrangement of glycans. (B) Low density arrangement of glycans. (C) Multivalent glycoconjugates to modulate glycan presentation on microarray surfaces.

immobilized on epoxide slides. This strategy permits immobilization of both synthetic carbohydrates as well as natural carbohydrates, presented on glyco-proteins. Important factors that affect the binding are the number of glycans on a neoglycopeptide, the linker length between the individual sugars, the distance between neoglycopeptide probes on the surface, and the type of protein.[113,146–148] These parameters can be tuned to affect the recognition process. Variations in the neoglycoprotein density revealed differences in specificity for antibodies that were not apparent at low density.[149]

Oligonucleotide hybridization permits to tailor spatial geometry. The rigidity of the double strand nucleic acid with well-defined nucleotide spacing permits to adjust the ligand presentation on this supramolecular scaffold.[150–152] Similarly, peptide nucleic acids (PNAs) have been used to tag glycans and evaluate their multivalent interactions with lectins. From an assembly stand-point, stable PNA–DNA duplexes can be achieved with shorter sequences than the corresponding DNA homoduplexes (10–14-mer PNA typically provides sufficient duplex stability).[153]

Chemical ligation of sugars at different positions within a PNA oligomer has been achieved[154] by using thiol moieties embedded in the backbone of the PNA, chemoselectively conjugated to a maleimide-glycan. DNA microarrays permitted the combinatorial pairing of diverse PNA-tagged glycan conjugates. The use of adjacent hybridization sites produced assemblies, emulating the diversity of di-, tri- and tetra-antennary glycans, mimicking the geometry of the HIV gp120 glycan epitope. The combinatorial synthesis of an extended library of PNA-encoded glycoconjugates represents the largest array of heteroglycan conjugates reported to date.[155–157]

Unnatural scaffolds, like dendrimers, were used for microarray analyses of CPIs.[158–161] Well-defined 3D saccharide arrangements on microarrays were constructed upon covalent binding of the dendrimers to the chip surface. Carbohydrates were attached to the dendrimer arm *via* "click" chemistry, prior or following the attachment of the whole construct to the chip. The multivalency can be precisely controlled with the structure of the glycodendrimer, with valencies ranging from one to eight sugars. Other unnatural alternatives used for multivalent presentation are glycoclusters. Calix[4]arenes are a suitable platform that can be easily derivatized at the upper and lower rims, resulting in well-organized three-dimensional architectures.[162,163] With such systems, the primary importance of the spatial arrangement, compared to the number of carbohydrate residues, was highlighted.[164]

The importance of spatial orientation was observed by 16 different fucosylated glycomimetics, bearing one to eight fucose moieties, synthesized with antenna-like, linear (or comb-like), or crown-like arrangements.[165] Binding properties using DNA directed immobilization (DDI)-based glycan microarrays showed that no chelate effect was present, with a one to one interaction between fucose and the lectin. Synthetic glycopolymers have been used to generate mucin-like structures which, as do natural ones, possess rigid extended structures.[124,166] Polymers of low polydispersity, displaying α-GalNAc residues, were produced by reversible addition-fragmentation chain transfer (RAFT) polymerization. This new class of orthogonally end-functionalized mucin-mimetics was printed on a microarray, where GalNAc valency and interligand spacing could be controlled. This system again proved that glycan valency and organization are critical parameters that determine the modes through which these interactions occur.

5.   Characterization and binding measurements

The readout of a glycan microarray is an important step to obtain precise and convincing data. To detect binding events of glycan binding proteins (GBPs) or successful enzymatic glycosylations on the array, different methods are available. The most frequently applied method is the detection of fluorescently-labeled binders, which directly or indirectly bind to the glycans on the microarray (see Fig. 8). The binding event can be visualized with a fluorescence scanner in several ways: either the GBP is fluorescently labelled or a fluorescently tagged secondary reagent (*e.g.* antibody) is used to bind to the GBP or to a tag (*e.g.* biotin, His tag) on the GBP. As discussed, multivalency plays a crucial role for CPIs and the glycan density on the microarray is essential to achieve differential binding. If the density is too low, the GBPs are sometimes unable to properly bind to the glycans, which results in a loss of signals and thus to misleading results.[167] Additional

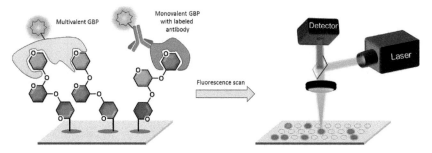

Fig. 8 Detection of directly or indirectly fluorescently labeled glycan binding proteins (GBPs) binding to specific glycans on a microarray by fluorescence scanning.

problems can be caused by the label, which can reduce the activity or influence the selectivity of the GBPs.[168] Unfortunately, indirect labeling of GBPs is often not possible, because fluorescently-labeled secondary reagents are not available.[9]

Mass spectrometry is a label-free method to monitor chemical or enzymatic glycosylations directly on an array. Thiol-linked sugars were deposited on a gold surface, whereby self-assembled monolayers are formed. Elongation reactions were then monitored by an on-slide mass spectrometry technique named SAMDI-TOF-MS.[21,24,169] With a similar non covalent approach, glycosylation of carbohydrates immobilized on modified gold surfaces using van der Waals forces between aliphatic[170,171] or perfluorinated[172] carbon chains was monitored. Multiple detection techniques could be used as proven by a multifunctional microarray platform consisting of a glass surface coated with an indium-tin oxide layer. Matrix-assisted laser desorption/ionization time-of-flight (MALDI-TOF) mass spectrometry, fluorescence spectroscopy, and optical microscopy can be employed on the same surface.[25]

Surface plasmon resonance (SPR) imaging is an alternative label-free method for the analysis of glycan microarrays that allows for determination of the thickness of layers on a metal surface in the nanometer range. SPR has the advantage of real-time monitoring of GBP binding events, which allows for measuring of kinetic and thermodynamic parameters. Metal surfaces (*e.g.* gold) are mandatory for this approach to excite surface plasmons within the metal by irradiation with polarized light. SPR was used to screen interactions between GBPs and glycans of the pathogen *Schistosoma mansoni*.[146] BSA–mannose-conjugates with different mannose substituents were attached to a gold surface and incubated with ConA to measure K_D values and relate it to multivalency.[173] Additionally, SPR permitted to identify ligand specificity of plant lectins[174] and to better understand siglec-8 (ref. 76) or ConA[175] binding specificities.

The above mentioned analysis technologies are the most common, but many others, such as evanescent-field fluorescence,[176–178] ellipsometry,[179] electrochemoluminescence,[180] detection of radioactivity,[181,182] oblique-incidence reflectivity (OI-RD) microscopy,[183] frontal affinity chromatography,[184,185] isothermal calorimetry,[186] and cantilever-based detection[187] exist. Nevertheless, multivalency cannot be detected directly with one of the analytical techniques. Experiments using multivalent scaffolds have to be compared to those with the monovalent analogue. New technologies to systematically vary the glycan density directly on the microarray are required, to understand multivalent events.

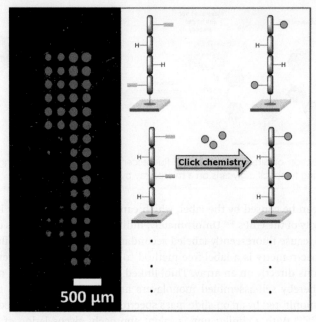

Fig. 9 Peptide array with peptide tetramers, synthesized *via* laser transfer, derivatized with up to four α-D-mannose azides clicked to the peptide backbone. Sixteen different peptides (quadruplicate spots) show differential lectin (ConA) binding, due to different multivalent display.

6.   Conclusions and outlook

The direct *in situ* synthesis on surfaces is already well-established and commercialized for oligonucleotides (*e.g.*, Affymetrix, Agilent, Illumina) and peptides (*e.g.*, Intavis, JPT, PEPperPRINT). A big part of this success is based on the enabling technologies, that had a major impact on high-throughput analysis and screening. To translate these technologies to *in situ* glycan synthesis will be far more challenging: to date, only the chemical synthesis of disaccharides on a "macroarray" surface has been shown,[21] whereas enzymatic synthesis is more promising.[22–25] Furthermore, multivalency is usually neglected in oligonucleotide and peptide synthesis.

In contrast, multivalency is essential for GBPs, because of the naturally weak ($K_D \sim \mu m$) protein–glycan affinity, compensated by multiple binding sites.[188] Since the advent of glycan microarrays, the main focus has been on the analysis of glycans on surfaces, with less interest in the control of molecular density and spacing. Yet, a defined way of presenting glycans on microarrays is the key step to strong GBP binding. Therefore, strategies are required to display glycans in a molecularly defined spatial order.

Different scaffolds or density variations have been proposed and quite successfully applied for multivalent glycan display.[166,167] Especially, the density variation presentation on surfaces leads to random and non-homogeneous systems, which lack reproducibility. Most scaffolds offer defined spacing, but lack flexibility, because the spacing cannot be changed easily. An elegant solution is DNA technology to display glycans. Only recently, this was shown for the display different molecules.[189,190] Using DNA-origami structures as scaffolds, multiple glycan structures can be placed in a wide variety of 2D and 3D configurations at exact positions in a controlled and reproducible way. Moreover, it may be used to exactly space the glycans to generate a perfectly matching template for the binding sites of multivalent GBPs. Thereby, control in the screening processes for pathogen interaction with a large variety of structures is possible.

The defined generation of many diverse scaffolds with defined glycan spacing will be one of the future research goals in glycan array technology. Progress in the field of *in situ* synthesis of scaffolds has been made. By growing brush-like gly-copolymers directly on the surface *via in situ* photo-polymerization, glycan microarrays with multivalent display were generated.[20] Different polymer lengths were produced with different amounts of sugar units on the polymer scaffold, by changing the irradiation time.

We recently employed a novel laser-based transfer setup[50] to generate peptide scaffolds for multivalent display. We synthesized arrays of peptide tetramers, containing all 16 possible sequences of L-glycine and L-propargylglycine. The propargylglycine offers an alkyne group for copper catalyzed click chemistry to attach up to four glycan azides to the peptide backbone. Depending on the amount and position of the α-D-mannose, we obtained differential binding of the lectin concanavalin A (see Fig. 9). This approach may serve as a basis to generate large and complex compound collections for the multivalent display of many different glycans in an orthogonal synthesis strategy. With our laser-based approach, molecules can be synthesized directly on surfaces step-by-step, by "printing" and stacking solid polymer nanolayers,[51,191] which embed all kinds of different chemicals and building blocks. Especially for peptide synthesis and applications in disease research, this offers a rapid strategy to generate diverse microarrays.[192–197] In the future, this technology may be exploited for the *in situ* synthesis of glycopeptides, glycans, and DNA in a microarray format.

A large gap remains in the multivalent display and analysis of complex glycans that needs to be filled. Advances accessing glycans and their synthesis and immobilization on surfaces show promising directions for future glycan micro-array research. Precisely defined multivalent arrangements on DNA or other structural scaffolds will enable the identification of cooperative effects between identical or diverse collections of glycans. Simultaneously, novel tools based on the presentation of single or multiple glycan molecules in specific arrangements and stoichiometry on the surfaces will be developed.

In the future, newly developed platforms will enable highly parallelized screenings, testing tens of thousands of combinations simultaneously in a microarray-based assay format. The field of protein–glycan interactions will benefit as researchers will be able to uncover the conformation of glycans in a biological environment and open new roads to develop efficient vaccines.

Conflicts of interest

P. H. S. declares a significant financial interest in GlycoUniverse GmbH & Co KGaA, the company that commercializes the synthesis instrument, building blocks and other reagents. F. F. L. is named on a pending patent application related to laser-based microarray synthesis.

Acknowledgements

We thank the Max-Planck Society, the Minerva Fast Track Program, and the MPG-FhG Cooperation Project Glyco3Dysplay, and the German Federal Ministry of Education and Research (BMBF, grant no. 13XP5050A) for generous financial support.

References

1 A. Varki, R. Cummings, J. Esko, H. Freeze, P. Stanley, G. Hart and P. H. Seeberger, *Essentials of glycobiology*, Cold Spring Harbor Laboratory Press, Cold Spring Harbor, N.Y., 2017.

2 Y. C. Lee and R. T. Lee, *Acc. Chem. Res.*, 1995, **28**, 321–327.

3 J. Rojo, J. C. Morales and S. Penadés, in *Host-Guest Chemistry: Mimetic Approaches to Study Carbohydrate Recognition*, ed. S. Penadés, Springer Berlin Heidelberg, Berlin, Heidelberg, 2002, vol. 218, pp. 45–92.

4 C. R. Bertozzi and L. L. Kiessling, *Science*, 2001, **291**, 2357–2364.

5 D. F. Smith and R. D. Cummings, *Curr. Opin. Virol.*, 2014, **7**, 79–87.

6 K. A. Kline, S. Fälker, S. Dahlberg, S. Normark and B. Henriques-Normark, *Cell Host Microbe*, 2009, **5**, 580–592.

7 Y. van Kooyk and G. A. Rabinovich, *Nat. Immunol.*, 2008, **9**, 593–601.

8 P. R. Crocker, J. C. Paulson and A. Varki, *Nat. Rev. Immunol.*, 2007, **7**, 255–266.

9 S. Park, J. C. Gildersleeve, O. Blixt and I. Shin, *Chem. Soc. Rev.*, 2013, **42**, 4310–4326.

10 C. D. Rillahan and J. C. Paulson, *Annu. Rev. Biochem.*, 2011, **80**, 797–823.

11 M. D. Disney and P. H. Seeberger, *Chem. Biol.*, 2004, **11**, 1701–1707.

12 J. J. Lundquist and E. J. Toone, *Chem. Rev.*, 2002, **102**, 555–578.

13 J. L. Jiménez Blanco, C. Ortiz Mellet and J. M. García Fernández, *Chem. Soc. Rev.*, 2013, **42**, 4518–4531.

14 C. Fasting, C. A. Schalley, M. Weber, O. Seitz, S. Hecht, B. Koksch, J. Dernedde, C. Graf, E.-W. Knapp and R. Haag, *Angew. Chem., Int. Ed.*, 2012, **51**, 10472–10498.

15 M. Delbianco, P. Bharate, S. Varela-Aramburu and P. H. Seeberger, *Chem. Rev.*, 2016, **116**, 1693–1752.

16 X. Han, Y. Zheng, C. J. Munro, Y. Ji and A. B. Braunschweig, *Curr. Opin. Biotechnol.*, 2015, **34**, 41–47.

17 L. Wang, R. D. Cummings, D. F. Smith, M. Huflejt, C. T. Campbell, J. C. Gildersleeve, J. Q. Gerlach, M. Kilcoyne, L. Joshi, S. Serna, N.-C. Reichardt, N. Parera Pera, R. J. Pieters, W. Eng and L. K. Mahal, *Glycobiology*, 2014, **24**, 507–517.

18 P.-H. Liang, S.-K. Wang and C.-H. Wong, *J. Am. Chem. Soc.*, 2007, **129**, 11177–11184.

19 F. Broecker and P. H. Seeberger, in *Small Molecule Microarrays: Methods and Protocols*, ed. M. Uttamchandani and S. Q. Yao, Springer New York, New York, NY, 2017, vol. 1518, pp. 227–240.

20 S. Bian, S. B. Zieba, W. Morris, X. Han, D. C. Richter, K. A. Brown, C. A. Mirkin and A. B. Braunschweig, *Chem. Sci.*, 2014, **5**, 2023–2030.

21 L. Ban and M. Mrksich, *Angew. Chem., Int. Ed.*, 2008, **47**, 3396–3399.

22 S. Serna, J. Etxebarria, N. Ruiz, M. Martin-Lomas and N.-C. Reichardt, *Chem.-Eur. J.*, 2010, **16**, 13163–13175.

23 O. Blixt, K. Allin, O. Bohorov, X. Liu, H. Andersson-Sand, J. Hoffmann and N. Razi, *Glycoconjugate J.*, 2008, **25**, 59–68.

24 L. Ban, N. Pettit, L. Li, A. D. Stuparu, L. Cai, W. L. Chen, W. Y. Guan, W. Q. Han, P. G. Wang and M. Mrksich, *Nat. Chem. Biol.*, 2012, **8**, 769–773.

25 A. Beloqui, J. Calvo, S. Serna, S. Yan, I. B. Wilson, M. Martin-Lomas and N. C. Reichardt, *Angew. Chem., Int. Ed.*, 2013, **52**, 7477–7481.

26 X. Song, J. Heimburg-Molinaro, R. D. Cummings and D. F. Smith, *Curr. Opin. Chem. Biol.*, 2014, **18**, 70–77.

27 R. B. Zheng, S. A. F. Jégouzo, M. Joe, Y. Bai, H.-A. Tran, K. Shen, J. Saupe, L. Xia, M. F. Ahmed, Y.-H. Liu, P. S. Patil, A. Tripathi, S.-C. Hung, M. E. Taylor, T. L. Lowary and K. Drickamer, *ACS Chem. Biol.*, 2017, **12**, 2990–3002.

28 J.-i. Kadokawa, *Chem. Rev.*, 2011, **111**, 4308–4345.

29 F. Pfrengle, *Curr. Opin. Chem. Biol.*, 2017, **40**, 145–151.

30 L. Wen, G. Edmunds, C. Gibbons, J. Zhang, M. R. Gadi, H. Zhu, J. Fang, X. Liu, Y. Kong and P. G. Wang, *Chem. Rev.*, 2018, **118**, 8151–8187.

31 T. Li, L. Liu, N. Wei, J.-Y. Yang, D. G. Chapla, K. W. Moremen and G.-J. Boons, *Nat. Chem.*, 2019, **11**, 229–236.

32 J. Zhang, C. Chen, M. R. Gadi, C. Gibbons, Y. Guo, X. Cao, G. Edmunds, S. Wang, D. Liu, J. Yu, L. Wen and P. G. Wang, *Angew. Chem., Int. Ed.*, 2018, **57**, 16638–16642.

33 P. H. Seeberger, *Perspect. Sci.*, 2017, **11**, 11–17.

34 M. Guberman and P. H. Seeberger, *J. Am. Chem. Soc.*, 2019, **141**, 5581–5592.

35 A. Pardo-Vargas, M. Delbianco and P. H. Seeberger, *Curr. Opin. Chem. Biol.*, 2018, **46**, 48–55.

36 Y. Yu, A. Kononov, M. Delbianco and P. H. Seeberger, *Chem.-Eur. J.*, 2018, **24**, 6075–6078.

37 M. Delbianco, A. Kononov, A. Poveda, Y. Yu, T. Diercks, J. Jiménez-Barbero and P. H. Seeberger, *J. Am. Chem. Soc.*, 2018, **140**, 5421–5426.

38 L. Krock, D. Esposito, B. Castagner, C.-C. Wang, P. Bindschadler and P. H. Seeberger, *Chem. Sci.*, 2012, **3**, 1617–1622.

39 M. Guberman, M. Bräutigam and P. Seeberger, *Chem. Sci.*, 2019, **10**, 5634.

40 B. Schumann, H. S. Hahm, S. G. Parameswarappa, K. Reppe, A. Wahlbrink, S. Govindan, P. Kaplonek, L. A. Pirofski, M. Witzenrath, C. Anish, C. L. Pereira and P. H. Seeberger, *Sci. Transl. Med.*, 2017, **9**, eaaf5347.

41 K. Naresh, F. Schumacher, H. S. Hahm and P. H. Seeberger, *Chem. Commun.*, 2017, **53**, 9085–9088.

42 Y. Yu, S. Gim, D. Kim, Z. A. Arnon, E. Gazit, P. H. Seeberger and M. Delbianco, *J. Am. Chem. Soc.*, 2019, **141**, 4833–4838.

43 M. W. Weishaupt, H. S. Hahm, A. Geissner and P. H. Seeberger, *Chem. Commun.*, 2017, **53**, 3591–3594.

44 H. S. Hahm, F. Broecker, F. Kawasaki, M. Mietzsch, R. Heilbronn, M. Fukuda and P. H. Seeberger, *Chem*, 2017, **2**, 114–124.

45 C. Ruprecht, P. Dallabernardina, P. J. Smith, B. R. Urbanowicz and F. Pfrengle, *ChemBioChem*, 2018, **19**, 793–798.

46 D. Schmidt, F. Schuhmacher, A. Geissner, P. H. Seeberger and F. Pfrengle, *Chem.–Eur. J.*, 2015, **21**, 5709–5713.

47 C. Ruprecht, M. P. Bartetzko, D. Senf, P. Dallabernadina, I. Boos, M. C. F. Andersen, T. Kotake, J. P. Knox, M. G. Hahn, M. H. Clausen and F. Pfrengle, *Plant Physiol.*, 2017, **175**, 1094–1104.

48 V. Romanov, S. N. Davidoff, A. R. Miles, D. W. Grainger, B. K. Gale and B. D. Brooks, *Analyst*, 2014, **139**, 1303–1326.

49 J. L. Wilbur, A. Kumar, H. A. Biebuyck, E. Kim and G. M. Whitesides, *Nanotechnology*, 1996, **7**, 452–457.

50 F. F. Loeffler, T. C. Foertsch, R. Popov, D. S. Mattes, M. Schlageter, M. Sedlmayr, B. Ridder, F. X. Dang, C. von Bojnicic-Kninski, L. K. Weber, A. Fischer, J. Greifenstein, V. Bykovskaya, I. Buliev, F. R. Bischoff, L. Hahn, M. A. Meier, S. Brase, A. K. Powell, T. S. Balaban, F. Breitling and A. Nesterov-Mueller, *Nat. Commun.*, 2016, **7**, 11844.

51 D. S. Mattes, B. Streit, D. R. Bhandari, J. Greifenstein, T. C. Foertsch, S. W. Munch, B. Ridder, C. Bojničić-Kninski, A. Nesterov-Mueller, B. Spengler, U. Schepers, S. Brase, F. F. Loeffler and F. Breitling, *Macromol. Rapid Commun.*, 2019, **40**, 1800533.

52 R. D. Piner, J. Zhu, F. Xu, S. Hong and C. A. Mirkin, *Science*, 1999, **283**, 661–663.

53 D. J. Valles, Y. Naeem, C. Carbonell, A. M. Wong, D. R. Mootoo and A. B. Braunschweig, *ACS Biomater. Sci. Eng.*, 2019, **5**, 3131.

54 J. Atwater, D. S. Mattes, B. Streit, C. von Bojnicic-Kninski, F. F. Loeffler, F. Breitling, H. Fuchs and M. Hirtz, *Adv. Mater.*, 2018, **30**, 1801632.

55 D. Wang, S. Liu, B. J. Trummer, C. Deng and A. Wang, *Nat. Biotechnol.*, 2002, **20**, 275–281.

56 V. I. Dyukova, N. V. Shilova, O. E. Galanina, A. Y. Rubina and N. V. Bovin, *Biochim. Biophys. Acta, Gen. Subj.*, 2006, **1760**, 603–609.

57 E. L. Shipp and L. C. Hsieh-Wilson, *Chem. Biol.*, 2007, **14**, 195–208.

58 C. J. Rogers, P. M. Clark, S. E. Tully, R. Abrol, K. C. Garcia, W. A. Goddard and L. C. Hsieh-Wilson, *Proc. Natl. Acad. Sci. U. S. A.*, 2011, **108**, 9747–9752.

59 S. Fukui, T. Feizi, C. Galustian, A. M. Lawson and W. Chai, *Nat. Biotechnol.*, 2002, **20**, 1011–1017.

60 T. Feizi and W. Chai, *Nat. Rev. Mol. Cell Biol.*, 2004, **5**, 582–588.

61 A. S. Palma, T. Feizi, Y. Zhang, M. S. Stoll, A. M. Lawson, E. Díaz-Rodríguez, M. A. Campanero-Rhodes, J. Costa, S. Gordon, G. D. Brown and W. Chai, *J. Biol. Chem.*, 2006, **281**, 5771–5779.

62 A. S. Palma, Y. Liu, H. Zhang, Y. Zhang, B. V. McCleary, G. Yu, Q. Huang, L. S. Guidolin, A. E. Ciocchini, A. Torosantucci, D. Wang, A. L. Carvalho, C. M. G. A. Fontes, B. Mulloy, R. A. Childs, T. Feizi and W. Chai, *Mol. Cell. Proteomics*, 2015, **14**, 974–988.

63 A. S. Palma, T. Feizi, R. A. Childs, W. Chai and Y. Liu, *Curr. Opin. Chem. Biol.*, 2014, **18**, 87–94.

64 Z. M. Khan, Y. Liu, U. Neu, M. Gilbert, B. Ehlers, T. Feizi and T. Stehle, *J. Virol.*, 2014, **88**, 6100–6111.

65 F. Klein, C. Gaebler, H. Mouquet, D. N. Sather, C. Lehmann, J. F. Scheid, Z. Kraft, Y. Liu, J. Pietzsch, A. Hurley, P. Poignard, T. Feizi, L. Morris, B. D. Walker, G. Fätkenheuer, M. S. Seaman, L. Stamatatos and M. C. Nussenzweig, *J. Exp. Med.*, 2012, **209**, 1469–1479.

		This journal is © The Royal Society of Chemistry 2019

66 C. Gao, Y. Liu, H. Zhang, Y. Zhang, M. N. Fukuda, A. S. Palma, R. P. Kozak, R. A. Childs, M. Nonaka, Z. Li, D. L. Siegel, P. Hanfland, D. M. Peehl, W. Chai, M. I. Greene and T. Feizi, *J. Biol. Chem.*, 2014, **289**, 16462–16477.

67 S. Hanashima, S. Götze, Y. Liu, A. Ikeda, K. Kojima-Aikawa, N. Taniguchi, D. Varón Silva, T. Feizi, P. H. Seeberger and Y. Yamaguchi, *ChemBioChem*, 2015, **16**, 1502–1511.

68 Y. Liu, R. A. Childs, A. S. Palma, M. A. Campanero-Rhodes, M. S. Stoll, W. Chai and T. Feizi, in *Carbohydrate Microarrays: Methods and Protocols*, ed. Y. Chevolot, Humana Press, Totowa, NJ, 2012, vol. 808, pp. 117–136.

69 K.-S. Ko, F. A. Jaipuri and N. L. Pohl, *J. Am. Chem. Soc.*, 2005, **127**, 13162–13163.

70 S. K. Mamidyala, K.-S. Ko, F. A. Jaipuri, G. Park and N. L. Pohl, *J. Fluorine Chem.*, 2006, **127**, 571–579.

71 G.-S. Chen and N. L. Pohl, *Org. Lett.*, 2008, **10**, 785–788.

72 S.-H. Chang, J.-L. Han, S. Y. Tseng, H.-Y. Lee, C.-W. Lin, Y.-C. Lin, W.-Y. Jeng, A. H. J. Wang, C.-Y. Wu and C.-H. Wong, *J. Am. Chem. Soc.*, 2010, **132**, 13371–13380.

73 Y. Li, E. Arigi, H. Eichert and S. B. Levery, *J. Mass Spectrom.*, 2010, **45**, 504–519.

74 O. E. Galanina, M. Mecklenburg, N. E. Nifantiev, G. V. Pazynina and N. V. Bovin, *Lab Chip*, 2003, **3**, 260–265.

75 Y. Guo, H. Feinberg, E. Conroy, D. A. Mitchell, R. Alvarez, O. Blixt, M. E. Taylor, W. I. Weis and K. Drickamer, *Nat. Struct. Mol. Biol.*, 2004, **11**, 591–598.

76 B. S. Bochner, R. A. Alvarez, P. Mehta, N. V. Bovin, O. Blixt, J. R. White and R. L. Schnaar, *J. Biol. Chem.*, 2005, **280**, 4307–4312.

77 K. Godula and C. R. Bertozzi, *J. Am. Chem. Soc.*, 2010, **132**, 9963–9965.

78 Y. Chevolot, C. Bouillon, S. Vidal, F. Morvan, A. Meyer, J.-P. Cloarec, A. Jochum, J.-P. Praly, J.-J. Vasseur and E. Souteyrand, *Angew. Chem., Int. Ed.*, 2007, **46**, 2398–2402.

79 F. Crevoisier, H. Gao, H. Sigrist, J. L. Ridet, N. Kusy, N. Sprenger, S. Angeloni, S. Guinchard and S. Kochhar, *Glycobiology*, 2004, **15**, 31–41.

80 Z. Pei, H. Yu, M. Theurer, A. Waldén, P. Nilsson, M. Yan and O. Ramström, *ChemBioChem*, 2007, **8**, 166–168.

81 G. T. Carroll, D. Wang, N. J. Turro and J. T. Koberstein, *Langmuir*, 2006, **22**, 2899–2905.

82 D. Wang, G. T. Carroll, N. J. Turro, J. T. Koberstein, P. Kováč, R. Saksena, R. Adamo, L. A. Herzenberg, L. A. Herzenberg and L. Steinman, *Proteomics*, 2007, **7**, 180–184.

83 H.-Y. Hsiao, M.-L. Chen, H.-T. Wu, L.-D. Huang, W.-T. Chien, C.-C. Yu, F.-D. Jan, S. Sahabuddin, T.-C. Chang and C.-C. Lin, *Chem. Commun.*, 2011, **47**, 1187–1189.

84 S. Park and I. Shin, *Angew. Chem., Int. Ed.*, 2002, **41**, 3180–3182.

85 S. Park, M.-r. Lee, S.-J. Pyo and I. Shin, *J. Am. Chem. Soc.*, 2004, **126**, 4812–4819.

86 I. Shin, A. D. Zamfir and B. Ye, *Methods Mol. Biol.*, 2008, **441**, 19–39.

87 B. T. Houseman, E. S. Gawalt and M. Mrksich, *Langmuir*, 2003, **19**, 1522–1531.

88 E. W. Adams, D. M. Ratner, H. R. Bokesch, J. B. McMahon, B. R. O'Keefe and P. H. Seeberger, *Chem. Biol.*, 2004, **11**, 875–881.

89 D. M. Ratner, E. W. Adams, J. Su, B. R. O'Keefe, M. Mrksich and P. H. Seeberger, *ChemBioChem*, 2004, **5**, 379–383.

90 M. A. Brun, M. D. Disney and P. H. Seeberger, *ChemBioChem*, 2006, **7**, 421–424.

91 D. M. Ratner and P. H. Seeberger, *Curr. Pharm. Des.*, 2007, **13**, 173–183.

92 J. H. Seo, K. Adachi, B. K. Lee, D. G. Kang, Y. K. Kim, K. R. Kim, H. Y. Lee, T. Kawai and H. J. Cha, *Bioconjugate Chem.*, 2007, **18**, 2197–2201.

93 S. Matthies, P. Stallforth and P. H. Seeberger, *J. Am. Chem. Soc.*, 2015, **137**, 2848–2851.

94 B. Schumann, R. Pragani, C. Anish, C. L. Pereira and P. H. Seeberger, *Chem. Sci.*, 2014, **5**, 1992–2002.

95 F. Kamena, M. Tamborrini, X. Liu, Y.-U. Kwon, F. Thompson, G. Pluschke and P. H. Seeberger, *Nat. Chem. Biol.*, 2008, **4**, 238–240.

96 M. Tamborrini, X. Liu, J. P. Mugasa, Y.-U. Kwon, F. Kamena, P. H. Seeberger and G. Pluschke, *Bioorg. Med. Chem.*, 2010, **18**, 3747–3752.

97 L. G. Harris, W. C. E. Schofield, K. J. Doores, B. G. Davis and J. P. S. Badyal, *J. Am. Chem. Soc.*, 2009, **131**, 7755–7761.

98 E. A. Smith, W. D. Thomas, L. L. Kiessling and R. M. Corn, *J. Am. Chem. Soc.*, 2003, **125**, 6140–6148.

99 O. Blixt, S. Head, T. Mondala, C. Scanlan, M. E. Huflejt, R. Alvarez, M. C. Bryan, F. Fazio, D. Calarese, J. Stevens, N. Razi, D. J. Stevens, J. J. Skehel, I. van Die, D. R. Burton, I. A. Wilson, R. Cummings, N. Bovin, C.-H. Wong and J. C. Paulson, *Proc. Natl. Acad. Sci. U. S. A.*, 2004, **101**, 17033–17038.

100 X. Song, Y. Lasanajak, B. Xia, J. Heimburg-Molinaro, J. M. Rhea, H. Ju, C. Zhao, R. J. Molinaro, R. D. Cummings and D. F. Smith, *Nat. Methods*, 2011, **8**, 85–90.

101 C. L. Pereira, A. Geissner, C. Anish and P. H. Seeberger, *Angew. Chem., Int. Ed.*, 2015, **54**, 10016–10019.

102 H.-Y. Lee, C.-Y. Chen, T.-I. Tsai, S.-T. Li, K.-H. Lin, Y.-Y. Cheng, C.-T. Ren, T.-J. R. Cheng, C.-Y. Wu and C.-H. Wong, *J. Am. Chem. Soc.*, 2014, **136**, 16844–16853.

103 O. Bohorov, H. Andersson-Sand, J. Hoffmann and O. Blixt, *Glycobiology*, 2006, **16**, 21C–27C.

104 M. Schwarz, L. Spector, A. Gargir, A. Shtevi, M. Gortler, R. T. Altstock, A. A. Dukler and N. Dotan, *Glycobiology*, 2003, **13**, 749–754.

105 A. Gargir, A. Shtevi, E. Fire, L. Nimrichter, M. Gortler, N. Dotan, O. Weisshaus, R. T. Altstock and R. L. Schnaar, *Glycobiology*, 2004, **14**, 197–203.

106 M.-r. Lee and I. Shin, *Org. Lett.*, 2005, **7**, 4269–4272.

107 A. Reinhardt, Y. Yang, H. Claus, C. L. Pereira, A. D. Cox, U. Vogel, C. Anish and P. H. Seeberger, *Chem. Biol.*, 2015, **22**, 38–49.

108 S. E. Tully, M. Rawat and L. C. Hsieh-Wilson, *J. Am. Chem. Soc.*, 2006, **128**, 7740–7741.

109 J. L. de Paz, D. Spillmann and P. H. Seeberger, *Chem. Commun.*, 2006, 3116–3118.

110 M.-r. Lee and I. Shin, *Angew. Chem., Int. Ed.*, 2005, **44**, 2881–2884.

111 S. Park and I. Shin, *Org. Lett.*, 2007, **9**, 1675–1678.

112 S. Park, M.-R. Lee and I. Shin, *Nat. Protoc.*, 2007, **2**, 2747–2758.

113 X. Tian, J. Pai and I. Shin, *Chem.–Asian J.*, 2012, **7**, 2052–2060.

114 S. Park, M.-R. Lee and I. Shin, in *Small Molecule Microarrays: Methods and Protocols*, ed. M. Uttamchandani and S. Q. Yao, Humana Press, Totowa, NJ, 2010, vol. 669, pp. 195–208.

115 M.-R. Lee, S. Park and I. Shin, in *Carbohydrate Microarrays: Methods and Protocols*, ed. Y. Chevolot, Humana Press, Totowa, NJ, 2012, vol. 808, pp. 103–116.

116 S. Götze, N. Azzouz, Y.-H. Tsai, U. Groß, A. Reinhardt, C. Anish, P. H. Seeberger and D. Varón Silva, *Angew. Chem., Int. Ed.*, 2014, **53**, 13701–13705.

117 S. Götze, A. Reinhardt, A. Geissner, N. Azzouz, Y.-H. Tsai, R. Kurucz, D. Varón Silva and P. H. Seeberger, *Glycobiology*, 2015, **25**, 984–991.

118 B. T. Houseman and M. Mrksich, *Chem. Biol.*, 2002, **9**, 443–454.

119 H. S. G. Beckmann, A. Niederwieser, M. Wiessler and V. Wittmann, *Chem.–Eur. J.*, 2012, **18**, 6548–6554.

120 X.-L. Sun, C. L. Stabler, C. S. Cazalis and E. L. Chaikof, *Bioconjugate Chem.*, 2006, **17**, 52–57.

121 C.-Y. Huang, D. A. Thayer, A. Y. Chang, M. D. Best, J. Hoffmann, S. Head and C.-H. Wong, *Proc. Natl. Acad. Sci. U. S. A.*, 2006, **103**, 15–20.

122 O. Michel and B. J. Ravoo, *Langmuir*, 2008, **24**, 12116–12118.

123 O. J. Barrett, A. Pushechnikov, M. Wu and M. D. Disney, *Carbohydr. Res.*, 2008, **343**, 2924–2931.

124 K. Godula, D. Rabuka, K. T. Nam and C. R. Bertozzi, *Angew. Chem., Int. Ed.*, 2009, **48**, 4973–4976.

125 M. Köhn, R. Wacker, C. Peters, H. Schröder, L. Soulère, R. Breinbauer, C. M. Niemeyer and H. Waldmann, *Angew. Chem., Int. Ed.*, 2003, **42**, 5830–5834.

126 A. Tyagi, X. Wang, L. Deng, O. Ramström and M. Yan, *Biosens. Bioelectron.*, 2010, **26**, 344–350.

127 K. J. Doores, D. P. Gamblin and B. G. Davis, *Chem.–Eur. J.*, 2006, **12**, 656–665.

128 A. Imberty, Y. M. Chabre and R. Roy, *Chem.–Eur. J.*, 2008, **14**, 7490–7499.

129 P. I. Kitov, J. M. Sadowska, G. Mulvey, G. D. Armstrong, H. Ling, N. S. Pannu, R. J. Read and D. R. Bundle, *Nature*, 2000, **403**, 669–672.

130 C.-H. Liang, S.-K. Wang, C.-W. Lin, C.-C. Wang, C.-H. Wong and C.-Y. Wu, *Angew. Chem.*, 2011, **123**, 1646–1650.

131 D. Valles, Y. Naeem, A. Rozenfeld, R. Aldasooky, A. Wong, C. Carbonell, D. R. Mootoo and A. Braunschweig, *Faraday Discuss.*, 2019, DOI: 10.1039/c9fd00028c.

132 R. J. Payne and C.-H. Wong, *Chem. Commun.*, 2010, **46**, 21–43.

133 P. M. Rendle, A. Seger, J. Rodrigues, N. J. Oldham, R. R. Bott, J. B. Jones, M. M. Cowan and B. G. Davis, *J. Am. Chem. Soc.*, 2004, **126**, 4750–4751.

134 I. Otsuka, B. Blanchard, R. Borsali, A. Imberty and T. Kakuchi, *ChemBioChem*, 2010, **11**, 2399–2408.

135 D. Ponader, F. Wojcik, F. Beceren-Braun, J. Dernedde and L. Hartmann, *Biomacromolecules*, 2012, **13**, 1845–1852.

136 J. Rieger, F. Stoffelbach, D. Cui, A. Imberty, E. Lameignere, J.-L. Putaux, R. Jérôme, C. Jérôme and R. Auzély-Velty, *Biomacromolecules*, 2007, **8**, 2717–2725.

137 L. Baldini, A. Casnati, F. Sansone and R. Ungaro, *Chem. Soc. Rev.*, 2007, **36**, 254–266.

138 S. Cecioni, R. Lalor, B. Blanchard, J.-P. Praly, A. Imberty, S. E. Matthews and S. Vidal, *Chem.–Eur. J.*, 2009, **15**, 13232–13240.

139 A. Dondoni and A. Marra, *Chem. Rev.*, 2010, **110**, 4949–4977.

140 S. André, R. J. Pieters, I. Vrasidas, H. Kaltner, I. Kuwabara, F.-T. Liu, R. M. J. Liskamp and H.-J. Gabius, *ChemBioChem*, 2001, **2**, 822–830.

141 M. A. Mintzer, E. L. Dane, G. A. O'Toole and M. W. Grinstaff, *Mol. Pharmaceutics*, 2012, **9**, 342–354.

142 C. D. Heidecke and T. K. Lindhorst, *Chem.–Eur. J.*, 2007, **13**, 9056–9067.

143 D. A. Fulton and J. F. Stoddart, *Bioconjugate Chem.*, 2001, **12**, 655–672.

144 A. Bernardi, J. Jiménez-Barbero, A. Casnati, C. De Castro, T. Darbre, F. Fieschi, J. Finne, H. Funken, K.-E. Jaeger, M. Lahmann, T. K. Lindhorst, M. Marradi, P. Messner, A. Molinaro, P. V. Murphy, C. Nativi, S. Oscarson, S. Penadés, F. Peri, R. J. Pieters, O. Renaudet, J.-L. Reymond, B. Richichi, J. Rojo, F. Sansone, C. Schäffer, W. B. Turnbull, T. Velasco-Torrijos, S. Vidal, S. Vincent, T. Wennekes, H. Zuilhof and A. Imberty, *Chem. Soc. Rev.*, 2013, **42**, 4709–4727.

145 M. Kilcoyne, J. Q. Gerlach, R. Gough, M. E. Gallagher, M. Kane, S. D. Carrington and L. Joshi, *Anal. Chem.*, 2012, **84**, 3330–3338.

146 A. R. de Boer, C. H. Hokke, A. M. Deelder and M. Wuhrer, *Glycoconjugate J.*, 2008, **25**, 75–84.

147 E. W. Adams, J. Ueberfeld, D. M. Ratner, B. R. O'Keefe, D. R. Walt and P. H. Seeberger, *Angew. Chem., Int. Ed.*, 2003, **42**, 5317–5320.

148 J. C. Manimala, Z. Li, A. Jain, S. VedBrat and J. C. Gildersleeve, *ChemBioChem*, 2005, **6**, 2229–2241.

149 Y. Zhang, C. Campbell, Q. Li and J. C. Gildersleeve, *Mol. BioSyst.*, 2010, **6**, 1583–1591.

150 F. Morvan, S. Vidal, E. Souteyrand, Y. Chevolot and J.-J. Vasseur, *RSC Adv.*, 2012, **2**, 12043–12068.

151 N. Spinelli, E. Defrancq and F. Morvan, *Chem. Soc. Rev.*, 2013, **42**, 4557–4573.

152 V. Wittmann and R. J. Pieters, *Chem. Soc. Rev.*, 2013, **42**, 4492–4503.

153 Z. L. Pianowski and N. Winssinger, *Chem. Soc. Rev.*, 2008, **37**, 1330–1336.

154 C. Scheibe, A. Bujotzek, J. Dernedde, M. Weber and O. Seitz, *Chem. Sci.*, 2011, **2**, 770–775.

155 K.-T. Huang, K. Gorska, S. Alvarez, S. Barluenga and N. Winssinger, *ChemBioChem*, 2011, **12**, 56–60.

156 K. Gorska, K.-T. Huang, O. Chaloin and N. Winssinger, *Angew. Chem., Int. Ed.*, 2009, **48**, 7695–7700.

157 A. Novoa, T. Machida, S. Barluenga, A. Imberty and N. Winssinger, *ChemBioChem*, 2014, **15**, 2058–2065.

158 X. Zhou, C. Turchi and D. Wang, *J. Proteome Res.*, 2009, **8**, 5031–5040.

159 H. M. Branderhorst, R. Ruijtenbeek, R. M. J. Liskamp and R. J. Pieters, *ChemBioChem*, 2008, **9**, 1836–1844.

160 T. Fukuda, S. Onogi and Y. Miura, *Thin Solid Films*, 2009, **518**, 880–888.

161 N. Parera Pera, H. M. Branderhorst, R. Kooij, C. Maierhofer, M. van der Kaaden, R. M. J. Liskamp, V. Wittmann, R. Ruijtenbeek and R. J. Pieters, *ChemBioChem*, 2010, **11**, 1896–1904.

162 V. Böhmer, *Angew. Chem., Int. Ed. Engl.*, 1995, **34**, 713–745.

163 A. Ikeda and S. Shinkai, *Chem. Rev.*, 1997, **97**, 1713–1734.

164 L. Moni, G. Pourceau, J. Zhang, A. Meyer, S. Vidal, E. Souteyrand, A. Dondoni, F. Morvan, Y. Chevolot, J.-J. Vasseur and A. Marra, *ChemBioChem*, 2009, **10**, 1369–1378.

165 B. Gerland, A. Goudot, G. Pourceau, A. Meyer, V. Dugas, S. Cecioni, S. Vidal, E. Souteyrand, J.-J. Vasseur, Y. Chevolot and F. Morvan, *Bioconjugate Chem.*, 2012, **23**, 1534–1547.

166 K. Godula and C. R. Bertozzi, *J. Am. Chem. Soc.*, 2012, **134**, 15732–15742.

167 H. S. Kim, J. Y. Hyun, S.-H. Park and I. Shin, *RSC Adv.*, 2018, **8**, 14898–14905.

168 Y. Fei, Y.-S. Sun, Y. Li, K. Lau, H. Yu, H. A. Chokhawala, S. Huang, J. P. Landry, X. Chen and X. Zhu, *Mol. BioSyst.*, 2011, **7**, 3343–3352.

169 W. Guan, L. Ban, L. Cai, L. Li, W. Chen, X. Liu, M. Mrksich and P. G. Wang, *Bioorg. Med. Chem. Lett.*, 2011, **21**, 5025–5028.

170 A. Sanchez-Ruiz, S. Serna, N. Ruiz, M. Martin-Lomas and N.-C. Reichardt, *Angew. Chem., Int. Ed.*, 2011, **50**, 1801–1804.

171 A. Beloqui, A. Sanchez-Ruiz, M. Martin-Lomas and N.-C. Reichardt, *Chem. Commun.*, 2012, **48**, 1701–1703.

172 T. R. Northen, J.-C. Lee, L. Hoang, J. Raymond, D.-R. Hwang, S. M. Yannone, C.-H. Wong and G. Siuzdak, *Proc. Natl. Acad. Sci. U. S. A.*, 2008, **105**, 3678–3683.

173 S. Tao, T.-W. Jia, Y. Yang and L.-Q. Chu, *ACS Sens.*, 2017, **2**, 57–60.

174 M. Fais, R. Karamanska, S. Allman, S. A. Fairhurst, P. Innocenti, A. J. Fairbanks, T. J. Donohoe, B. G. Davis, D. A. Russell and R. A. Field, *Chem. Sci.*, 2011, **2**, 1952–1959.

175 M. Dhayal and D. M. Ratner, *Langmuir*, 2009, **25**, 2181–2187.

176 H. Tateno, A. Mori, N. Uchiyama, R. Yabe, J. Iwaki, T. Shikanai, T. Angata, H. Narimatsu and J. Hirabayashi, *Glycobiology*, 2008, **18**, 789–798.

177 K. Takahara, T. Arita, S. Tokieda, N. Shibata, Y. Okawa, H. Tateno, J. Hirabayashi and K. Inaba, *Infect. Immun.*, 2012, **80**, 1699–1706.

178 A. Kuno, N. Uchiyama, S. Koseki-Kuno, Y. Ebe, S. Takashima, M. Yamada and J. Hirabayashi, *Nat. Methods*, 2005, **2**, 851–856.

179 Y. Fei, Y.-S. Sun, Y. Li, H. Yu, K. Lau, J. Landry, Z. Luo, N. Baumgarth, X. Chen and X. Zhu, *Biomolecules*, 2015, **5**, 1480–1498.

180 E. Han, L. Ding, S. Jin and H. Ju, *Biosens. Bioelectron.*, 2011, **26**, 2500–2505.

181 M. Shipp, R. Nadella, H. Gao, V. Farkas, H. Sigrist and A. Faik, *Glycoconjugate J.*, 2008, **25**, 49–58.

182 H. L. Pedersen, J. U. Fangel, B. McCleary, C. Ruzanski, M. G. Rydahl, M.-C. Ralet, V. Farkas, L. von Schantz, S. E. Marcus, M. C. F. Andersen, R. Field, M. Ohlin, J. P. Knox, M. H. Clausen and W. G. T. Willats, *J. Biol. Chem.*, 2012, **287**, 39429–39438.

183 Y. Y. Fei, A. Schmidt, G. Bylund, D. X. Johansson, S. Henriksson, C. Lebrilla, J. V. Solnick, T. Borén and X. D. Zhu, *Anal. Chem.*, 2011, **83**, 6336–6341.

184 H. Tateno, S. Nakamura-Tsuruta and J. Hirabayashi, *Nat. Protoc.*, 2007, **2**, 2529–2537.

185 J. Iwaki and J. Hirabayashi, *Trends Glycosci. Glycotechnol.*, 2018, **30**, SE137–SE153.

186 Y. Takeda and I. Matsuo, in *Lectins: Methods and Protocols*, ed. J. Hirabayashi, Springer New York, New York, NY, 2014, vol. 1200, pp. 207–214.

187 K. Gruber, T. Horlacher, R. Castelli, A. Mader, P. H. Seeberger and B. A. Hermann, *ACS Nano*, 2011, **5**, 3670–3678.

188 A. Geissner and P. H. Seeberger, *Annu. Rev. Anal. Chem.*, 2016, **9**, 223–247.

189 C. Kielar, F. V. Reddavide, S. Tubbenhauer, M. Cui, X. Xu, G. Grundmeier, Y. Zhang and A. Keller, *Angew. Chem., Int. Ed.*, 2018, **57**, 14873–14877.

190 W. Hawkes, D. Huang, P. Reynolds, L. Hammond, M. Ward, N. Gadegaard, J. F. Marshall, T. Iskratch and M. Palma, *Faraday Discuss.*, 2019, DOI: 10.1039/c9fd00023b.

191 G. Paris, J. Heidepriem, A. Tsouka, M. Mende, S. Eickelmann and F. F. Loeffler, Proc. SPIE 10875, *Microfluidics, BioMEMS, and Medical Microsystems XVII*, 2019, vol. 10875, p. 108750C.

192 A. Palermo, L. K. Weber, S. Rentschler, A. Isse, M. Sedlmayr, K. Herbster, V. List, J. Hubbuch, F. F. Loffler, A. Nesterov-Muller and F. Breitling, *Biotechnol. J.*, 2017, **12**, 1700197.

193 L. K. Weber, A. Isse, S. Rentschler, R. E. Kneusel, A. Palermo, J. Hubbuch, A. Nesterov-Mueller, F. Breitling and F. F. Loeffler, *Eng. Life Sci.*, 2017, **17**, 1078–1087.

194 L. K. Weber, A. Palermo, J. Kugler, O. Armant, A. Isse, S. Rentschler, T. Jaenisch, J. Hubbuch, S. Dubel, A. Nesterov-Mueller, F. Breitling and F. F. Loeffler, *J. Immunol. Methods*, 2017, **443**, 45–54.

195 M. C. L. C. Freire, L. Pol-Fachin, D. F. Coelho, I. F. T. Viana, T. Magalhaes, M. T. Cordeiro, N. Fischer, F. F. Loeffler, T. Jaenisch, R. F. Franca, E. T. A. Marques and R. D. Lins, *ACS Omega*, 2017, **2**, 3913–3920.

196 F. F. Loeffler, J. Pfeil and K. Heiss, *Methods Mol. Biol.*, 2016, **1403**, 569–582.

197 T. Jaenisch, K. Heiss, N. Fischer, C. Geiger, F. R. Bischoff, G. Moldenhauer, L. Rychlewski, A. Sie, B. Coulibaly, P. H. Seeberger, L. S. Wyrwicz, F. Breitling and F. F. Loeffler, *Mol. Cell. Proteomics*, 2019, **18**, 642.

Faraday Discussions

PAPER

High-throughput protein nanopatterning†

Xiangyu Liu, [ID] [a] Mohit Kumar, [ID] [b] Annalisa Calo', [ID] [a]
Edoardo Albisetti, [ac] Xiaouri Zheng, [a] Kylie B. Manning, [d]
Elisabeth Elacqua, [ID] [d] Marcus Weck, [ID] [d] Rein V. Ulijn [b]
and Elisa Riedo [ID] *[a]

Received 2nd March 2019, Accepted 11th March 2019

DOI: 10.1039/c9fd00025a

High-throughput and large-scale patterning of enzymes with sub-10 nm resolution, the size range of individual protein molecules, is crucial for propelling advancement in a variety of areas, from the development of chip-based biomolecular nano-devices to molecular-level studies of cell biology. Despite recent developments in bio-nanofabrication technology, combining 10 nm resolution with high-throughput and large-scale patterning of enzymes is still an open challenge. Here, we demonstrate a high resolution and high-throughput patterning method to generate enzyme nanopatterns with sub-10 nm resolution by using thermochemical scanning probe lithography (tc-SPL). First, tc-SPL is used to generate amine patterns on a methacrylate copolymer film. Thermolysin enzymes functionalized with sulfonate-containing fluorescent labels (Alexa-488) are then directly immobilized onto the amine patterns through electrostatic interaction. Enzyme patterns with sub-10 nm line width are obtained as evidenced by atomic force microscopy (AFM) and fluorescence microscopy. Moreover, we demonstrate large-scale and high throughput (0.13 × 0.1 mm^2 at a throughput of 5.2 × 10^4 μm^2 h^{-1}) patterning of enzymes incorporating 10 nm detailed pattern features. This straightforward and high-throughput method of fabricating enzyme nanopatterns will have a significant impact on future bio-nanotechnology applications and molecular-level biological studies. By scaling up using parallel probes, tc-SPL is promising for implementation to scale up the fabrication of nano-biodevices.

1. Introduction

Enzymes are a class of biomolecules which are known to catalyze a variety of biochemical reactions with great specificity under mild conditions[1] and have

[a]Tandon School of Engineering, New York University, Brooklyn, NY, USA

[b]Advanced Science Research Center (ASRC), CUNY Graduate Center, New York, NY, USA

[c]Dipartimento di Fisica, Politecnico di Milano, Milano, Italy

[d]Department of Chemistry, New York University, New York, NY, USA. E-mail: elisa.riedo@nyu.edu

† Electronic supplementary information (ESI) available. See DOI: 10.1039/c9fd00025a

been exploited for decades in various fields, such as biological studies, biomedicine, bio-nanotechnology and in the food industry. Enzyme immobilization on surface patterns with predefined configurations and spatial control has garnered significant interest and shows significant importance in many biological and biotechnological applications, such as biosensors,[2,3] enzyme-assisted lithography,[4] bio-nanoreactors,[5,6] as well as in the systematic studies of biomolecule interactions.[7] Enzyme immobilized on solid supports also constitute functional surfaces which may exhibit the advantages of improved enzyme stability and reusability after multiple exposures to reaction mixtures.

To fully realize the potential of enzymes for applications in bio-nanotechnology, the ability to generate enzyme surface patterns at the nanoscale is critical. Since miniaturized size and greater detection sensitivity are desired for future biosensors, highly defined enzyme nanopatterns and single-enzyme positioning capability could significantly facilitate the fabrication of biosensors with densely packed arrays.[4,8,9] Also, with the capability of catalyzing the degradation of certain macromolecules, enzyme nanopatterns can also be used as a lithography tools to generate templates for nanoreactors, photonic structures and electrical circuits.[8,10,11] Furthermore, patterning enzymes in quasi 2D, soft polymer landscapes and using feature sizes that are in the same (<10 nm) range as these enzymes, is advantageous as it allows a more precise engineering of the enzymes' microenvironment, potentially allowing for this environment to be designed to more closely mimic the biological context and thereby to enhance enzyme function and stability.

Fabrication of enzyme surface patterns relies on advanced patterning techniques which were initially developed for the semiconductor industry and now have been translated to a variety of fields, such as biomedicine and bio-nanotechnology. These patterns have been successfully generated using a variety of techniques, including micro-contacting printing,[11-13] inkjet printing[14] and optical lithography[15-17] which can readily generate large-scale patterns but lack the capability of patterning at the nanoscale. Electron beam lithography (EBL),[18-21] dip pen nanolithography (DPN),[10,22,23] nanoimprint lithography (NIL),[24] scanning probe lithography (SPL)[25-28] and bottom-up self-assembly[29-32] can generate patterns at the nanometer scale and several works report enzyme patterns with sub-100 nm resolution using these methods. However, it is still challenging to fabricate enzyme patterns with sub-10 nm resolution using a high-throughput top-down method in a facile and effective manner, avoiding complex and multi-step fabrications. Only a few works[22,29] reported enzyme immobilization onto patterns with 10 nm resolution. For example, Chai et al. reported the only top-down approach that generated an array of 10 nm Au nanoparticles with avidin–horseradish peroxidase attached using DPN.[22] However, multiple steps of fabrication are required to produce the 10 nm Au dots and protein immobilization is needed to finally immobilize the enzymes. As a bottom-up approach, self-assembly has been utilized to achieve enzyme patterns of sub-10 nm resolution.[29] However, the lack of specific control over enzyme density, pattern configuration and pattern topography hinders its further implementation.[33] Additionally, a direct proof of enzyme immobilization on the patterns by AFM or fluorescence microscopy is missing. Moreover, for future industrial-scale applications, one of the most important criteria is large-scale manufacturing using high-throughput and cost-effective fabrication processes. Therefore, to fulfil the potential of

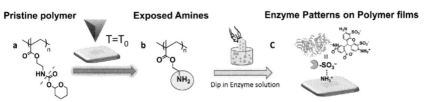

Fig. 1 Schematic illustration of the general process of generating enzyme patterns by tc-SPL. (a and b) Generating amine patterns by local deprotection of the amine groups using a heated probe. (b and c) Enzyme immobilization onto amine patterns through electro-static interactions.

enzyme nanopatterns for large-scale bio-nanotechnology applications, it is important to develop a fabrication process that can achieve high-throughput and large-scale enzyme patterning together with ultrahigh resolution.

In our previous works,[34–37] protein and DNA patterns were demonstrated by using a home-built thermochemical SPL (tc-SPL) system, where the heated scanning probe induced local amine deprotection on a synthesized methacrylate copolymer film followed by the immobilization of the proteins and DNAs through appropriate linker chemistry. Although those works demonstrated the great versatility of this method to pattern biomolecules, the smallest object achieved was only around 30 nm in the lateral dimension. Also, they required intermediate linkers to bridge the proteins or DNAs to the amine patterns with extra steps and increased cost. Furthermore, due to the limited throughput, no large-scale patterning was demonstrated.

Here, we report a SPL method for generating robust enzyme nanopatterns with sub-10 nm resolution and with large-scale and high-throughput patterning capability $(0.13 \times 0.1 \text{ mm}^2$ at a throughput of $5.2 \times 10^4 \text{ } \mu\text{m}^2 \text{ h}^{-1})$. This is enabled by using a tc-SPL system (see the Experimental section) featuring simultaneous patterning and *in situ* imaging, and the immobilization of enzymes onto the amine patterns through electrostatic interactions between negatively charged enzymes and positively charged amine patterns (Fig. 1c). This electrostatic immobilization approach greatly simplifies the fabrication process and opens up a door for generating nanopatterns of charged biomolecules using our fabrication method. This method of generating high-throughput and high-resolution enzyme nanopatterns may significantly impact the future fabrication of highly sensitive biosensors and the studies of cell behavior on bio-surfaces.

2. Experimental

Materials

The polymer used in this work is a non-commercial diblock polymethacrylate copolymer, poly((tetrahydropyran-2-yl-*N*-(2-methacryloxyethyl)carbamate)-*co*-(methyl 4-(3-methacryloyloxy-propoxy)cinnamate)) with 1 : 4 ratio of cinnamate to carbamate, which was synthesized using uncontrolled free-radical polymerization according to previous reports.[38] The polymer contains thermally labile tetrahy-dropyran (THP) carbamate groups. It deprotects and exposes the primary amines upon heating at the deprotection temperature $(T_0 \sim 150 \text{ °C})$.[34,35,37] The fluores-cence fluorophore, Alexa-488 NHS ester was purchased from ThermoFisher and dissolved in dimethyl sulfoxide (DMSO) to obtain a 2 mM stock solution. It was further diluted by water to 100 nM for incubating patterned films. The enzyme, thermolysin, was purchased from *Bacillus thermoproteolytics Rokko* and conjugated with Alexa-488 NHS ester fluorophore for enzyme immobilization and optical detection.

Polymer film preparation

The polymers were dissolved in chloroform to obtain solution with the concentration of 2.5 mg mL^{-1} followed by the filtration with 0.45 μm PTFE filters (Thermofisher). The polymer solution was spin coated onto Si substrate (1500 rpm, 60 s) followed by a quick baking (50 °C, 1 min) to remove the residual solvent. The resulting film thickness ranges from 20 to 30 nm.

Synthesis of Alexa-488 thermolysin conjugate

4 mg of thermolysin was dissolved in 3 mL water (pH adjusted to 8) followed by adding 60 μL Alexa-NHS ester solution with a concentration of 2 mM in DMSO. The solution was then shaken for two hours. After dialysis of the solution to wash off unreacted dye (using a dialysis membrane with a molecular weight cut-off of 3500 Dalton), the remaining solution was lyophilized to obtain Alexa-488 conjugated thermolysin (Alexa-thermolysin). The product formation was confirmed by mass-spectrometry (MALDI).

tc-SPL and pattern fabrication

Thermal patterning of the polymer films was conducted in ambient environment using a commercial tc-SPL system (NanoFrazor, SwissLitho AG, Switzerland). It is equipped with a silicon thermal cantilever comprising a resistive microheater for thermal patterning and a separate resistor used as a thermal sensor for topography imaging. The temperature of the microheater on the cantilever is calibrated by the NanoFrazor software by fitting the knee point in the current–voltage curve with the theoretical values for doped silicon.[39] During the patterning, the cantilever is in contact with the sample surface. The cantilever is bent down by the electrostatic force between the cantilever and Si substrate, and the bimorph bending due to the temperature gradient in the cantilever.[40] Another major feature of this lithography system is the capability of simultaneous patterning and *in situ* imaging. After each patterning line, the thermal reading sensor immediately maps the topography of the patterned structure when retracing back in contact mode. This also enables closed feedback loop correction for 3D patterning.

In this work, patterning was performed with a pixel dwell time of 140 μs and a pixel pitch of 30 nm for patterns in Fig. 2–4. For the NYU Tandon pattern in Fig. 5, pixel dwell time and pixel pitch were 50 μs and 50 nm (1 mm s^{-1}), respectively. In detail, the NanoFrazor system was operated in a pulsed heating mode by which the probe was only heated shortly before and in contact with samples. This operating mode extends its lifetime. Typical settings of the heating pulse and electrostatic force pulse used in this work were 120 μs and 100 μs, respectively (for pattern in Fig. 5, 30 μs and 15 μs were used). The patterning temperature was typical above the amine deprotection temperature ($T_0 \sim 150$ °C). It should be noted that the patterning temperature at the probe–sample interface cannot be directly measured due to a reduced heat transport from the micro-heater to the sample surface. To estimate the contact temperature at the probe–sample interface, a heating efficiency of 0.24 between the temperature of the microheater and the probe–sample interface was derived and shown to follow the equation $T_{contact} = 0.24(T_{heater} - T_{RT}) + T_{RT}$.[41–43]

Fluorophore and enzyme immobilization

Alexa 488 was immobilized onto the amine patterns by incubating the polymer film with 100 nM Alexa 488 NHS ester water solution for 45 minutes. After incubation, the surface was rinsed with Milli-Q water (ultrapure and filtered) and phosphate buffer solution (PBS), and then dried with N_2. The thermolysin enzyme was immobilized by incubating the polymer film in 30 μL of 1 mg mL^{-1} Alexa 488-thermolysin conjugate solution in water for one hour. The above-mentioned washing and drying procedures were performed.

Pattern characterization

Fluorophore-labelled patterns were imaged using laser scanning confocal fluorescence microscopy (Zeiss LSM 880 Airyscan) with excitation laser wavelength $\lambda_{exc} = 488$ nm and emission detection window $\lambda_{em} = 510–600$ nm. The fluorescence images were processed using ImageJ and the optical contrast was tuned for better visibility. All AFM measurements were conducted on a Bruker MultiMode 8 AFM. Gwyddion software was used to flatten and adjust the z-scale of the AFM images.

3.   Results & discussion

Amine patterns and enzyme immobilization

The critical step for achieving enzyme patterns is to fabricate active amine patterns. As reported in our previous works,[34,35,37] the heated probe induces local deprotection of amines on the surface of the polymer films. Fig. 2a shows several square patterns generated by the tc-SPL system with different temperatures (141 °C to 191 °C). To confirm the deprotection of primary amines on the patterned surface, a green-emitting fluorophore, Alexa-488 NHS ester, was used to label these amines. From Fig. 2b, the enhanced emission intensity from the square patterns confirms that amine groups are exposed on the patterned surface.

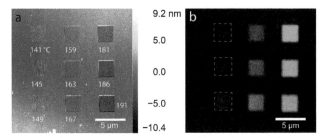

Fig. 2 Amine patterns generated by tc-SPL. (a) AFM topography image of square patterns using different temperatures. (b) Corresponding fluorescence image of the amine patterns labelled by Alexa-488 NHS ester.

It is shown that different emission intensities are obtained from different squares. This is due to the fact that the amine deprotection is a thermally-driven reaction and therefore the final amine concentration strongly depends on the patterning temperature. The patterns shown here are only used to confirm the presence of amines. Detailed discussion on the temperature dependence of the amine deprotection can be found in our previous work.[37]

Following from the successful fabrication of active amine patterns, we seek to immobilize enzymes onto amine patterns through a simple and effective approach. This is based on the specific electrostatic interactions between positively charged amines and negatively charged Alexa-488 sulfonate groups. Thus, thermolysin enzyme was modified by covalently conjugating with Alexa-488 which contains sulfonate groups (see the Experimental section) such that it can strongly adhere onto positively charged amine surfaces created by tc-SPL. In this way, the enzyme patterns can be also easily imaged by using fluorescence microscopy. To test our hypothesis, both AFM and fluorescence microscopy were used to characterize the patterns after incubation with thermolysin enzyme solution, and the results are shown in Fig. 3. Here, an array of $1 \times 1 \ \mu m^2$ square-patterns were generated by tc-SPL with a temperature above the amine deprotection temperature. The AFM topography images of a single square pattern before and after incubation (Fig. 3a and b) show that after incubation, the pattern is filled up with the immobilized proteins. The fluorescence image of this array of square patterns shows the enhanced emission intensity from the square patterns (Fig. 3c), which confirms the selective immobilization of Alexa-thermolysin molecules onto the patterns. The cross-sectional profiles (Fig. 3d) shows that the height difference of

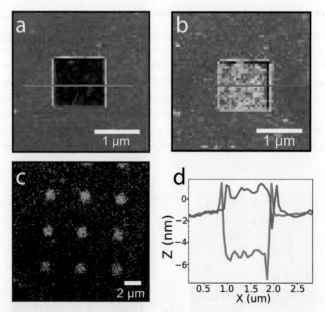

Fig. 3 $1 \times 1 \ \mu m^2$ square amine patterns filled with electrostatically immobilized thermolysin. (a and b) AFM topography images collected before (a) and after (b) thermolysin immobilization (z-scale is 16 nm). (c) Confocal fluorescence images of an array of square patterns. (d) Cross-sectional profiles from green line in (a) and red line in (b).

the pattern before and after enzyme immobilization is around 6 nm. This indicates the immobilization of a single layer of thermolysin enzyme molecules onto the pattern (thermolysin enzyme size ~ 5.75 nm, see ESI† for the size characterization of the thermolysin molecules).

Sub-10 nm enzyme patterning with single-enzyme resolution

With thermolysin enzymes successfully immobilized onto the micro-scale patterns, we investigate the limit of the patterning resolution in x, y. Since features smaller than 100 nm cannot be easily imaged by fluorescence microscopy, in order to probe and visualize the enzyme immobilization onto such small features, we patterned a 1 × 5 μm² rectangular area consisting of multiple single-line patterns with amines exposed on the surface (Fig. 4a). From the cross-sectional profile (Fig. 4c, purple curve), it is shown that each individual line has a full width at half maximum (FWHM) of 9 nm comparable to the size of a single enzyme molecule. This ultrahigh resolution benefits from the sharp temperature gradient produced by the ultra-sharp tip integrated within this lithography system.[44] After enzyme immobilization, similarly, those patterned lines are filled up by the enzyme molecules (Fig. 4b and red cross-sectional profile in Fig. 4c). From the fluorescence emission of this area (Fig. 4d), the selective immobilization of enzymes onto these single-line patterns is clearly shown, which also confirms the patterning of single-enzyme lines. Using this method, sub-10 nm enzyme patterning with remarkable control of pattern position and configuration can be readily achieved, which may greatly benefit the scaling of future biosensor and bio-nanoreactors.

Combination of large-scale patterning of enzymes and high resolution

After demonstrating the smallest feature of enzyme patterning by this new tc-SPL system, we explore its capability for generating large-scale enzyme patterns. The

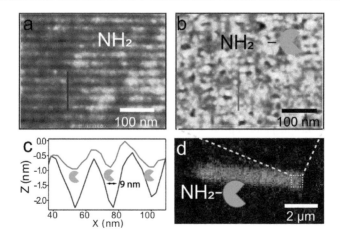

Fig. 4 Sub-10 nm thermolysin enzyme patterns. (a and b) AFM topography images of the single-line amine patterns before (a) and after (b) thermolysin immobilization (z-scale is 10 nm). (c) Cross-sectional profile from purple line in (a) and red line in (b), showing 9 nm FWHM. (d) Fluorescence image of the entire 1 × 5 μm² rectangular pattern consisting of closely-packed single lines with thermolysin enzyme immobilized.

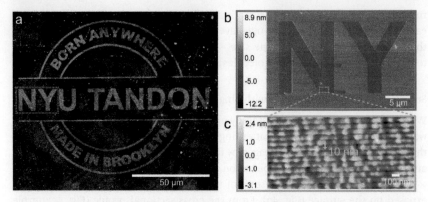

Fig. 5 Combination of high resolution and large-scale patterning of enzymes. (a) 0.13 × 0.1 mm^2 enzyme pattern of the NYU Tandon School of Engineering logo. (b) Thermal topography image of patterned "NY", as shown in panel (a), obtained during the simultaneous tc-SPL patterning and *in situ* thermal imaging. (c) AFM topography image of the zoomed-in area in (b), showing high resolution single-line patterning of 10 nm FWHM. Permission to use logo was granted by the New York University Trademark Licensing.

logo for NYU Tandon School of Engineering was selected for the large-scale enzyme patterning experiments. Fig. 5a shows the fluorescence image of a 0.13 × 0.1 mm^2 thermolysin enzyme pattern of the NYU Tandon logo generated in 15 min (5.2×10^4 µm^2 h^{-1}). This straightforward and large-scale patterning is enabled by the high-throughput capability of this system (up to 5.2×10^4 µm^2 h^{-1} per single probe). To date, this is the largest enzyme pattern achieved using scanning probe lithography techniques. Beside the high throughput, simultaneously *in situ* thermal imaging empowers this lithography system to perform real-time metrology and monitor the patterning process, which greatly enhances the fabrication efficiency. This is a significant advantage over other mainstream lithography techniques, such as optical lithography, EBL and DPN. Fig. 5b shows an example of *in situ* thermal topography image of "NY" obtained simultaneously during patterning. Furthermore, when we zoom in the patterned region, it is shown that the smallest component of this large logo pattern, *i.e.* a single patterning line, has a FWHM around 10 nm. Although this hasn't reached the resolution limit of this technique, it is still among the highest patterning resolutions of any other lithography technique. This demonstrates the capability of generating large-scale enzyme patterns with high -resolution detail. When scaled up by using multiple parallel probes, this method of patterning enzymes can have remarkable potential for fabricating industrial-scale biochips and will also facilitate the studies of cell behavior on bio-surfaces.

4. Conclusion

In this contribution, we present a high-throughput and high resolution patterning method to generate enzyme patterns using thermal scanning probe lithography. With amine patterns first generated by tc-SPL, thermolysin enzymes functionalized with a sulfonate group were directly immobilized onto positively-charged amine patterns under aqueous conditions through electrostatic interaction without any intermediate linker, improving greatly the efficiency of the fabrication process. We demonstrate that, by using our new tc-SPL system, sub-10 nm (single-enzyme level) enzyme patterns can be achieved. In comparison to the work achieving 10 nm enzyme patterns using top-down dip pen nanolithography,[22] our method is more cost-effective and has a significantly easier fabrication process. Together with the advantage over bottom-up self-assembly methods of controlling the spatial arrangement and configuration of the patterns, our method introduces a facile and highly effective way of generating arbitrary enzyme patterns with single-enzyme resolution (sub-10 nm). We also demonstrate the high-throughput ($5.2 \times 10^4 \ \mu m^2 \ h^{-1}$) and large-scale ($0.13 \times 0.1$ mm^2) patterning of enzymes combined with 10 nm pattern details using a single probe. This is shown to be very promising for future fabrication of industrial-scale biochips and provides an alternative method to generate bio-patterned surfaces for studies of cell behavior.

Conflicts of interest

There are no conflicts to declare.

Acknowledgements

The authors acknowledge support from the US Army Research Office (proposal number 69180-CH), the Office of Basic Energy Sciences of the US Department of Energy, and the National Science Foundation.

References

1 I. Schomburg, A. Chang, S. Placzek, C. Söhngen, M. Rother, M. Lang, C. Munaretto, S. Ulas, M. Stelzer and A. Grote, *Nucleic Acids Res.*, 2012, **41**, D764–D772.

2 A. Sassolas, L. J. Blum and B. D. Leca-Bouvier, *Biotechnol. Adv.*, 2012, **30**, 489–511.

3 Z. Li, X. Liu and Z. Zhang, *IEEE Trans. NanoBiosci.*, 2008, **7**, 194–199.

4 L. Riemenschneider, S. Blank and M. Radmacher, *Nano Lett.*, 2005, **5**, 1643–1646.

5 A. Küchler, M. Yoshimoto, S. Luginbühl, F. Mavelli and P. Walde, *Nat. Nanotechnol.*, 2016, **11**, 409.

6 R. De La Rica and H. Matsui, *Angew. Chem., Int. Ed.*, 2008, **47**, 5415–5417.

7 E. W. Miles, S. Rhee and D. R. Davies, *J. Biol. Chem.*, 1999, **274**, 12193–12196.

8 K. I. Fabijanic, R. Perez-Castillejos and H. Matsui, *J. Mater. Chem.*, 2011, **21**, 16877–16879.

9 K.-B. Lee, J.-H. Lim and C. A. Mirkin, *J. Am. Chem. Soc.*, 2003, **125**, 5588–5589.

10 H. Li, Q. He, X. Wang, G. Lu, C. Liusman, B. Li, F. Boey, S. S. Venkatraman and H. Zhang, *Small*, 2011, **7**, 226–229.

11 R. Kargl, T. Mohan, S. Köstler, S. Spirk, A. Doliška, K. Stana-Kleinschek and V. Ribitsch, *Adv. Funct. Mater.*, 2013, **23**, 308–315.

12 M. Buhl, B. Vonhören and B. J. Ravoo, *Bioconjugate Chem.*, 2015, **26**, 1017–1020.

13 A. Guyomard-Lack, N. Delorme, C. Moreau, J.-F. o. Bardeau and B. Cathala, *Langmuir*, 2011, **27**, 7629–7634.

14 E. Gdor, S. Shemesh, S. Magdassi and D. Mandler, *ACS Appl. Mater. Interfaces*, 2015, **7**, 17985–17992.

15 Y.-N. Lee, T. Horio, K. Okumura, T. Iwata, K. Takahashi, M. Ishida and K. Sawada, *Patterning an enzyme-membrane of bio-image sensor using lithography technique*, SENSORS 2015 IEEE, Busan, South Korea, 2015.

16 S. Marchesan, C. D. Easton, K. E. Styan, P. Leech, T. R. Gengenbach, J. S. Forsythe and P. G. Hartley, *Colloids Surf., B*, 2013, **108**, 313–321.

17 R. E. Alvarado, H. T. Nguyen, B. Pepin-Donat, C. Lombard, Y. Roupioz and L. Leroy, *Langmuir*, 2017, **33**, 10511–10516.

18 S. Kim, B. Marelli, M. A. Brenckle, A. N. Mitropoulos, E. S. Gil, K. Tsioris, H. Tao, D. L. Kaplan and F. G. Omenetto, *Nat. Nanotechnol.*, 2014, **9**, 306–310.

19 E. Bat, J. Lee, U. Y. Lau and H. D. Maynard, *Nat. Commun.*, 2015, **6**, 6654.

20 J. N. Lockhart, A. B. Hmelo and E. Harth, *Polym. Chem.*, 2018, **9**, 637–645.

21 R. J. Mancini, S. J. Paluck, E. Bat and H. D. Maynard, *Langmuir*, 2016, **32**, 4043–4051.

22 J. Chai, L. S. Wong, L. Giam and C. A. Mirkin, *Proc. Natl. Acad. Sci. U. S. A.*, 2011, **108**, 19521–19525.

23 Z. Xie, C. Chen, X. Zhou, T. Gao, D. Liu, Q. Miao and Z. Zheng, *ACS Appl. Mater. Interfaces*, 2014, **6**, 11955–11964.

24 S. Merino, A. Retolaza, V. Trabadelo, A. Cruz, P. Heredia, J. A. Alduncín, D. Mecerreyes, I. Fernández-Cuesta, X. Borrisé and F. Pérez-Murano, *J. Vac. Sci. Technol., B: Microelectron. Nanometer Struct.-Process., Meas., Phenom.*, 2009, **27**, 2439.

25 R. De La Rica and H. Matsui, *J. Am. Chem. Soc.*, 2009, **131**, 14180–14181.

26 J. Shi, J. Chen and P. S. Cremer, *J. Am. Chem. Soc.*, 2008, **130**, 2718–2719.

27 R. V. Martinez, J. Martinez, M. Chiesa, R. Garcia, E. Coronado, E. Pinilla-Cienfuegos and S. Tatay, *Adv. Mater.*, 2010, **22**, 588–591.

28 R. Garcia, A. W. Knoll and E. Riedo, *Nat. Nanotechnol.*, 2014, **9**, 577.

29 J. Gajdzik, J. Lenz, H. Natter, R. Hempelmann, G. W. Kohring, F. Giffhorn, M. Manolova and D. M. Kolb, *Phys. Chem. Chem. Phys.*, 2010, **12**, 12604–12607.

30 D. Samanta and A. Sarkar, *Chem. Soc. Rev.*, 2011, **40**, 2567–2592.

31 J. Fu, M. Liu, Y. Liu, N. W. Woodbury and H. Yan, *J. Am. Chem. Soc.*, 2012, **134**, 5516–5519.

32 L. A. Wollenberg, J. E. Jett, Y. Wu, D. R. Flora, N. Wu, T. S. Tracy and P. M. Gannett, *Nanotechnology*, 2012, **23**, 385101.

33 M. J. Dalby, N. Gadegaard and R. O. Oreffo, *Nat. Mater.*, 2014, **13**, 558.

34 E. Albisetti, K. Carroll, X. Lu, J. Curtis, D. Petti, R. Bertacco and E. Riedo, *Nanotechnology*, 2016, **27**, 315302.

35 D. Wang, V. K. Kodali, W. D. Underwood II, J. E. Jarvholm, T. Okada, S. C. Jones, M. Rumi, Z. Dai, W. P. King and S. R. Marder, *Adv. Funct. Mater.*, 2009, **19**, 3696–3702.

36 K. M. Carroll, M. Desai, A. J. Giordano, J. Scrimgeour, W. P. King, E. Riedo and J. E. Curtis, *ChemPhysChem*, 2014, **15**, 2530–2535.

37 K. M. Carroll, A. J. Giordano, D. Wang, V. K. Kodali, J. Scrimgeour, W. P. King, S. R. Marder, E. Riedo and J. E. Curtis, *Langmuir*, 2013, **29**, 8675–8682.

38 W. D. Underwood, Master thesis, Georgia Institute of Technology, 2009.

39 U. Dürig, *J. Appl. Phys.*, 2005, **98**, 044906.

40 D. Pires, J. L. Hedrick, A. De Silva, J. Frommer, B. Gotsmann, H. Wolf, M. Despont, U. Duerig and A. W. Knoll, *Science*, 2010, 1187851.

41 S. T. Zimmermann, D. W. Balkenende, A. Lavrenova, C. Weder and J. r. Brugger, *ACS Appl. Mater. Interfaces*, 2017, **9**, 41454–41461.

42 B. Gotsmann, M. A. Lantz, A. Knoll and U. Dürig, *Nanotechnology*, 2010, 121–169.

43 F. Holzner, PhD thesis, ETH Zurich, 2013.

44 NanoFrazor Cantilevers, https://swisslitho.com/nanofrazor-cantilevers/, accessed February 2019.

Faraday Discussions

PAPER

UV-responsive cyclic peptide progelator bioinks†

Andrea S. Carlini,[abf] Mollie A. Touve,[bf] Héctor Fernández-Caro,[bi]
Matthew P. Thompson,[bcdfgh] Mary F. Cassidy,[e] Wei Cao[bcdfgh]
and Nathan C. Gianneschi ⓘ *[abcdfgh]

Received 5th March 2019, Accepted 12th April 2019
DOI: 10.1039/c9fd00026g

We describe cyclic peptide progelators which cleave in response to UV light to generate linearized peptides which then self-assemble into gel networks. Cyclic peptide progelators were synthesized, where the peptides were sterically constrained, but upon UV irradiation, predictable cleavage products were generated. Amino acid sequences and formulation conditions were altered to tune the mechanical properties of the resulting gels. Characterization of the resulting morphologies and chemistry was achieved through liquid phase and standard TEM methods, combined with matrix assisted laser desorption ionization imaging mass spectrometry (MALDI-IMS).

1   Introduction

UV-responsive biomaterials are commonly employed in printing technologies to make artificial tissues and biomimetic substrates because of the high spatial resolution and tunable wavelength of activation lasers. Such bioinks have been employed in projection-based,[1] sterolithography,[2] and laser-assisted[3] three-dimensional (3D) printing. Rigorous control over assembly crosslinking density and feature sizes can enable the exploration of nL-scale reactions, either within a defined boundary or between reagents across a diffusion-controlled barrier.

[a]Department of Chemistry & Biochemistry, University of California San Diego, La Jolla, California 92093, USA. E-mail: nathan.gianneschi@northwestern.edu

[b]Department of Chemistry, Northwestern University, Evanston, Illinois 60208, USA

[c]Department of Materials Science & Engineering, Northwestern University, Evanston, Illinois 60208, USA

[d]Department of Biomedical Engineering, Northwestern University, Evanston, Illinois 60208, USA

[e]Department of Chemical and Biological Engineering, Northwestern University, Evanston, Illinois 60208, USA

[f]International Institute for Nanotechnology, Northwestern University, Evanston, Illinois 60208, USA

[g]Chemistry of Life Processes Institute, Northwestern University, Evanston, Illinois 60208, USA

[h]Simpson Querrey Institute, Northwestern University, Evanston, Illinois 60208, USA

[i]Centro Singular de Investigación en Química Biolóxica e Materiais Moleculares (CIQUS), Departamento de Química Orgánica, Universidade de Santiago de Compostela, Rúa Jenaro de La Fuente s/n, 15782, Santiago de Compostela, Spain

† Electronic supplementary information (ESI) available. See DOI: 10.1039/c9fd00026g

However, current strategies generally rely on inks that produce feature sizes controlled entirely by the laser aperture. Thus, 3D printed tissue scaffolds[4–7] do not recapitulate the architectural complexity exhibited by most biological scaffolds, such as the fibrous (10–300 nm diameter) extracellular matrix (ECM).[8,9] Notably, the ECM exists as a porous, delicate network of peptide nanofibers entangled to generate a weak physical gel.

More recently, four-dimensional (4D) printing technologies, in which position (x, y, z) and chemical composition are controlled using shape-memory and/or transforming polymers that respond to various stimuli (*e.g.* irradiation,[10] temperature,[11] humidity,[12] electricity,[13] magnetism,[14] and solvation[15]), have expanded the behavioral diversity of printed materials. The use of UV-responsive bioinks for 4D printing of patterned self-assembling scaffolds with nanoscale resolution are of particular interest for designing microfluidic devices, bioarrays, and mimetic biological interfaces.[16]

We propose a strategy to reverse-engineer a biomimetic ECM through controllable printing of soft material inks. Peptides were selected as the base component of our ink material due to their inherent biocompatibility, absolute sequence control lending reproducible manufacturing, and a toolbox of commercially available amino acids which allows the expansion of sequence modularity. Furthermore, we were inspired by synthetic and processing efforts to control peptide morphologies[17–19] and their spatial organization[20–22] because of their inherent capacity to self-assemble through ionic crosslinking, amphiphilic, and/or π–π stacking interactions. As such, a UV-responsive ink that produces a local and patterned array of these active peptides, which themselves produce hierarchical assemblies, will enable "4D" printing of a biomimetic ECM.

Herein, we present the development of a photocleavable peptide that is sterically constrained to prevent gelation prior to treatment with UV light (Fig. 1). This stimuli-responsive peptide ink selectively generates macromolecular complexes with nanoscale features (Fig. 1a–c). We show that liquid cell transmission electron microscopy (LCTEM) can be used to simultaneously activate the UV-responsive nitrophenyl substrate and probe self-assembly in real time. Matrix-assisted laser desorption ionization imaging mass spectrometry (MALDI-IMS) characterization provides a further means for chemically verifying that the e$^-$ beam can be used as a source of inducing cleavage without initiating damage (Fig. 1d). Our method for printing complex scaffolds by design, as opposed to homogenous traces in conventional practices, may provide nanoscale features necessary for the study and construction of biological systems.

2   Results and discussion

Design of amphiphilic peptide as gelling ink template

A variety of short peptide sequences are reported to assemble as nanofibers such as rP11, KLD-12, MAX1 RADA16, FF dipeptide, KFE8, and OVA$_{257–264}$.[23–25] Many of these sequences possess primary amines from an unprotected N-terminus or Lys R-group. This moiety can react with UV active *o*-nitrobenzyl groups,[26] causing potential side reactions that disrupt the overall capacity to self-assemble into ordered structures. Therefore, the peptide-based inks generated in this study utilized a core gelator sequence free of Lys residues and N-capped with acetyl groups. We designed a new gelling peptide sequence that blends the amino acid

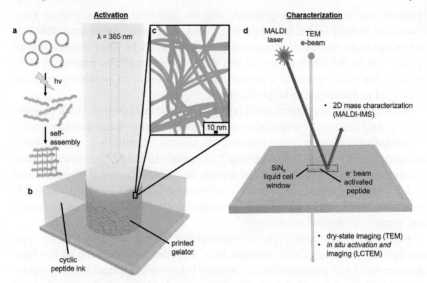

Fig. 1 Schematic of UV-responsive ink activation and characterization. (a and b) Sterically constrained cyclic peptide linearize in response to UV activation, and self-assemble through amphiphilic interactions. (b) Bulk scale activation from a solution of cyclic peptide ink provides a means for 4D printing of an ECM-like gel comprised of (c) entangled nanofiber networks. (d) Characterization of silicon nitride (SiN$_x$) chip surfaces to explore morphology by transmission electron microscopy (TEM) and chemical signatures by matrix-assisted laser desorption ionization imaging mass spectrometry (MALDI-IMS).

sequences possessed by small hydrophobic Fmoc-FF dipeptide[27] and more hydrophilic stimuli-responsive PhAc-FFAGLDD,[18] to generate an amphiphilic gelator composed of nonionic R-groups, Ac-GFFFLGSGS (Fig. 2 and S1†). The GSGS repeat has been used in our lab to provide a neutral linker for increasing peptide–polymer amphiphile solubility in aqueous solutions.[28] This unique nonionic, amphiphilic gelator sequence formed a physical gel in which unimer packing was influenced by both peptide and salt concentration, as well as applied shear forces (Fig. 2a). TEM under dilute conditions revealed a fibrous morphology akin to native ECM (Fig. 2b) but concentrated samples in aqueous mixtures of H$_2$O/acetonitrile or PBS/acetonitrile (90 : 10% v/v) formed viscoelastic gels, exemplified by larger storage (G′) than loss (G″) moduli (Fig. 2c and d). Additionally, greater angular frequency independence in the salt-containing sample demonstrated an increased resistance to flow compared to salt-free peptide solutions. Finally, this gelator exhibited a shear-thinning capacity, a characteristic property of self-assembling peptides.

Identification of a responsive moiety with predictable UV cleavage products

A multitude of UV cleavable amino acids have been reported in the literature, such as 2-nitrophenylglycine (NPG),[29] expanded o-nitrobenzyl linker,[30] o-nitrobenzyl caged phenol,[31] o-nitrobenzyl caged thiol,[32] nitroveratryloxycarbonyl (NVOC) caged aniline,[33] o-nitrobenzyl caged selenides,[34] bis-azobenzene,[35] coumarin,[36] cinnamyl,[37] spiropyran,[38] 2-nitrophenylalanine (2-nF),[39] and 3-amino-3(2-nitrophenyl)propionic acid (ANP).[40,41] For the purpose of simplicity and

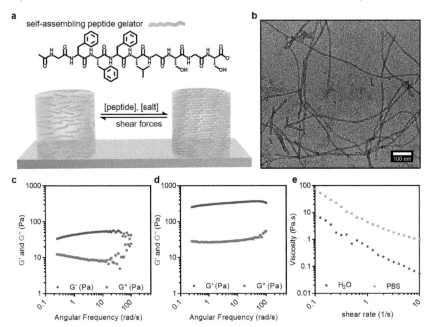

Fig. 2 Design of the self-assembling peptide gelator. (a) Schematic of the self-assembling peptide gelator, with amino acid sequence Ac-GFFFLGSGS, showing that the peptide and salt concentration are directly correlated to unimer packing and shear forces induce disassembly. (b) Fibrous morphology of the gelator (100 µM) by dry state TEM. (c–e) Bulk scale rheological characterization of the gelator (15 mg mL^{-1}). Frequency sweeps showing storage (G′) and loss (G″) moduli of gelator prepared in (c) H$_2$O (pH 6.4) and (d) 1× PBS (pH 7.4). (e) Corresponding complex viscosity as a function of shear rate.

scalability, we focused on commercially available nitrophenyl-based amino acids, as they are easily incorporated into the peptide sequence using SPPS without further modification. Initial systems utilized 2-nitrophenylalanine (2-nF), due to its low cost and sensitivity to UV light at 365 nm. Linear peptide sequences were synthesized and cyclized *via* N-to-C amidation through the N-terminal lysine amine and C-terminal carboxylic acid, to create **cyc(2-nF)** (Fig. S2†). To eliminate the presence of free amines and their subsequent reactivity with the activated nitrobenzyl group, the N-terminus was acetylated. Photoactivation of **cyc(2-nF)** yielded a viscous gel, demonstrating that self-assembly because of UV activation is possible. However, LCMS revealed a multitude of hydrophilic chemical species over time. Furthermore, the mass of these chemical species did not correlate with the predicted cleavage products proposed by Peters *et al.* for 2-nF amino acid bearing peptides.[39] Given the need for self-complementary peptide sequences to generate discrete morphologies and bulk physical gels, predictable and reliable photocleavage products are necessary.[25]

ANP was then selected as an alternative commercially available responsive moiety which achieves specific, single site peptide cleavage for reliable and reproducible macrocycle linearization.[42] Given the similarity in mass and expected HPLC peak retention time inherent to the cyclic precursor ink and its cleaved, linearized product, we developed a linear ink analogue, **lin(ANP)**, in which photocleavage was expected to release a short hydrophilic peptide linker and change the product peptide hydrophobicity (Fig. 3). The linear analogue, **lin(ANP)**, with sequence Ac-GG(ANP)GFFFLGSGS (Fig. 3a), was dissolved in an aqueous mixture of H_2O and ACN, and submitted to photocleavage at 365 nm for 4 h. The proposed mechanism of photocleavage and side product formation in Fig. 3b was based on the mechanism reported by Liang et. al.[42] Activation of the starting material (1) nitrophenyl moiety is accompanied by cleavage of the N-terminal amine, releasing the soluble Ac-GG linker, with subsequent rearrangement into an N-terminal nitroso compound (2). The dehydration product (3) and methanol adduct (4) were then generated. Notably, addition of methanol to the solution has been reported to push equilibrium towards a methyl ester compound for simpler product characterization. As such, all photocleavage reactions in this study were performed in the presence of a small quantity of methanol (~5% v/v). HPLC analysis of the solution revealed a mixture of compounds 1–4 after 1 h, with convergence upon the methanol adduct after 3 hours (Fig. 3c). Decreased solubility of cleaved **lin(ANP)** exhibited self-assembly into a cloudy aggregate. Additionally, a distinct color change from clear to yellow was observed, indicative of the indolin-3-one cleavage product. The observed peak masses by electrospray ionization mass spectrometry (ESI-MS) agree with expected values (Fig. 3d and S3†).

Peptide macrocyclization to generate sterically constrained inks

Gelator macrocyclization presents a rapid, clean, and synthetically simple method for generating a structurally dynamic peptide ink (Fig. 4). Peptides synthesized through SPPS were cleaved from 2-CTC resin with simultaneous deprotection of the Lys(Mtt) to generate the semi-protected peptide, Ac-K(ANP)GFFFLGS(otBu)GS(otBu) (Fig. 4a). Amide bond formation between the C-terminal carboxylate and Lys R-group amine occurred under dilute conditions (500 μM in DMF) for

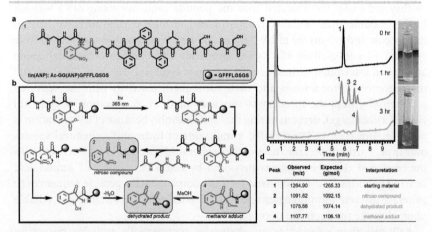

Fig. 3 Photoactivation of a linear ink analogue produces reliable cleavage products. (a) Linear ink analogue with photocleavable ANP residue. (b) Proposed photocleavage mechanism showing conversion of the (1) starting material to a (2) nitroso compound, (3) dehydrated product, and (4) methanol adduct.[42] (c) HPLC spectra vial pictures of linear ink analogue during cleavage at 0, 1, and 3 h. Peaks labeled according to (b). (d) Corresponding observed and expected mass spectra of peaks.

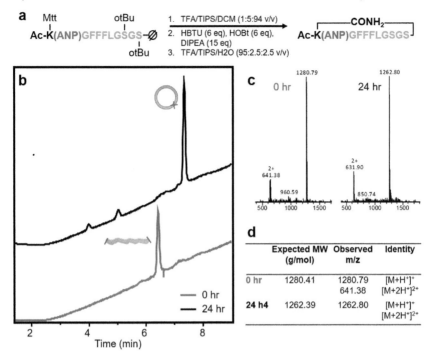

Fig. 4 Peptide macrocyclization to generate photocleavable inks. (a) Synthetic scheme for the generation of macrocycle, cyc(ANP) from the resin-bound peptide. Synthesis involved (1) cleavage from resin and Lys(Mtt) deprotection, (2) amide bond formation between the Lys R-group amine and C-terminal carboxylate under dilute conditions (500 µM in DMF), and (3) serine R-group deprotection. (b) LCMS of fully deprotected peptides at 0 h and 24 h of cyclization. (c) Corresponding ESI mass spectra. (d) Table of expected and observed masses with corresponding mass identities.

24 h to generate the semi-protected macrocyclic peptide. Final amino acid deprotection yielded **cyc(ANP)**. LCMS analysis shows a characteristic shift in peak retention from 6.4 min to 7.3 min, indicative of an increase in peptide hydrophobicity (Fig. 4b). Additionally, no dimerization was observed. Corresponding ESI mass spectra of peptide peaks at 0 h and 24 h reveal a loss of H_2O [M−18] upon cyclization, which corresponds to the expected masses (Fig. 4c and d). Finally, **cyc(ANP)** was purified by preparatory phase HPLC.

Photocleavage of a UV-responsive cyclic peptide ink

To test the mechanism of photocleavage, our ink was dissolved in a high concentration of organic solvent to ensure that the starting material and products remained dispersed in solution during UV treatment, and that self-assembly did not slow the reaction (Fig. 5). A bulk solution of **cyc(ANP)** in H_2O/ACN/MeOH (50 : 45 : 5% v/v) was submitted to photocleavage (365 nm) for 4 h and monitored for cleavage products 2, 3, and 4 (Fig. 5a). HPLC coupled with ESI-MI analysis of peaks revealed that all three compounds were present during UV treatment, demonstrating that synthetic and structural differences between **lin(ANP)** and **cyc(ANP)** do not affect the reaction (Fig. 5b). After 4 hours, the

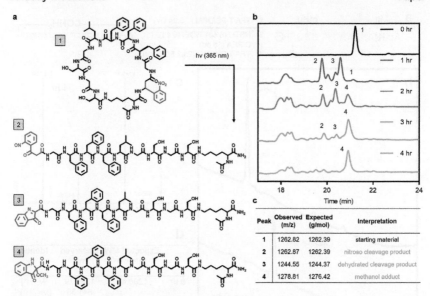

Fig. 5 UV cleavage of cyclic peptide ink. (a) Chemical structures of cyclic peptide (1) proposed cleavage products (2, 3, and 4). (b) HPLC spectra of UV-treated ink after 0, 1, 2, 3, and 4 h of UV (365 nm) treatment, with relevant peaks labeled. (c) Corresponding ESI mass spectra of collected peaks from (b).

methanol adduct (4) was the primary photocleavage product present in solution. Experimentally observed peak masses agree with expected values (Fig. 5c). Notably, linearized photocleavage products were more hydrophilic than the cyclic precursor ink, as shown by their decreased retention times. We hypothesized that increased solvation in aqueous solution will enable gel assembly of linearized photocleavage products.

Given the high hydrophobicity of **cyc(ANP)**, dissolution directly into H_2O was not possible. Various conditions were explored, with formulations 1 and 2 shown in Fig. 6. In formulation 1, peptide was dissolved in a mixture of H_2O/ACN (66 : 33% v/v), heated to evaporate-off ACN, and cooled to room temperature. Slow aggregation of the peptide in solution resulted in a slurry (Fig. 6a). Dry state TEM revealed the presence of micron scale amorphous **cyc(ANP)** aggregates (Fig. 6b). Treatment of the slurry solution with UV light and subsequent linearization increased peptide solubility, yielding a visible gel in solution. The resulting TEM morphology shows large elongated ribbon structures (~100 nm thick), significantly thicker than those of **linGelator** (~12 nm thick). To improve solubility, formulation 2 involved dissolving **cyc(ANP)** in a mixture of H_2O/ACN/AcOH (85 : 10 : 5% v/v) with heating, which produced a clear free-flowing solution (Fig. 6c). The size of **cyc(ANP)** aggregates was reduced from micro- to nano-scale structures (~200 nm) as observed by TEM (Fig. 6d). In this improved formulation, UV treatment also produced a gel, but with fibrous morphology and thickness (~10–15 nm) akin to that of **linGelator**. These results demonstrate that ink formulation significantly impacts the product morphologies, but it does not disrupt the overall capacity to self-assemble as a gel in response to photocleavage. Furthermore, we show that resulting gel morphology can be manipulated solely

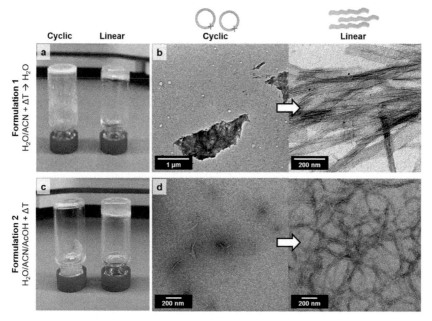

Fig. 6 Morphological analysis of **cyc(ANP)** ink with and without UV treatment. (a and b) Formulation 1 conditions: dissolution in H_2O/ACN (66 : 33% v/v) with heating (60 °C), and heating (80 °C) to remove acetonitrile. (a) Picture of vial flips from (left) untreated and (right) treated peptide sample (10 mg mL^{-1}). (b) Corresponding dry state TEM of (left) untreated and (right) treated samples (500 μM). (c and d) Formulation 2 conditions: dissolution in $H_2O/ACN/AcOH$ (85 : 10 : 5% v/v) with heating (60 °C). Neutralization to pH 6.4 with NH_4OH. (c) Picture of vial flips from (left) untreated and (right) treated peptide sample (10 mg mL^{-1}). (d) Corresponding dry state TEM of (left) untreated and (right) treated samples (500 μM).

from ink formulation alone. As the linearized photocleavage product is more hydrophilic than the cyclic precursor, we suspected that solvent exchange into pure H_2O or PBS after gel formation may be a viable means for 4D printing of ECM mimics for biological applications. Alternatively, modified peptide sequences which produce more water-soluble inks could eliminate the need for these harsh formulation conditions.

A modified gelator (**linGelator-II: GFFFLGSGSGS**) was generated for increased solubility by removing the N-terminal acetyl group and added extra GS sequence (Fig. S4†). Given the zwitterionic nature of this new gelator, which contains a cationic N-terminal amine and anionic C-terminal carboxylate, overall solubility was increased in water after using a pH-switch from pH 10 to pH 6.4, negating the need for ACN with this gelator. Conventional dry-state TEM confirmed that **lin-Gelator-II** forms well-defined fibrous networks (Fig. 7).

Utilizing MALDI-IMS to verify chemical identities of dry-state TEM samples

Validating chemical composition is an important analytical component in the generation of a novel nanomaterial. For materials that are generated in solution, various standard analytical techniques are available for probing chemical

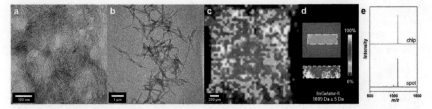

Fig. 7 Dry-state TEM imaging and MALDI-IMS mapping of **linGelator-II**. (a) Conventional dry-state, uranyl acetate stained TEM image of **linGelator-II** fibers. (b) Dry-state, unstained image of **linGelator-II** fibers on a SiN$_x$ chip. (c) MALDI-IMS map of the surface of the chip shown in (b) with mass filter for **linGelator-II** (1089 Da $\pm$ 5 Da) applied. (d) Optical picture of **linGelator-II** spot on ITO slide with MALDI-IMS measurement region outlined (top) and corresponding MALDI-IMS map with mass filter for **linGelator-II** applied (bottom). (e) Overall MALDI mass spectra for **linGelator-II** spot (bottom) and chip surface (top).

identities (*e.g.*, nuclear magnetic resonance, size exclusion chromatography). However, few techniques exist for probing the chemistry of materials that are dried-on or covalently bound to a surface, especially at low analyte concentrations. We previously demonstrated the use of MALDI-IMS to directly probe the chemical identities of substances dried on silicon nitride (SiN$_x$) chip surfaces.[43] Specifically, using MALDI-IMS, thousands of mass spectra could be generated across a SiN$_x$ TEM chip surface to determine chemical identities of materials generated *in situ* during LCTEM experiments. With this technique, we can verify whether the observed structures possess a claimed chemical identity. Analogous to energy-dispersive X-ray (EDX) and electron energy loss spectroscopy (EELS) analysis of S/TEM samples, here we were interested in directly analyzing the chemical signatures of macromolecular assemblies that had been imaged by TEM, either in the dry state by conventional TEM, or in a hydrated state by LCTEM.

Similar to conventional dry-state TEM sample preparation on copper grids with carbon films, an aliquot of **linGelator-II** was deposited onto a SiN$_x$ chip, then the solution was allowed to dry. Dry-state TEM imaging of the SiN$_x$ chip showed well-defined fibers, without staining, identical to those observed by conventional dry-state, uranyl acetate stained TEM (Fig. 7a and b). After imaging, the chip was attached to an indium-tin-oxide (ITO) slide, evenly coated with matrix, then analyzed by MALDI-IMS. The mass spectra verified that the fiber structures observed on the chips possessed the expected mass: charge signature (1089 Da) (Fig. 7c–e). By comparing the relative intensities of a specific *m/z* value across the generated MALDI-IMS map, we can gain information on the abundance of the material, which can be important when generating materials with questionable purity.

Use of an electron beam for nanoscale printing *in situ*

To analyze real-time morphology changes of the peptide inks, we hypothesized that the electron beam of a transmission electron microscope (TEM) could act as a stimulus to activate the more labile components of the **cyc(ANP)** peptide (*i.e.*, the ANP UV-responsive moiety) and, as a result, promote the growth of nanoscale assemblies observable by TEM.[44] For this imaging experiment, a solution of **cyc(ANP)** was irradiated under low electron flux conditions, where the amount of electrons irradiating a unit area over time (e⁻/Å² s) is kept to a minimum to mitigate beam-induced damage to the peptide structures and the solvent (Fig. 8).[44–46] All LCTEM imaging was conducted at the corners of the liquid cell, where the window thickness is minimal, to detect and observe the formation of these structures which possess minimal TEM contrast. Immediately upon exposure of the electron beam to the peptide solution under low flux conditions (0.1 e⁻/Å² s), low-contrast structures were observed to form and grow with increasing contrast for 2–3 min within the irradiated regions of the window (Fig. 8a–d). As macromolecular structures assembled and grew during imaging, they were not observed to move along the surface of the window region. We suspect that limited

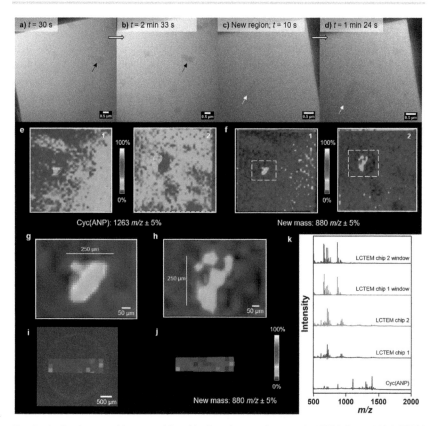

Fig. 8 Activating peptide assembly with the electron beam of a TEM. (a and b) LCTEM snapshots of a region of a liquid cell containing **cyc(ANP)**, where $t = 0$ is when the region imaged was first irradiated with the electron beam. Images at (a) 30 s and (b) 2 min 33 s are shown. Arrows point out one structure change over time. Electron flux = 0.1 e⁻/Å² s. (c and d) Snapshots of another region of the same liquid cell at (c) 10 s and (d) 1 min 24 s. (e and f) MALDI-IMS 2D mapping of the two chip surfaces from the liquid cell experiment shown in (a–d). Applied mass filter for (e) **cyc(ANP)** (1263 ± 5%) and (f) a new previously unobserved mass, (880 ± 5%). (g) Zoom-in of a region of interest on the chip surface shown in (f, left). (h) Zoom-in of a region of interest on the chip surface shown in (f, right). (i) Optical picture and MALDI-IMS 2D map overlay of a dried spot of UV peptide. (j) 2D map overlay only of the dried spot of **cyc(ANP)** shown in (i). (k) MALDI spectra for **cyc(ANP)** (purple), the entire surface of chip 1 (red), the entire surface of chip 2 (green), the window of chip 1 (teal), the window of chip 2 (blue).

diffusion in the confined environment of the liquid cell, as well as potential interaction with the SiN$_x$ surface, provided an immobilizing environment.

Given that we were able to observe the formation of structures from a peptide solution by LCTEM, we wanted to probe the chemistry of these structures to elucidate whether the formed architectures are from the same cleavage event which occurs during UV irradiation, or if the architectures are instead forming as a result of a different route of cleavage, as might be expected because of a change in the type of radiation utilized. To do this, we chemically analyzed the surface of the LCTEM chips after the *in situ* experiment by MALDI-IMS. After LCTEM, the two chips were gently pried apart and the sample solution was dried. The chips were mounted onto an ITO slide, evenly coated with matrix, then analyzed by MALDI-IMS. The mass for **cyc(ANP)** was present across the surface of the chips (1263 *m/z*) (Fig. 8e). Interestingly, we did not observe the UV cleavage product, as discussed above (Fig. 8e–k). Rather, the new masses obtained were approximately 400 *m/z* lower than that of the starting material, suggesting an altered mechanism of chemical cleavage when stimulated with an electron beam. Colorized mapping of these new masses (880 ± 5% *m/z*) reveal two small localized regions (~250 nm × 250 nm) of this material near the center of the chip on both chip surfaces of the liquid cell assembly. We suspect that this new mass corresponds to an electron beam-activated peptide within the window region, and that the displacement of this material away from the centered window regions of the chips occurred during chip separation prior to MALDI-IMS analysis. Altogether, we observe that the electron beam can successfully and locally activate peptides bearing the ANP-responsive moieties, but the mechanism of activation by a TEM electron beam is clearly different from when UV light is utilized as a stimulus.

3 Summary and conclusion

The results presented herein validate our hypothesis for conformational steric hindrance as a method of triggering a bioink. We demonstrate with UV activation that bulk scale and localized self-assembly, respectively, were possible, with predictable cleavage products. The required adjustments in peptide chemistry highlight the importance that controlled cleavage can have on resulting morphologies. In addition to novel ink generation, after verifying fiber morphologies by dry-state TEM, MALDI-IMS can be utilized to probe and confirm the chemical identities of the self-assembled architectures at high spatial resolution and sensitivity, showing that MALDI-IMS can be extended as a more routine method of verification for nanoscale assemblies on surfaces and surface localized patterns more generally. We suggest that MALDI-IMS mapping is a powerful method for adding chemical information, or "color" to the otherwise purely morphological information of grayscale TEM images. Further, current 3D printing technologies to generate tissue scaffolds, such as porous gel networks, are traditionally defined by the user-defined scaffold design and printer specifications. Our strategy introduces a new method for printing of complex features without the need for increasing 3D print model complexity or printers with higher UV laser resolution. This may be an interesting avenue to explore multilayered 4D printing of variable morphologies using the **cyc(ANP)** ink or synthetically modified analogues. We envision these materials will be useful in nanolithography applications for generating microfluidic devices, and artificial tissue scaffolds for studying single cell or protein reactions.

Conflicts of interest

The authors declare no conflicts of interest.

Acknowledgements

This research was conducted with Government support under and awarded by DoD through an AFOSR MURI (FA9550-16-1-0150) and a National Defense Science and Engineering Graduate Fellowship to M. A. T. (32 CFR 168a). A. C. would like to thank the NSF for a Graduate Research Fellowship (DGE-1144086). H. F.-C. thanks Xunta de Galicia (Predoctoral fellowship: ED481A-2017/047). The authors are also grateful for the support of an NIH Director's Transformative Research Award (R01HL117326), part of the NIH Common Fund, and the NHLBI (R01HL139001). This work made use of the EPIC facility of Northwestern University's NUANCE Center, which has received support from the Soft and Hybrid Nanotechnology Experimental (SHyNE) Resource (NSF ECCS-1542205); the MRSEC program (NSF DMR-1720139) at the Materials Research Center; the International Institute for Nanotechnology (IIN); the Keck Foundation; and the State of Illinois, through the IIN. This work also made use of the IMSERC at Northwestern University, which has received support from the Soft and Hybrid Nanotechnology Experimental (SHyNE) Resource (NSF ECCS-1542205); the State of Illinois and International Institute for Nanotechnology (IIN).

References

1 D. Xue, Y. Wang, J. Zhang, D. Mei, Y. Wang and S. Chen, *ACS Appl. Mater. Interfaces*, 2018, **10**(23), 19428–19435.
2 X. Zhou, W. Zhu, M. Nowicki, S. Miao, H. Cui, B. Holmes, R. I. Glazer and L. G. Zhang, *ACS Appl. Mater. Interfaces*, 2016, **8**(44), 30017–30026.
3 K. S. Lim, B. S. Schon, N. V. Mekhileri, G. C. J. Brown, C. M. Chia, S. Prabakar, G. J. Hooper and T. B. F. Woodfield, *ACS Biomater. Sci. Eng.*, 2016, **2**(10), 1752–1762.
4 Y. Yan, H. Chen, H. Zhang, C. Guo, K. Yang, K. Chen, R. Cheng, N. Qian, N. Sandler, Y. S. Zhang, H. Shen, J. Qi, W. Cui and L. Deng, *Biomaterials*, 2019, **190–191**, 97–110.
5 B. I. Oladapo, S. A. Zahedi and A. O. M. Adeoye, *Composites, Part B*, 2019, **158**, 428–436.
6 N. Contessi Negrini, L. Bonetti, L. Contili and S. Farè, *Bioprinting*, 2018, **10**, e00024.
7 Y. He, F. Yang, H. Zhao, Q. Gao, B. Xia and J. Fu, *Sci. Rep.*, 2016, **6**, 29977.
8 X. Guo, A. E. K. Hutcheon, S. A. Melotti, J. D. Zieske, V. Trinkaus-Randall and J. W. Ruberti, *Invest. Ophthalmol. Visual Sci.*, 2007, **48**(9), 4050–4060.
9 S. Bancelin, C. Aimé, I. Gusachenko, L. Kowalczuk, G. Latour, T. Coradin and M.-C. Schanne-Klein, *Nat. Commun.*, 2014, **5**, 4920.
10 M. López-Valdeolivas, D. Liu, D. J. Broer and C. Sánchez-Somolinos, *Macromol. Rapid Commun.*, 2018, **39**(5), 1700710.

11 D. Han, Z. Lu, S. A. Chester and H. Lee, *Sci. Rep.*, 2018, **8**(1), 1963.

12 A. Sydney Gladman, E. A. Matsumoto, R. G. Nuzzo, L. Mahadevan and J. A. Lewis, *Nat. Mater.*, 2016, **15**, 413.

13 S. R. Shin, Y.-C. Li, H. L. Jang, P. Khoshakhlagh, M. Akbari, A. Nasajpour, Y. S. Zhang, A. Tamayol and A. Khademhosseini, *Adv. Drug Delivery Rev.*, 2016, **105**, 255–274.

14 X. Zhao, J. Kim, C. A. Cezar, N. Huebsch, K. Lee, K. Bouhadir and D. J. Mooney, *Proc. Natl. Acad. Sci. U. S. A.*, 2011, **108**(1), 67–72.

15 D.-G. Shin, T.-H. Kim and D.-E. Kim, *International Journal of Precision Engineering and Manufacturing-Green Technology*, 2017, **4**(3), 349–357.

16 C. Carbonell and A. B. Braunschweig, *Acc. Chem. Res.*, 2017, **50**(2), 190–198.

17 A. S. Carlini, R. Gaetani, R. L. Braden, C. Luo, K. L. Christman and N. C. Gianneschi, *Nat. Commun.*, 2019, **10**, 1735.

18 D. Kalafatovic, M. Nobis, N. Javid, P. W. J. M. Frederix, K. I. Anderson, B. R. Saunders and R. V. Ulijn, *Biomater. Sci.*, 2015, **3**(2), 246–249.

19 C. J. Bowerman and B. L. Nilsson, *J. Am. Chem. Soc.*, 2010, **132**(28), 9526–9527.

20 L. Adler-Abramovich, D. Aronov, P. Beker, M. Yevnin, S. Stempler, L. Buzhansky, G. Rosenman and E. Gazit, *Nat. Nanotechnol.*, 2009, **4**, 849.

21 S. Zhang, M. A. Greenfield, A. Mata, L. C. Palmer, R. Bitton, J. R. Mantei, C. Aparicio, M. O. de la Cruz and S. I. Stupp, *Nat. Mater.*, 2010, **9**, 594.

22 L. Adler-Abramovich, P. Marco, Z. A. Arnon, R. C. G. Creasey, T. C. T. Michaels, A. Levin, D. J. Scurr, C. J. Roberts, T. P. J. Knowles, S. J. B. Tendler and E. Gazit, *ACS Nano*, 2016, **10**(8), 7436–7442.

23 N. Habibi, N. Kamaly, A. Memic and H. Shafiee, *Nano Today*, 2016, **11**(1), 41–60.

24 J. Chen and X. Zou, *Bioact. Mater.*, 2019, **4**, 120–131.

25 H. Zhang, J. Park, Y. Jiang and K. A. Woodrow, *Acta Biomater.*, 2017, **55**, 183–193.

26 N. De Alwis Watuthanthrige, P. N. Kurek and D. Konkolewicz, *Polym. Chem.*, 2018, **9**(13), 1557–1561.

27 J. Raeburn, G. Pont, L. Chen, Y. Cesbron, R. Lévy and D. J. Adams, *Soft Matter*, 2012, **8**(4), 1168–1174.

28 A. P. Blum, J. K. Kammeyer and N. C. Gianneschi, *Chem. Sci.*, 2016, **7**(2), 989–994.

29 P. M. England, H. A. Lester, N. Davidson and D. A. Dougherty, *Proc. Natl. Acad. Sci. U. S. A.*, 1997, **94**(20), 11025–11030.

30 J.-P. Pellois and T. W. Muir, *Angew. Chem., Int. Ed.*, 2005, **44**(35), 5713–5717.

31 A. Shigenaga, D. Tsuji, N. Nishioka, S. Tsuda, K. Itoh and A. Otaka, *ChemBioChem*, 2007, **8**(16), 1929–1931.

32 G. Liu and C.-M. Dong, *Biomacromolecules*, 2012, **13**(5), 1573–1583.

33 M. E. Roth-Konforti, M. Comune, M. Halperin-Sternfeld, I. Grigoriants, D. Shabat and L. Adler-Abramovich, *Macromol. Rapid Commun.*, 2018, **39**(24), 1800588.

34 A. L. Eastwood, A. P. Blum, N. M. Zacharias and D. A. Dougherty, *J. Org. Chem.*, 2009, **74**(23), 9241–9244.

35 M. Mba, D. Mazzier, S. Silvestrini, C. Toniolo, P. Fatás, A. I. Jiménez, C. Cativiela and A. Moretto, *Chem.–Eur. J.*, 2013, **19**(47), 15841–15846.

36 Y. Shao, C. Shi, G. Xu, D. Guo and J. Luo, *ACS Appl. Mater. Interfaces*, 2014, **6**(13), 10381–10392.

37 L. Yan, L. Yang, H. He, X. Hu, Z. Xie, Y. Huang and X. Jing, *Polym. Chem.*, 2012, **3**(5), 1300–1307.

38 V. K. Kotharangannagari, A. Sánchez-Ferrer, J. Ruokolainen and R. Mezzenga, *Macromolecules*, 2011, **44**(12), 4569–4573.

39 F. B. Peters, A. Brock, J. Wang and P. G. Schultz, *Chem. Biol.*, 2009, **16**(2), 148–152.

40 N. Umezawa, Y. Noro, K. Ukai, N. Kato and T. Higuchi, *ChemBioChem*, 2011, **12**(11), 1694–1698.

41 G. M. Grotenbreg, M. J. Nicholson, K. D. Fowler, K. Wilbuer, L. Octavio, M. Yang, A. K. Chakraborty, H. L. Ploegh and K. W. Wucherpfennig, *J. Biol. Chem.*, 2007, **282**(29), 21425–21436.

42 X. Liang, S. Vézina-Dawod, F. Bédard, K. Porte and E. Biron, *Org. Lett.*, 2016, **18**(5), 1174–1177.

43 M. A. Touve, A. S. Carlini and N. C. Gianneschi, *Nat. Commun.*, 2019, in press.

44 M. A. Touve, C. A. Figg, D. B. Wright, C. Park, J. Cantlon, B. S. Sumerlin and N. C. Gianneschi, *ACS Cent. Sci.*, 2018, **4**(5), 543–547.

45 T. J. Woehl and P. Abellan, *J. Microsc.*, 2017, **265**(2), 135–147.

46 N. M. Schneider, M. M. Norton, B. J. Mendel, J. M. Grogan, F. M. Ross and H. H. Bau, *J. Phys. Chem. C*, 2014, **118**(38), 22373–22382.

PAPER

Poly(alkyl glycidyl ether) hydrogels for harnessing the bioactivity of engineered microbes†

Trevor G. Johnston, [ID] ‡[a] Christopher R. Fellin, [ID] ‡[a] Alberto Carignano[b] and Alshakim Nelson [ID] *[a]

Received 14th February 2019, Accepted 5th March 2019

DOI: 10.1039/c9fd00019d

Herein, we describe a method to produce yeast-laden hydrogel inks for the direct-write 3D printing of cuboidal lattices for immobilized whole-cell catalysis. A poly(alkyl glycidyl ether)-based triblock copolymer was designed to have three important features for this application: (1) a temperature response, which allowed for facile processing of the material; (2) a shear response, which facilitated the extrusion of the material through a nozzle; and (3) UV light induced polymerization, which enabled the post-extrusion chemical crosslinking of network chains, and the fabrication of robust printed objects. These three key stimuli responses were confirmed *via* rheometrical characterization. A genetically-engineered yeast strain with an upregulated α-factor production pathway was incorporated into the hydrogel ink and 3D printed. The immobilized yeast cells exhibited adequate viability of 87.5% within the hydrogel. The production of the upregulated α-factor was detected using a detecting yeast strain and quantified at 268 nM (s = 34.6 nM) over 72 h. The reusability of these bioreactors was demonstrated *via* immersion of the yeast-laden hydrogel lattice in fresh SC media and confirmed by the detection of similar amounts of upregulated α-factor at 259 nM (s = 45.1 nM). These yeast-laden materials represent an attractive opportunity for whole-cell catalysis of other high-value products in a sustainable and continuous manner.

Introduction

Whole-cell biocatalysis is a standard practice across a wide range of industries wherein cells are used to transform molecular precursors into a product of interest (antibiotics, drugs, vitamins, insulin, vaccines, *etc.*). These reactions are generally employed as a batch process, wherein the cells and the necessary molecular precursors are combined into a single reaction vessel, stirred for

[a]Department of Chemistry, University of Washington, Seattle, WA 98195, USA. E-mail: alshakim@uw.edu

[b]Department of Electrical Engineering, University of Washington, Seattle, WA 98195, USA

† Electronic supplementary information (ESI) available. See DOI: 10.1039/c9fd00019d

‡ These authors contributed equally to this work.

several days, and then the product is isolated from the complex mixture.[1-10] However, batch cell reactors are costly and time consuming because they require sterile conditions, can have low yields, and purification protocols are labor and cost intensive. Immobilized-cell bioreactors, wherein metabolically active cells are trapped within a material such as a hydrogel, offer an alternative method that can simplify the isolation of the product, minimize product inhibition or toxicity, and allow recycling of the cells.[11]

We have previously reported additive-manufactured catalytically-active living materials (AMCALM) as a platform for immobilized-cell bioreactors.[12] In contrast to previous reports, wherein calcium alginate beads[13-16] or electrospun fibers[17-19] were used to encapsulate microbes for fermentation, our approach utilized customized lattices of yeast-laden hydrogels that were printed using a direct-write 3D printer. Additive manufacturing (or 3D printing) is a fabrication process that utilizes layer-by-layer material deposition to construct three-dimensional geometries according to a computer-aided design (CAD) model.[20-27] As such, 3D printing is well suited for manufacturing living materials[28-33] with spatial and geometrical organization of cells. Hydrogel-based materials are particularly attractive as cell-laden inks for direct-write 3D printing because these materials can recapitulate some of the chemical and physical attributes of the extracellular matrix that exists in biofilms, living tissue, and other naturally occurring microenvironments.[31,34-36] These features could include the presence of ligands or functional groups,[37-40] or the stiffness of the hydrogel.[41,42]

There were three important features of the yeast-laden hydrogel ink that were developed for AMCALMs: (1) a temperature response, wherein the material exhibited a reversible gel-to-sol transition upon cooling, which enabled homogeneous dispersion of the yeast cells within the gel and facile loading of the hydrogel into a syringe; (2) a shear-thinning response that facilitated the extrusion of the hydrogel ink from a nozzle to enable the layer-by-layer formation of three-dimensional objects; and (3) an irreversible photo-response wherein polymerizable groups chemically cross-linked the hydrogel into a robust structure. Yeast-laden hydrogel lattices were printed and then utilized in a continuous batch process for the fermentation of glucose to produce ethanol.

Herein, we demonstrate a new polymer hydrogel for the encapsulation and direct-write 3D printing of yeast-laden hydrogel lattices that can be used for the production of a polypeptide. The poly(glycidyl ether)-based ABA triblock copolymer (polymer 2) has similar features (temperature-, shear-, and photo-responses, Fig. 1) to the Pluronic-based triblock copolymer that was reported previously.[12] Furthermore, a genetically-engineered yeast strain with an upregulated α-factor production pathway was incorporated into the hydrogel to demonstrate the production of an extracellularly excreted polypeptide from the yeast-laden hydrogel. While α-factor is a yeast hormone that does not have any known therapeutic effects, the polypeptide is commonly utilized in the expression and recovery of recombinant proteins.[43,44] When the α-factor leader sequence is appended to the genetic code for an engineered protein, the yeast cell utilizes its native cellular machinery to secrete the protein beyond the cell membrane.[45-50] In our experiments, we not only demonstrate that the secreted α-factor can be detected in the liquid media surrounding the hydrogel lattice, but we also demonstrate the reusability of the yeast-laden hydrogel lattices.

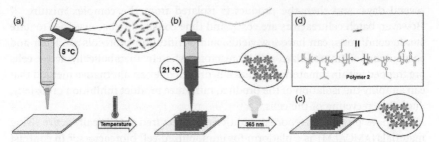

Fig. 1   A graphical overview of the three key stimuli responses of the polymer **2** hydrogels necessary for AMCALM applications. (a) The temperature responsive feature of the hydrogels enabled facile loading of the hydrogel material at 5 °C. (b) The shear-stress response facilitated the formation of complex three-dimensional objects at 21 °C. (c) UV light (365 nm) initiated the polymerization of the methacrylate end-groups and chemical crosslinking of the polymer **2** hydrogel. (d) The chemical structure of polymer **2**. The letter designations ($z = 6.4$, $y = 7.4$, and $x = 182$) refer to the average number of isopropyl glycidyl ether, ethyl glycidyl ether, and ethylene oxide repeat units, respectively.

Experimental

Materials

All chemicals and solvents were purchased from Sigma-Aldrich or Fisher Scientific and used without further purification unless noted otherwise. Isopropyl glycidyl ether (iPGE, 98%) and ethyl glycidyl ether (EGE, 98%) were dried over CaH_2 for 24 h, distilled into a flask containing butyl magnesium chloride (2 M in tetrahydrofuran, THF), re-distilled, and stored under a N_2 atmosphere. Poly(-ethylene oxide) (PEO, $M_n = 8000$ g mol^{-1}) was dried under vacuum overnight prior to use. Dry THF was obtained using neutral alumina and a Pure Process Technology solvent purification system. A potassium naphthalenide solution (1 M) was prepared by dissolving naphthalene (3.2 g) in THF (25 mL), adding potassium (0.975 g), and storing under a N_2 atmosphere. ^{1}H NMR spectra were obtained on a Bruker Advance 300 or 500 MHz spectrometer. Gel permeation chromatography was performed using a Waters chromatograph equipped with two 10 µm Malvern columns (300 mm × 7.8 mm) connected in series with increasing pore size (1000 and 10 000 Å), using chloroform (Optima, 0.1% v/v trimethylamine) as the eluent, and calibrated with poly(ethylene oxide) standards (400 to 40 000 g mol^{-1}). The relative molecular weights were measured in chloroform using poly(ethylene oxide) standards and a refractive index detector (flow rate: 1 mL min^{-1}). Drop-out Mix Complete without Yeast Nitrogen Base (D9515) and Yeast Nitrogen Base (Y2025) were purchased from US Biological Life Sciences. Sytox green nucleic acid strain (5 mM solution in DMSO; Invitrogen) was purchased from Thermo Fisher Scientific.

Yeast strains

For the microscopy and live/dead viability tests, the strain yJS001 was used (genotype SO992 mfa2::pTEF1_mCherry(kanR)). This strain constitutively expressed the mCherry protein, which facilitated characterization using fluorescence microscopy. The *Saccharomyces cerevisiae* strain SO992 was used as the control strain in the α-factor production experiments. This strain is a derivative of W303 (genotype MATa ura3 leu2 trp1 his3 can1R ade).

As the α factor-producing strain, we employed a genetically modified MAT-α yeast strain ('secreting strain') expressing the *S. cerevisiae* gene MFα1 (YPL187W) on a constitutively-expressed promoter pGPD (natively expressing YPL197W). The MFα1 gene expresses 3 copies of the α-factor peptide, a 13-amino acid peptide natively used as a mating hormone from the MAT-α to the MAT-A strains.

In order to quantify the α-factor secreted from the yeast immobilized in the hydrogels, we employed a genetically modified MAT-A yeast strain that expresses the fluorescence protein yeVenus, driven by the pFUS1 promoter, which is downstream in the MAPK signaling pathway, activated by α-factor detection ('detecting strain'). The MAT-A native strain was engineered by deleting BAR1 (native α-factor protease) and integrating the POG1 gene on a constitutively-expressed pGPD promoter to avoid α-factor-induced growth arrest.

Yeast transformations were carried out using a standard lithium acetate protocol. The yeast cells were made competent by growing 50 mL cultures in rich media to log the growth phase, and then spinning down the cells and washing with H_2O. Next, linearized DNA, salmon sperm donor DNA, 50% polyethylene glycol, and 1 M LiOAc were combined with 50 mL of competent cells and the mixture was heat shocked at 42 °C for 15 min. The cells were then spun down, the supernatant was removed, and they were resuspended in H_2O and then plated on selective agar media. Transformations were done into MATa W303-1A and MATα W303-1B background strains.

Synthesis of polymer 1

The ABA triblock copolymer was synthesized *via* anionic ring-opening polymerization. PEO ($M_n = 8000$ g mol^{-1}, 20 g, 2.5 mmol) was added to the reaction vessel and dried under vacuum overnight. Dry THF (250 mL) was added under an Ar atmosphere and heated to 50 °C to facilitate dissolution of the macroinitiator. Once sufficiently dissolved, a potassium naphthalenide solution (1 M in THF) was titrated into the flask until the solution remained a slight green color, indicating full deprotonation of the PEO hydroxyl end groups. Isopropyl glycidyl ether (4.94 g, 42.5 mmol) and ethyl glycidyl ether (4.34 g, 42.5 mmol) were added to begin polymerization. The reaction continued for 24 h at 65 °C and was subsequently quenched with a degassed solution of 1% v/v AcOH in MeOH. The reaction mixture was then precipitated into cold hexane. The polymer was collected *via* centrifugation (4400 rpm, 10 min) and the supernatant decanted. The product was washed twice with additional hexane and collected again in the same manner. The isolated polymer solution was dried in a vacuum oven for at least 24 h to afford polymer **1** as an off-white solid (27.5 g). ^{1}H NMR (500 MHz, CDCl$_3$): $\delta = 1.15$–1.17 (m, –O–CH–(C$\underline{H}_3$)$_2$), 1.17–1.23 (t, –O–CH$_2$–C$\underline{H}_3$, $J = 7.0$ Hz), 3.47–3.81 (m, –O–C$\underline{H}_2$–C$\underline{H}_2$–O– and –O–C$\underline{H}_2$–C$\underline{H}$(C$\underline{H}_2$–O–C$\underline{H}_2$–CH$_3$)–O– and –O–C$\underline{H}_2$–C$\underline{H}$(C$\underline{H}_2$–O–C$\underline{H}$–(CH$_3$)$_2$)–O–).

Synthesis of polymer 2

Polymer **1** (20 g, 1.81 mmol) was dissolved in dry THF (250 mL) under a nitrogen atmosphere until complete dissolution of the polymer occurred. Triethylamine

(2.45 mL, 18.1 mmol) was added to increase the reactivity of the polymer hydroxyl chain ends and the mixture was heated at 65 °C for 30 min. Methacrylic anhydride (26.9 mL, 181 mmol) was then added and the reaction mixture was stirred for 16 h at 65 °C. After this, the reaction was quenched with a degassed solution of 1% v/v AcOH in MeOH. The reaction mixture was then precipitated into cold ether. The polymer was collected *via* centrifugation (4400 rpm, 10 min) and the supernatant decanted. The product was washed twice with additional ether, once with hexane, and collected again by centrifugation. The isolated polymer was dried in a vacuum oven for 24 h to afford polymer **2** as an off-white solid (16.7 g). The degree of functionalization (f_n) was determined by comparing the integrations of the methacrylate vinyl (6.12 and 5.55 ppm) and methyl (1.94 ppm) protons, as well as the PEO chain-end methylene protons (5.08–5.15 ppm), to their theoretical values. For example, 1.99 (vinyl, actual)/2 (vinyl, theoretical) × 100 ≈ 100% functionalization of chain ends. These integration values were referenced to the total alkyl glycidyl ether protons for each polymer chain (1.12–1.20 ppm, 121 H). ^{1}H NMR (300 MHz, CDCl$_3$): δ = 1.12–1.20 (m, –O–CH–(C$\underline{H}_3$)$_2$ and –O–CH$_2$–C$\underline{H}_3$), 1.94 (s, C$\underline{H}_3$C(CO$_2$)=CH$_2$), 3.39–3.89 (m, –O–C$\underline{H}_2$–C$\underline{H}_2$–O– and –O–C$\underline{H}_2$–C$\underline{H}$(CH$_2$–O–C$\underline{H}_2$–CH$_3$)–O–, and –O–C$\underline{H}_2$–C$\underline{H}$(C$\underline{H}_2$–O–C$\underline{H}$–(CH$_3$)$_2$)–O–), 5.08–5.15 (m, –CH$_2$–C$\underline{H}$–O–(C=O)), 5.48 (s, $\underline{H}$–CH=C), 6.18 (s, $\underline{H}$–CH=C).

Preparation of synthetic complete media

The SC media (1 L) were prepared by dissolving drop-out mix (2 g), yeast nitrogen base (6.7 g), and glucose (20 g) in Milli-Q water. The resulting solution was sterilized by filtration through a 0.2 μm nylon filter.

Preparation of hydrogel solution

Polymer **2** was dissolved in sterile deionized water at a concentration of 20 wt% polymer. The resulting polymer solution was cooled overnight at 5 °C to facilitate hydrogel formation *via* LCST behavior. After homogenization of the solution at low temperature, the solution was warmed to 21 °C to induce the formation of a gel state. The hydrogels used to print the proof-of-concept models in Fig. 3 were mixed with the photo-radical initiator 2-hydroxy-2-methylpropiophenone (10 μL) and centrifuged (4400 rpm, 10 min) to remove bubbles.

Preparation of yeast-laden hydrogel ink

In order to prepare yeast-laden hydrogels, the aforementioned hydrogel solution was cooled to 5 °C in a refrigerator. At this temperature, the hydrogel solution underwent gel-to-sol transition, affording a free-flowing liquid. Yeast cells were added from an overnight liquid culture at a concentration of 10^7 cells per gram of hydrogel while the gel was in its solution state. The resulting solution was mixed thoroughly and allowed to equilibrate at 5 °C until all of the bubbles present in the solution were removed. Finally, the hydrogel was warmed to 21 °C to undergo a sol-to-gel transition, resulting in a shear-responsive gel.

Rheometrical characterization

For the rheometrical characterization of the hydrogels, dynamic oscillatory experiments were performed on a TA Instruments Discovery Hybrid Rheometer-2

(DHR-2) equipped with a Peltier temperature controller. The instrument operates by applying a known displacement (strain) and measuring the material's resistance (stress) to the force. Rheometry experiments were conducted by depositing the hydrogel between the rheometer base plate and 20 mm parallel plate geometry with a final gap of 1 mm. The samples were equilibrated in an ice bath for at least 30 min and then were carefully loaded onto the Peltier plate at 5 °C and a preshear experiment was applied to eliminate the bubbles from the sample. The sample was equilibrated at 21 °C for 8 min before each run. The gel yield stress values were measured under oscillatory strain (frequency: 1 Hz, 21 °C), starting with an initial strain of 0.01%, and converted to applied oscillatory stress. Viscosity *versus* shear rate experiments were performed in a range of shear rates between 0.01 and 100 Hz. Cyclic shear-thinning experiments (frequency = 1 Hz) were performed at 21 °C using alternating strains of 1% for 5 min and 100% for 3 min per cycle, in order to investigate the shear-thinning and recovery behavior of the hydrogels. Temperature ramp experiments were performed at 1 Hz from 5–50 °C at 2 °C min^{-1}. Photocuring was performed using a fully integrated smart swap LED photocuring accessory. A 120 s dwell time (frequency: 1 Hz) elapsed before the UV lamp (365 nm LED with an irradiation intensity of 5 mW cm^{-2}) was turned on for 420 s at a constant oscillatory strain (1%).

Additive manufacturing (direct-write 3D printing) of a cuboidal lattice

A modified pneumatic direct-write 3D printer was assembled based on a Tronxy P802E 3D Printer kit from Shenzhen Tronxy Technology Co. The printer was controlled with an Arduino using Marlin firmware. All CAD models were designed in OpenSCAD. The G-code commands for the printer were generated using Slic3r. The resulting G-code was modified using Python to introduce the required commands for the dispensing of the hydrogel *via* pneumatic pressure. All printing was performed using a 20 wt% hydrogel ink with an extrusion air pressure of 20 psi, a print speed of 5 mm s^{-1}, and a 0.41 mm inner diameter CML supply conical nozzle attachment.

The dimensions for the 3D-printed lattices were 1.9 cm by 1.9 cm by 1.2 cm, and each cube weighed between 1.8 and 2.0 g. Upon completion of the 3D printing, the cubes were irradiated under 365 nm light (at 3.4 mW cm^{-2}) for 3 min to cure and chemically fix the structures.

Microscopy and imaging

Optical imaging: images of the 3D-printed hydrogels were captured using an iPhone XS. Confocal imaging: confocal microscopy images were taken using a Leica TCS SP5 II laser scanning confocal microscope. All images were taken with a dry 20× objective. mCherry protein fluorescence was excited with a 561 nm laser at 5% laser power, and emission was scanned from 569 to 700 nm. Sytox green viability dye was excited with a 488 nm laser at 5% laser power, and emission was scanned from 500 to 550 nm. The samples were sequentially scanned and the output images were processed using ImageJ Java software.

Cell viability assay

The yeast-laden hydrogel was prepared as described above. This gel was cooled to 5 °C to induce a gel-to-sol transition, and was then poured and spread into a sterile Petri dish to produce a thin film (~0.5 mm) of the yeast-laden hydrogel. The sample was irradiated with 365 nm UV light for 3 min. This film was then incubated in fresh SC media at 30 °C and periodically agitated. The medium was exchanged every 24 h to ensure fresh nutrient delivery to the embedded cells. At the imaging intervals described in this work, a small square of the film (5 mm × 5 mm) was cut and removed for staining. Sytox green stain stock solution was diluted to 5 μM, and 20 μL of the dye solution was exposed to the hydrogel sample for 5 min. The sample was then washed with SC media and imaged using the Leica SP5 confocal microscope. Constitutively-expressed mCherry protein fluorescence from the yeast cells was measured to indicate live cells, while the Sytox dye fluorescence indicated the presence of dead cells.

α-Factor production

The α-factor production from the 3D-printed yeast-laden cuboidal lattices was quantified as follows. A yeast-laden hydrogel ink containing 10^7 yeast cells per gram of hydrogel was direct-write 3D printed and photocured (3 min) with 365 nm irradiation. The lattices were then washed with SC media and placed into 50 mL falcon tubes. The tubes were then filled with 10 mL of SC media for fermentation and placed in a 30 °C shaker (225 rpm) for incubation. After a period of 72 h, the lattices were removed from the reactors and the media were collected. The fermentation media were filtered with a 0.2 μm nylon filter. The up-regulated yeast strain experiments were performed in triplicate alongside a wild-type yeast-laden lattice.

In order to determine the reusability of the lattices in subsequent fermentation cycles, the same lattices from the first round of α-factor production were again washed individually in sterile SC media. The lattices were then placed into new falcon tubes containing 10 mL of SC media, and incubated under the same conditions for an additional 72 h. After incubation, the samples were collected and prepared for characterization in the same manner as mentioned above.

α-Factor detection and quantification

In order to quantify the α-factor synthesis from within the yeast-laden hydrogel, we collected the SC media that were used to submerge the hydrogel samples after 72 h. We collected 9 mL of media for each experimental condition and replicate; each sample was split into six 1.5 mL Eppendorf tubes and then dehydrated in a Savant SpeedVac Plus SC110A Concentrator for 12 h at medium dehydration speed. Subsequently, we added 100 μL of molecular grade water into each dehydrated sample, resuspended the samples, and collected the resulting 600 μL into a single sample. The final sample was run again through the Speed-Vac for 12 h at medium speed and resuspended into 100 μL of molecular grade water. This process concentrated the initial sample to 90-fold its original concentration.

We experimentally tested for α-factor synthesis using detecting strains with a standard cytometer assay protocol. The detecting strains were grown overnight for 20 h in SC media, then diluted to 30 events per μL, again in SC media. After 3 h, we inoculated the concentrated samples from the yeast-laden hydrogels into different vials, plus two extra vials: one kept as a control (3 μL of 0.1 M sodium acetate) and one inoculated with 10 μM 98% HPLC-pure α-factor peptide in 0.1 M sodium acetate obtained from Zymo Research (Irvine, CA, USA). The samples were collected for cytometer measurement 8 h after induction, and median fluorescence values were computed from the resulting histogram.

The fluorescence intensity of the detecting strain was measured with a BD Accuri C6 flow cytometer equipped with a CSampler plate adapter using excitation wavelengths of 488 and 640 nm and an emission detection filter at 533 nm (FL1 channel). A total of 10 000 events above a 400 000 FSC-H threshold (to exclude debris) were recorded for each sample with a core size of 22 mm using Accuri C6 CFlow Sampler software. Cytometry data were exported as FCS 3.0 files and processed using custom Python scripts to obtain the median FL1-A value at each data point.

We used a model fitted to the data to estimate the amount of α-factor detected by the detecting strain, starting from its median fluorescence value response. The model interpolates the titration curve of an α-factor-detecting strain that is equivalent to the one used in this study, except it synthesizes the fluorescent protein yeGFP instead of yeVenus. The interpolation follows a standard Hill activation function:

$$y = K \frac{u^n}{\varphi + u^n} + y_0$$

where u represents the α-factor input concentration in nM, and y the median fluorescence output. To account for the different fluorescent protein, we used the control and the 10 μM α-factor median fluorescent outputs from this experiment to re-fit K and y_0.

In order to estimate the amount of α-factor synthesized by our strains in the hydrogel, we inverted the formula above to get:

$$u = \sqrt[n]{\varphi \frac{y - y_0}{K + y_0 - y}}$$

Results and discussion

Synthesis and functionalization of the triblock copolymer

ABA triblock copolymers of poly(isopropyl glycidyl ether-*stat*-ethyl glycidyl ether)-*block*-poly(ethylene oxide)-*block*-poly(isopropyl glycidyl ether-*stat*-ethyl glycidyl ether) afford shear-thinning hydrogels that have a tunable sol–gel transition temperature based on the ratio of ethyl and isopropyl glycidyl ether monomers (EGE and iPGE, respectively), 'A' block chain length, and concentration.[51] When these polymers were dissolved in aqueous media, flower-like micelles[52] were expected to form based on the design of this ABA triblock copolymer's hydrophobic 'A' blocks flanking a hydrophilic 'B' block. Hydrogels based on this ABA triblock copolymer platform (20 wt%) were suitable inks for direct-write 3D printing and were able to create self-supporting hydrogel structures.[52] However, the physical cross-links present in this system were insufficient to maintain the integrity of the 3D-printed object when subjected to either excess aqueous media or mechanical loads. Thus, chemically cross-linked hydrogels are necessary to improve the robustness of the 3D-printed objects.

Polymer **1** was synthesized (Scheme 1) *via* living anionic ring-opening polymerization[53–56] from a poly(ethylene oxide) (PEO) macroinitiator (M_n = 8000 g mol^{-1}) with an EGE : iPGE ratio of 1.15 : 1 to afford triblock copolymers with

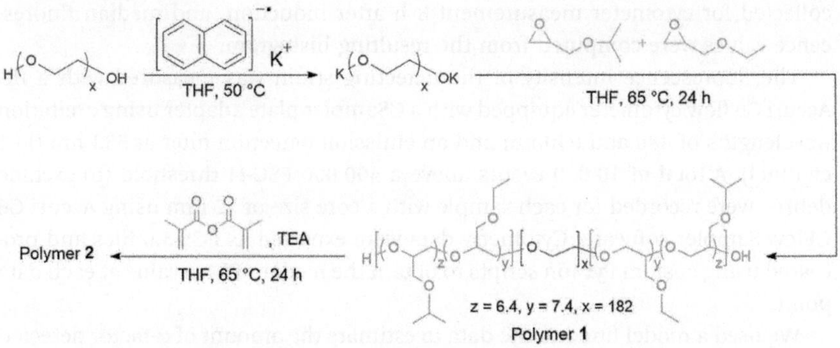

Scheme 1 Synthesis of the ABA triblock copolymer (polymer 1) and functionalization with methacrylic anhydride (polymer 2). The letter designations ($z = 6.4$, $y = 7.4$, and $x = 182$) refer to the degree of polymerization for isopropyl glycidyl ether, ethyl glycidyl ether, and ethylene oxide, respectively.

narrow dispersity ($Đ = 1.11$). Polymerizable methacrylate groups were introduced onto the chain-ends of polymer 1 to afford polymer 2 (Scheme 1). We hypothesized that the resulting polymer hydrogel would maintain its thermal- and shear-responses pre-extrusion, and then photochemically cross-link post-extrusion. The degree of chain-end functionalization (f_n) was determined by comparing the integrations of the methacrylate vinyl (6.12 and 5.55 ppm) and methyl (1.94 ppm) protons, as well as the PEO chain-end methylene protons (5.08–5.15 ppm), to their theoretical values and was found to be quantitative. The polymer was dissolved in water (20 wt%) and 2-hydroxy-2-methylpropiophenone was added as the photoradical initiator (0.1 wt%) for polymerization of the methacrylate groups.

Rheology of the functionalized triblock copolymer hydrogel

The viscoelastic properties of a 20 wt% solution of polymer 2 in water were characterized using a rheometer. The temperature-dependent viscoelastic behavior of the gel was confirmed by the presence of a sol–gel transition as defined by the intersection of the elastic (G') and viscous (G'') moduli (15.61 °C, Fig. 2a). The solution maintained a non-viscous liquid-like morphology at 5 °C and became a self-supporting hydrogel network by 21 °C. The temperature-dependent sol–gel transition was also confirmed visually, as shown in the photographs in Fig. S1.†

Hydrogels based on polymer 2 also exhibited shear-thinning behavior. The viscosity of the material decreased with increasing applied shear rate, which is indicative of a shear-thinning hydrogel. An oscillatory strain sweep experiment afforded a yield stress value of 1.33 kPa (Fig. S2 and S3†). A dynamic oscillatory strain experiment (Fig. 2b) demonstrated the strain dependent viscoelastic behavior of the gel upon repeated cycles of high (100%) and low (1%) strain. Initially, under low strain, the material exhibited a gel-like morphology as indicated by greater values for G' relative to G''. During periods of high strain, G'' exceeded G', which indicated that the gel had a higher viscous character, consistent with the material in its sol state. The material rapidly recovered to its gel state when the strain was reduced to 1% and exhibited very little mechanical hysteresis.

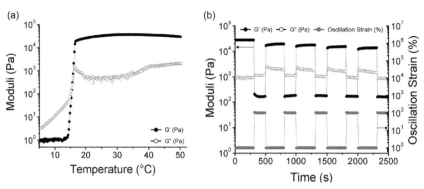

Fig. 2 Rheometrical experiments for a 20 wt% formulation of polymer 2. (a) Dynamic oscillatory temperature ramp displaying the elastic (G', filled) and viscous (G'', open) moduli. $T_{gel} = 15.61\,°C$. (b) Cyclic strain experiment demonstrating the rapid recovery of the hydrogel elastic modulus (black circles) from periods of high (100%) to low (1%) oscillatory strain (red circles). Arrows indicate the reference axis: elastic/viscous moduli (left axis) and oscillatory strain (right axis).

The post-extrusion UV cure of the gel was simulated under constant strain (1%) and frequency (1 Hz). After 120 s of equilibration time, the hydrogel was subjected to 5 mW cm^{-2} of 365 nm UV light for 420 s. The G' value of the hydrogel increased from 21.56 kPa to 94.23 kPa within 75 s of UV exposure, and the G'' value decreased from 1.88 kPa to 0.47 kPa in the same time frame. These changes to G' and G'' were indicative of the rapid chemical crosslinking of the hydrogel network *via* photo-induced radical polymerization of the methacrylated chain ends (Fig. 3a).

Direct-write 3D printing of triblock copolymer dimethacrylate hydrogels

Utilizing these three stimuli responses, a proof-of-concept cuboidal lattice was 3D printed using a pneumatic direct-write 3D printer at 5 mm s^{-1} and 20 psi with a 0.41 mm inner diameter conical nozzle. The temperature response allowed for facile processing of the hydrogel into the syringe at 5 °C in its liquid state. The

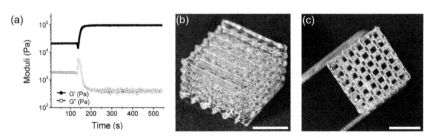

Fig. 3 (a) A rheological UV-cure experiment using a 20 mm parallel plate geometry. The hydrogel was equilibrated for 120 s before being subjected to 5 mW cm^{-2} of 365 nm UV light for 420 s under constant strain (1%) and frequency (1 Hz). (b and c) A 3D-printed proof-of-concept cuboid structure (1.9 cm by 1.9 cm by 1.2 cm). This structure was printed from a pneumatic direct-write 3D printer at 5 mm s^{-1} and 20 psi using a 0.41 mm inner diameter nozzle (scale bar: 1 cm).

hydrogel was then warmed to ambient temperature and extruded. The shear response facilitated the extrusion of the hydrogel as rod-like filaments. A cuboidal lattice structure, with dimensions of 1.9 cm by 1.9 cm by 1.2 cm, was cured post-extrusion using a custom-made UV box equipped with two 365 nm A19 UV lamps for 180 s at 3.4 mW cm^{-2} (Fig. 3b and c). The cuboidal lattice structure was chosen to simultaneously demonstrate a high-surface-area structure that promotes media circulation around the bioreactor in application, while also demonstrating the structural integrity of the printed filaments that span unsupported overhangs.

Incorporation of yeast cells and cell viability

In order to ensure the viability of the encapsulated yeast cells within the polymer **2** hydrogel, *Saccharomyces cerevisiae* constitutively expressing mCherry fluorescent proteins was inoculated into a solution of polymer **2** at 5 °C and mixed to make a homogenous mixture. A film of the resulting yeast-laden solution was cast at the same temperature, subsequently warmed to 21 °C to induce gelation, and cured to create a physically robust hydrogel film. This yeast-laden film appeared transparent after processing due to the relatively low loading concentration of the yeast cells. A small sample of the yeast-laden hydrogel film was extracted, stained, and imaged periodically over seven days. Sytox green staining results showed that cells remained 87.5% viable within the cast hydrogel at the end of the week of imaging, as seen in Fig. 4a and S4–S6.† Significant cell colony growth was observed by both confocal microscopy and optical imaging. This growth was also observed visually, as indicated by the opacity of the 3D-printed hydrogel structure in Fig. 4b and c. These results suggest that the poly(alkyl glycidyl ether)-based hydrogels are comparable in their ability to house and promote yeast cell viability over time to our previously-employed[12] F127-based hydrogels.

α-Factor production with 3D printed AMCALMs

An engineered yeast strain with upregulated α-factor production was used to demonstrate AMCALM lattices that produced a polypeptide. The pGPD promoter present in the secreting yeast strain allowed for the constitutive expression of the MFα1 gene, and thus, continuous production of the α-factor polypeptide. After the reaction, the aqueous reaction media were exposed to the detecting strain that fluoresced in the presence of α-factor.

Fig. 4 (a) A sample composite image of live cells (green channel) and dead cells (red channel) after 7 days of incubation (scale bar: 200 μm). (b) Side and (c) top-down views of the 3D-printed yeast-laden hydrogel after two rounds of α-factor synthesis in SC media (scale bar: 1 cm).

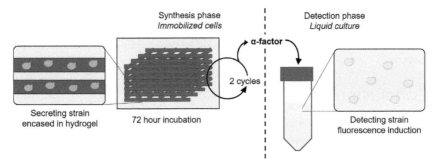

Fig. 5 Immobilized yeast within the hydrogel matrix was incubated in SC media for 72 h at 30 °C with automated shaker agitation. The media were collected at the end of the incubation period and exposed to the receiver strain in a separate tube, and the resulting fluorescence was measured. The cuboidal lattice was reused in a subsequent incubation with fresh media to produce an additional batch of α-factor.

These lattices were incubated in SC media for a period of 72 h (Fig. 5), producing an average of 268 nM ($s = 34.6$ nM) α-factor (Fig. 6). The induced fluorescence response from the cuboidal lattice with the secreting strain was 4.40 times greater than that of the control strain cuboidal lattice (without upregulated α-factor), which indicated that the upregulation of the α-factor pathway was successful in the secreting strain. These results provide evidence that the AMCALMs were suitable for the production of polypeptides, and that the products could readily diffuse out from the hydrogel matrix into the surrounding media.

One advantage of using an immobilized yeast platform for polypeptide synthesis is the potential to reuse the printed cuboidal lattices. We investigated the reusability of the poly(alkyl glycidyl ether)-based hydrogel living materials for the continued production of α-factor in subsequent batch reactions. When the printed AMCALMs were removed from their first 72 h incubation and placed into fresh media, they continued to produce an average of 259 nM ($s = 45.1$ nM) α-

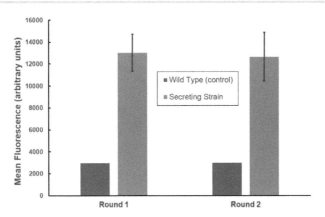

Fig. 6 Fluorescence values obtained from the quantification of α-factor produced from the wild type (control) AMCALMs and the secreting strain AMCALMs. Round 1 indicates the first incubation period of 72 h, while round 2 indicates the re-employment of the same printed materials in a second, subsequent 72 h batch reaction in fresh SC media.

factor after a second 72 h cycle (Fig. 6). The second batch production was directly comparable to the output achieved during the first batch reaction, which suggests that these immobilized cell bioreactors could be useful for continuous whole-cell catalysis.

Conclusion

In conclusion, we developed an ABA triblock copolymer based on a poly(alkyl glycidyl ether) that is suitable for 3D printing yeast-laden hydrogels for whole-cell catalysis. These polymer hydrogels exhibit temperature, shear, and UV light responsive behavior, which are integral to the preparation of the hydrogel ink and subsequent printing. The post-extrusion chemical crosslinking of the polymer micelles induced by UV light is particularly important to fabricate robust forms that do not degrade or dissolve over time.

The poly(alkyl glycidyl ether) based hydrogels also proved to have a negligible effect on the viability and biological activity of the encapsulated yeast cells. The engineered yeast-laden living materials were shown to be capable of producing an average of 263.5 nM α-factor during two consecutive batch reactions, exhibiting no significant reduction in efficiency between the two cycles.

In this work, we demonstrated that polypeptides can be produced using an AMCALM platform. This result, combined with the prominence of the α-factor leader sequence in recombinant protein design and production, suggests that these materials could be employed in the whole-cell catalysis of other higher-value molecules in a sustainable and continuous manner.

Conflicts of interest

We have no conflicts to declare.

Acknowledgements

The authors thank Dr Eric Klavins of the University of Washington for continued consultation and expertise on yeast and biological systems. A. N. gratefully acknowledges the support of this research by the National Science Foundation CAREER grant (DMR 1752972). This work was also partially funded by the Army Research Office (W911NF-17-1-0595). The authors thank The Biology Imaging Facility at the University of Washington for the use of the SP5 confocal microscope.

References

1 Y.-H. P. Zhang, J. Sun and Y. Ma, *J. Ind. Microbiol. Biotechnol.*, 2017, **44**, 773–784.

2 Y. Li and C. D. Smolke, *Nat. Commun.*, 2016, **7**, 12137.

3 K. Thodey, S. Galanie and C. D. Smolke, *Nat. Chem. Biol.*, 2014, **10**, 837–844.

4 A. R. Awan, B. A. Blount, D. J. Bell, W. M. Shaw, J. C. H. Ho, R. M. McKiernan and T. Ellis, *Nat. Commun.*, 2017, **8**, 15202.

5 T. U. Gerngross, *Nat. Biotechnol.*, 2004, **22**, 1409–1414.

6 C. J. Paddon, P. J. Westfall, D. J. Pitera, K. Benjamin, K. Fisher, D. McPhee, M. D. Leavell, A. Tai, A. Main, D. Eng, D. R. Polichuk, K. H. Teoh,

D. W. Reed, T. Treynor, J. Lenihan, H. Jiang, M. Fleck, S. Bajad, G. Dang, D. Dengrove, D. Diola, G. Dorin, K. W. Ellens, S. Fickes, J. Galazzo, S. P. Gaucher, T. Geistlinger, R. Henry, M. Hepp, T. Horning, T. Iqbal, L. Kizer, B. Lieu, D. Melis, N. Moss, R. Regentin, S. Secrest, H. Tsuruta, R. Vazquez, L. F. Westblade, L. Xu, M. Yu, Y. Zhang, L. Zhao, J. Lievense, P. S. Covello, J. D. Keasling, K. K. Reiling, N. S. Renninger and J. D. Newman, *Nature*, 2013, **496**, 528–532.

7 W. C. DeLoache, Z. N. Russ, L. Narcross, A. M. Gonzales, V. J. J. Martin and J. E. Dueber, *Nat. Chem. Biol.*, 2015, **11**, 465–471.

8 S. Yang, Q. Fei, Y. Zhang, L. M. Contreras, S. M. Utturkar, S. D. Brown, M. E. Himmel and M. Zhang, *Microb. Biotechnol.*, 2016, **9**, 699–717.

9 A. M. Davy, H. F. Kildegaard and M. R. Andersen, *Cell Syst.*, 2017, **4**, 262–275.

10 P. P. Peralta-Yahya, F. Zhang, S. B. del Cardayre and J. D. Keasling, *Nature*, 2012, **488**, 320–328.

11 P. J. Verbelen, D. P. De Schutter, F. Delvaux, K. J. Verstrepen and F. R. Delvaux, *Biotechnol. Lett.*, 2006, **28**, 1515–1525.

12 A. Saha, T. G. Johnston, R. T. Shafranek, C. J. Goodman, J. G. Zalatan, D. W. Storti, M. A. Ganter and A. Nelson, *ACS Appl. Mater. Interfaces*, 2018, **10**, 13373–13380.

13 P. S. J. Cheetham, K. W. Blunt and C. Bocke, *Biotechnol. Bioeng.*, 1979, **21**, 2155–2168.

14 S. Nagarajan, A. L. Kruckeberg, K. H. Schmidt, E. Kroll, M. Hamilton, K. McInnerney, R. Summers, T. Taylor and F. Rosenzweig, *Proc. Natl. Acad. Sci. U. S. A.*, 2014, **111**, E1538–E1547.

15 B. A. E. Lehner, D. T. Schmieden and A. S. Meyer, *ACS Synth. Biol.*, 2017, **6**, 1124–1130.

16 A. Lode, F. Krujatz, S. Brüggemeier, M. Quade, K. Schütz, S. Knaack, J. Weber, T. Bley and M. Gelinsky, *Eng. Life Sci.*, 2015, **15**, 177–183.

17 A. Townsend-Nicholson and S. N. Jayasinghe, *Biomacromolecules*, 2006, **7**, 3364–3369.

18 Y. Liu, M. H. Rafailovich, R. Malal, D. Cohn and D. Chidambaram, *Proc. Natl. Acad. Sci. U. S. A.*, 2009, **106**, 14201–14206.

19 I. Letnik, R. Avrahami, J. S. Rokem, A. Greiner, E. Zussman and C. Greenblatt, *Biomacromolecules*, 2015, **16**, 3322–3328.

20 A. Brimmo, P. A. Goyette, R. Alnemari, T. Gervais and M. A. Qasaimeh, *Sci. Rep.*, 2018, **8**, 10995.

21 T. Dahlberg, T. Stangner, H. Zhang, K. Wiklund, P. Lundberg, L. Edman and M. Andersson, *Sci. Rep.*, 2018, **8**, 3372.

22 S. Mamatha, P. Biswas, P. Ramavath, D. Das and R. Johnson, *Ceram. Int.*, 2018, **44**, 19278–19281.

23 R. Melnikova, A. Ehrmann and K. Finsterbusch, *IOP Conf. Ser.: Mater. Sci. Eng.*, 2014, **62**, 012018.

24 M.-W. Sa, B.-N. B. Nguyen, R. A. Moriarty, T. Kamalitdinov, J. P. Fisher and J. Y. Kim, *Biotechnol. Bioeng.*, 2018, **115**, 989–999.

25 R. U. Kiran, S. Malferrari, A. Van Haver, F. Verstreken, S. N. Rath and D. M. Kalaskar, *Mater. Des.*, 2018, **162**, 263–270.

26 S. C. Ligon, R. Liska, J. Stampfl, M. Gurr and R. Mülhaupt, *Chem. Rev.*, 2017, **117**, 10212–10290.

27 M. J. Kimlinger and R. S. Martin, *Electroanalysis*, 2018, **30**, 2241–2249.

28 M. Müller, J. Becher, M. Schnabelrauch and M. Zenobi-Wong, *Biofabrication*, 2015, **7**, 035006.

29 L. E. Bertassoni, J. C. Cardoso, V. Manoharan, A. L. Cristino, N. S. Bhise, W. A. Araujo, P. Zorlutuna, N. E. Vrana, A. M. Ghaemmaghami, M. R. Dokmeci and A. Khademhosseini, *Biofabrication*, 2014, **6**, 024105.

30 Y.-C. Li, Y. S. Zhang, A. Akpek, S. R. Shin and A. Khademhosseini, *Biofabrication*, 2016, **9**, 012001.

31 S. Patra and V. Young, *Cell Biochem. Biophys.*, 2016, **74**, 93–98.

32 M. Schaffner, P. A. Rühs, F. Coulter, S. Kilcher and A. R. Studart, *Sci. Adv.*, 2017, **3**, eaao6804.

33 C. Bader, W. G. Patrick, D. Kolb, S. G. Hays, S. Keating, S. Sharma, D. Dikovsky, B. Belocon, J. C. Weaver, P. A. Silver and N. Oxman, *3D Print. Addit. Manuf.*, 2016, **3**, 79–89.

34 J. M. Lee and W. Y. Yeong, *Adv. Healthcare Mater.*, 2016, **5**, 2856–2865.

35 A. S. Hoffman, *Adv. Drug Delivery Rev.*, 2002, **54**, 3–12.

36 K. Y. Lee and D. J. Mooney, *Chem. Rev.*, 2001, **101**, 1869–1880.

37 K. Smetana, *Biomaterials*, 1993, **14**, 1046–1050.

38 K. Smetana, J. Vacík, D. Součková, Z. Krčová and J. Šulc, *J. Biomed. Mater. Res.*, 1990, **24**, 463–470.

39 D. S. W. Benoit, M. P. Schwartz, A. R. Durney and K. S. Anseth, *Nat. Mater.*, 2008, **7**, 816–823.

40 J. H. Lee, H. W. Jung, I.-K. Kang and H. B. Lee, *Biomaterials*, 1994, **15**, 705–711.

41 D. E. Discher, P. Janmey and Y. L. Wang, *Science*, 2005, **310**, 1139–1143.

42 A. Buxboim, K. Rajagopal, A. E. X. Brown and D. E. Discher, *J. Phys.: Condens. Matter*, 2010, **22**, 194116.

43 Z. Liu, K. E. J. Tyo, J. L. Martínez, D. Petranovic and J. Nielsen, *Biotechnol. Bioeng.*, 2012, **109**, 1259–1268.

44 T. Achstetter, *Mol. Cell. Biol.*, 1989, **9**, 4507–4514.

45 E. H. Baba and I. Berkower, *Biochem. Biophys. Res. Commun.*, 1992, **184**, 50–59.

46 J. A. Rakestraw, S. L. Sazinsky, A. Piatesi, E. Antipov and K. D. Wittrup, *Biotechnol. Bioeng.*, 2009, **103**, 1192–1201.

47 A. S. Robinson, V. Hines and K. D. Wittrup, *Nat. Biotechnol.*, 1994, **12**, 381–384.

48 J. Staniulis, *Biologija*, 2006, **3**, 49–53.

49 Y. Chigira, T. Oka, T. Okajima and Y. Jigami, *Glycobiology*, 2008, **18**, 303–314.

50 T. Kjeldsen, A. Frost Pettersson and M. Hach, *J. Biotechnol.*, 1999, **75**, 195–208.

51 C. R. Fellin, S. M. Adelmund, D. G. Karis, R. T. Shafranek, R. J. Ono, C. G. Martin, T. G. Johnston, C. A. DeForest and A. Nelson, *Polym. Int.*, 2018, **68**, 1238–1246.

52 M. A. Winnik and A. Yekta, *Curr. Opin. Colloid Interface Sci.*, 1997, **2**, 424–436.

53 B. F. Lee, M. J. Kade, J. A. Chute, N. Gupta, L. M. Campos, G. H. Fredrickson, E. J. Kramer, N. A. Lynd and C. J. Hawker, *J. Polym. Sci., Part A: Polym. Chem.*, 2011, **49**, 4498–4504.

54 S. Heinen, S. Rackow, A. Schäfer and M. Weinhart, *Macromolecules*, 2017, **50**, 44–53.

55 Y. Satoh, K. Miyachi, H. Matsuno, T. Isono, K. Tajima, T. Kakuchi and T. Satoh, *Macromolecules*, 2016, **49**, 499–509.

56 K. P. Barteau, M. Wolffs, N. A. Lynd, G. H. Fredrickson, E. J. Kramer and C. J. Hawker, *Macromolecules*, 2013, **46**, 8988–8994.

Faraday Discussions

Multidimensional micro- and nano-printing technologies: general discussion

Helena S. Azevedo, Adam Braunschweig, Joseph P. Byrne,
Yuri Diaz Fernandez, Jeff Gildersleeve, Laura Hartmann, Mia Huang,
Alshakim Nelson, Bart Jan Ravoo, Stephan Schmidt, Tekla Tammelin,
W. Bruce Turnbull, Zijian Zheng and Dejian Zhou

DOI: 10.1039/C9FD90061F

Adam Braunschweig opened the discussion of the paper by Peter Seeberger: In many systems that involve binding in the glycocalyx, there are often multiple different protein–receptor interactions that occur. These may involve multivalent interactions from a single protein to a simple type of sugar, or may involve binding to different types of sugars to achieve a desired response. How do you imagine that the next generation of glycan microarrays can be built to interrogate these interactions by possibly printing different materials with programmed spacings and ratios within the same feature of an array?

Peter Seeberger responded: In the future, glycan arrays can be built on surfaces that will be more three dimensional. Such surfaces my incorporate polymers to expose the glycan chains in ways that make them look more like they do on the cell surface.

Adam Braunschweig asked: Multivalency and its role in glycan–glycan binding protein interactions is a major topic in the field and something that remains poorly understood. As yet, there is no accepted or broadly adopted model for taking binding data – either or arrays or in solution – and using these data for anticipating how multivalent interactions may affect quantitatively the binding data. Given your work in the field, how do you feel we can begin resolving this challenge in the field?

Peter Seeberger answered: There is virtually no data that can be used for comparison. Ideally, we would all agree on a standard way to print arrays and to measure binding. That approach could be used as a metric to compare data. There are some efforts underway in the European Union but so far there is no agreement as to how one should go about it.

Helena S. Azevedo remarked: In the movie showing the growth of carbohydrate fibres, what is the role of the hexafluoroisopropanol?

Peter Seeberger replied: Hexafluoroisopropanol is the solvent in which the disaccharide is dissolved. The remainder is water, where the disaccharide is poorly soluble and where the fibrils are being formed.

W. Bruce Turnbull commented: I think Peter highlights a very important challenge in glycobiology, which is defining what is the structure of the glycocalyx that we might want to mimic. We know what are the components in the glycocalyx, but not its three-dimensional structure. We also need new tools to probe the 3D structure of a natural glycocalyx to help guide the reconstruction of model glycocalyxes for biological studies.

Bart Jan Ravoo asked: You have emphasized the power of organic chemistry but do you see a role for biomimetic/bioinspired enzyme catalyzed glycosylation reactions in automated glycan synthesis?

Peter Seeberger commented: Automated glycan assembly and enzymatic syntheses are complementary and not in competition. If one or the other is better for making a structure, then use that method. I strongly believe that in many cases a combination of both may be the most powerful method. For example, you can make a backbone polysaccharide including non-natural monosaccharides by automated synthesis and then you modify it with the help of enzymes.

Adam Braunschweig remarked: Peter, your automated synthesizer is a revolutionary advance in carbohydrate chemistry. The synthesis of carbohydrates, even on an automated machine, is far more complex than the automated synthesis of oligonucleotides and oligopeptides. To get this machine into the hands of those who are not experts in carbohydrate synthesis, you will have to simplify the synthetic planning so that non-experts can use the tool for synthesizing carbohydrates. How do you plan to do this?

Peter Seeberger responded: We are already on a program that can break down a target oligosaccharide and provides a suggestion for the user as to what building blocks (s)he should use on the synthesizer. I hope that we can reveal that program in 2020.

Adam Braunschweig opened the discussion of the paper by Nathan Gianneschi: You have shown new organic materials that can be used to make complex 3D materials with properties important for interfacing with cells and other biological systems. If you were given an imaginary lithography system that could print your system, and the resolution of the features in 3D were limited by the material and not the printer, what are the inherent attributes of your materials that might limit feature dimensions?

Nathan Gianneschi answered: If feature dimensions are limited to structures that give rise to specific properties, then those would define the useful limit. For example, spacing receptor functionalities to sparingly, or too tightly, would and does change how they engage cells, for example. I think this would depend, in other words, on the type of function you are seeking. The easy answer is that you really want arbitrary chemistry and resolution to be possible, so that you can explore the design and ultimately engineering of customized materials in 3D.

Laura Hartmann continued: Could you attach your cyclic peptides onto polymers, *e.g.* at the chain ends, to reduce the amount of peptide you need to form your hydrogel?

Nathan Gianneschi replied: Very good question. We have done this to some degree in work published several years ago in *Chemical Communications*. In that work, we used a lot of peptide, attached as a brush of cyclic oligomers along a backbone of a polymer. Used properly this could be a significant advance in how to program gelation, but use far less peptide.

Yuri Diaz Fernandez asked: Did you observe a different metabolic activity for the yeast cells extruded within your gel compared to planktonic cells? If so, what is the reason for that?

Alshakim Nelson answered: We did not compare the metabolic activity of the immobilized and non-immobilized yeast in the production of alpha factor. However, in other systems that we have investigated, we observed that there is a difference in the growth rates of the cells, in that the hydrogels appear to slow their growth. Over time, the immobilized yeast visually appears to fully take over the space within the hydrogel.

Tekla Tammelin then asked: What is the reason to select 3D architecture in this application?
No speakers have yet replied.

Zijian Zheng began a general discussion by addressing all three speakers: Funding available for lithography is being reduced – will there be the time and funding to develop complex structures?

Nathan Gianneschi commented: Complexity that gives rise to transformational science and technology will always be funded. It has to be about more than seeking complexity and more about seeking function. In addition, the bandwidth for funding tool development needs to be maintained and battled for.

Mia Huang then spoke: The technologies you all have created are really fascinating, but I wonder if there is such a thing as "too much engineering" that veers us away from what a biological interface looks like? Do you think there is a theoretical limit?
No speakers have yet replied.

W. Bruce Turnbull addressed Alshakim Nelson: Regarding the processibility of cellulose – is this not a solved problem? The regenerated cellulose industry has very effective ways of solubilising cellulose with sodium hydroxide and carbon disulfide to form viscose as a liquid which can be dispensed and returned to crystalline cellulose upon acidification. Could this be (or has this been) used for 3D printing applications?

Alshakim Nelson responded: Not to my knowledge. Bio-sourced and biodegradable materials for 3D printing represent a significant gap in the field. Cellulose-based materials could represent a great opportunity to fill this gap if there is a way to make these materials compatible with 3D printing processes.

Adam Braunschweig concluded the discussion by addressing Alshakim Nelson: Multivalency and how to study and quantify multivalent interactions remains an unresolved open chemistry in glycobiology. One of the reasons is that sugars, proteins, and cells exist within a 3D environment, whereas most of our methods for studying multivalency are in solution or on 2D substrates. Your materials can control structure in 3D. Can your materials be used to build systems for studying the multivalency with structured 3D probes?

Alshakim Nelson responded: I agree that the investigation of multivalency in natural systems is complex, and requires model systems that provide 3D topography to present ligands and receptors in a controlled fashion. Our approach utilizes a hierarchically organized system wherein triblock copolymers self-assembled in aqueous solution to form nanostructures that afford a hydrogel. We then use a direct-write 3D printer to print microscale structures of these hydrogels. Multivalency occurs at the nanoscale, and therefore, we would have to alter the structure of the polymers to present carbohydrate ligands in a defined manner. Perhaps this could be possible by utilizing some of the polymer-based approaches for creating multivalent ligands.

Conflicts of interest

Peter Seeberger holds significant shares in GlycoUniverse, the company that produces the automated glycan synthesizer.

Faraday Discussions

PAPER

Multivalent binding of concanavalin A on variable-density mannoside microarrays

Daniel J. Valles, ⓘ *[abc] Yasir Naeem, ⓘ [bc] Angelica Y. Rozenfeld,[bc] Rawan W. Aldasooky,[bd] Alexa M. Wong, ⓘ [bc] Carlos Carbonell, ⓘ [bc] David R. Mootoo ⓘ [ac] and Adam B. Braunschweig ⓘ [abce]

Received 31st March 2019, Accepted 18th April 2019

DOI: 10.1039/c9fd00028c

Interactions between cell surface glycans and glycan binding proteins (GBPs) have a central role in the immune response, pathogen–host recognition, cell–cell communication, and a myriad other biological processes. Because of the weak association between GBPs and glycans in solution, multivalent and cooperative interactions in the dense glycocalyx have an outsized role in directing binding affinity and selectivity. However, a major challenge in glycobiology is that few experimental approaches exist for examining and understanding quantitatively how glycan density affects avidity with GBPs, and there is a need for new tools that can fabricate glycan arrays with the ability to vary their density controllably and systematically in each feature. Here, we use thiol−ene reactions to fabricate glycan arrays using a recently developed photochemical printer that leverages a digital micromirror device and microfluidics to create multiplexed patterns of immobilized mannosides, where the density of mannosides in each feature was varied by dilution with an inert spacer allyl alcohol. The association between these immobilized glycans and FITC-labeled concanavalin A (ConA) – a tetrameric GBP that binds to mannosides multivalently – was measured by fluorescence microscopy. We observed that the fluorescence decreased nonlinearly with increasing spacer concentration in the features, and we present a model that relates the average mannoside–mannoside spacing to the abrupt drop-off in ConA binding. Applying these recent advances in microscale photolithography to the challenge of mimicking the architecture of the glycocalyx could lead to a rapid understanding of how information is trafficked on the cell surface.

[a]The PhD Program in Chemistry, Graduate Center of the City University of New York, 365 5th Ave, New York, NY 10016, USA. E-mail: Daniel.Valles@asrc.cuny.edu

[b]The Advanced Science Research Center at the Graduate Center of the City University of New York, 85 St. Nicholas Terrace, New York, NY 10031, USA

[c]Department of Chemistry, Hunter College, 695 Park Ave, New York, NY 10065, USA

[d]Department of Chemistry, Lehman College, 250 Bedford Park Blvd W, Bronx, NY 10468, USA

[e]The PhD Program in Biochemistry, Graduate Center of the City University of New York, 365 5th Ave, New York, NY 10016, USA

Introduction

Many biological processes, from the immune response to host–pathogen inter-actions, are mediated by recognition between glycan binding proteins (GBPs) and glycans within the glycocalyx – the dense layer of glycolipids, glycoproteins, and glycopolymers on the surface of every eukaryotic cell.[1-4] In this environment, multivalency, whereby contacts form between a GBP and multiple glycans simultaneously, plays an outsized role in determining substrate specificity and binding avidity.[5-9] As a result of multivalency, increases in avidity up to 10^6 M^{-1} are commonly observed between GBPs and substrates that present multiple glycans in close enough proximity to allow for multipoint binding, a phenomenon termed the "cluster glycoside effect".[10-13] Despite its central role in glycobiology, quantitative measures of multivalency, particularly in substrates that are designed to mimic the presentation of glycans on the cell surface, remain a significant and unresolved experimental challenge because of difficulties associated with surface immobilization chemistry, the paucity of lithographic methods that can print glycans with control over density, and unsatisfactory models for anticipating the impact of multivalent interactions on the surface.

Creating substrates that can systematically vary glycan presentation requires control over the surface immobilization chemistry, while employing printing technologies that are compatible with delicate organic materials, like glycans. As such, strategies for controlling the glycan surface presentation must consider both the printing method and the immobilization chemistry. The surface chemistry used to prepare the Consortium for Functional Glycomics glycan array, for example, involves pin-printing amino-functionalized glycans onto NHS-activated surfaces.[14,15] Other common immobilization chemistries involve: func-tionalizing a substrate with a monolayer of N-hydroxysuccinimidyl ester to covalently immobilize 3'-amine oligonucleotides conjugated with glycans;[16] epoxy-activated substrates that react with amino-functionalized glycans;[17,18] thiol–ene photochemical click reactions;[19-23] patterning fluorous derivatized glycans onto commercially available Teflon/epoxy coated microscope slides;[24] and immobilizing lipid-linked oligosaccharides onto nitrocellulose substrates.[25,26] Alternatively, the Gildersleeve group conjugated glycans to bovine serum albumin (BSA), creating "neoglycoproteins" that were then themselves immobilized to epoxide-coated glass substrates, where the glycan density could be manipulated by either mixing with non-conjugated BSA or varying the density of glycans within each neoglycoprotein.[27-29] The patterning of glycans into arrays is typically accomplished by creating spots using an inkjet or pin printer.[14] With these technologies, solutions of the appropriately functionalized glycans are deposited onto substrates functionalized with the complementary reactivity, resulting in features of covalently immobilized glycans with typical diameters of ~500 μm. These printing technologies are popular because they are non-destructive towards delicate glycans and are compatible with common immobilization reactions. However, other techniques could substantially reduce the feature diameter, and thereby increase the number of spots per array and limit the amount of expensive glycan or GBP needed for microarray analysis. Some of these next-generation printing techniques that have been explored recently in the context of glycan microarrays include: scanning-probe lithography,[22,30] microcontact printing,[21,31,32] and photochemical patterning enabled by a digital micromirror device (DMD).[20] Ideally, a glycan arraying platform – which constitutes immobilization chemistry and the lithography method for patterning – should be compatible with the immobilization chemistry, create features with <100 μm diameters, and possess the ability to control the glycan density within each feature.

Varying the density of glycans is particularly important for glycan microarrays because doing so is necessary for investigating how multipoint binding between GBPs and a glycan-coated surface affects avidity. To this end, diverse approaches have been adopted to vary the glycan density within the features of a microarray. Liang and coworkers, for example, studied the binding of concanavalin A (ConA) to mannosides and oligomannosides that were immobilized onto a NHS-activated glass substrate prepared *via* robotic pin-printing.[33] In this study, they varied the concentration of mannose in the printing solutions from 0.6 to 100 μM, and studied the binding to fluorophore-labeled lectin. They measured increased ConA avidity to the substrate at spots printed with higher concentration mannose solutions. The authors concluded that when the printing solutions dropped to 1 μM and 0.6 μM, the mannosides were too widely spaced for ConA to achieve multivalent binding, which was necessary to obtain avidity sufficiently high enough for the GBP to remain on the surface. Alternatively, Oyelaran and coworkers fabricated variable glycan-density arrays with BSA–neoglycoproteins in order to investigate how multivalency affects GBP avidity.[28] The binding of fluorescently labeled ConA to neoglycoproteins with systematically varying glycan (monomannose and oligomannose) concentration was assessed. The results of this experiment showed that, as the ratio between the mannose-conjugated neoglycoprotein and unmodified BSA changed from 1 : 0 to 1 : 7 in the printing solution, the spacing between glycans increased. The increased spacing between the glycans resulted in a decrease in the overall multivalent opportunities on the surface, thus decreasing the overall fluorescence of the feature, which they concluded was a consequence of the inability of ConA to bind multivalently as the monomannose density was reduced. However, at low concentrations of the oligomannose, ConA was still able to bind the features, as multivalent opportunities persisted. Concerned that the glycan monolayers typically used in glycan arrays bind GBPs weakly as a result of surface roughness, Godula and coworkers prepared glycopolymer microarrays. Attempting to mimic the natural presentation of mucins, which are highly glycosylated cell-surface proteins, they created heavily glycosylated synthetic brush polymers using reversible addition–fragmentation chain-transfer (RAFT) polymerization and printed them into microarrays.[34,35] Aminooxy-labeled glycans were grafted at different densities onto polymer chains of different lengths in order to examine how density affected avidity. The affinities of four lectins (soybean agglutinin (SBA), *Wisteria floribunda* lectin (WFL), *Vicia villosa*-B-4 agglutinin (VVA), and *Helix pomatia* agglutinin (HPA)) to the glycosylated brush polymers were assessed. SBA, WFL, and VVA all displayed multivalent binding, as the dissociation constants (K_d) decreased with increasing glycan grafting density. HPA, however, did not show any change in K_d despite the changes in glycan density, thus indicating that it did not associate to the brush polymer with avidity. Thus, glycan arrays are powerful platforms to assess the impact of multivalency on GBP binding, but despite these important contributions, there is still a need for techniques that can reduce the printing areas and systematically vary the surface concentration.

Another challenge that arises in understanding these data is scaling the changes in avidity to a molecular level understanding of multivalency, and, to this end, several models have been proposed. In a review on multivalency, Mammen *et al.* suggested[36] that the enthalpy of the system is dominated by the strain on the multivalent ligands, while entropic changes are dependent upon perturbations to the lectin conformation. Brewer and coworkers used isothermal titration calorimetry to investigate the thermodynamics of the binding between ConA and multivalent mannosides in solution.[37] The results showed that, as the number of mannosides on a ligand increases, the avidity to ConA also increases, and that the enthalpy of the system increased linearly with the number of mannosides. Brewer also emphasized the importance of the entropy of the system by showing how the conformation of the lectin, SBA in this case, affects the multivalency on a mucin. Because of the structure of SBA and its conformation when it binds to mucin galactose, SBA can "bind and jump" from one residue to the next.[38] Houseman and coworkers argued that it is particularly important to study multivalent interactions on surfaces because of the increased control over glycan spacing in monolayers, and the fact that immobilized ligands are in an environment that reduces nonspecific binding.[39] Here, we build upon these investigations to study ConA–mannoside multivalency by employing a new method for printing glycan microarrays,[20] which allows for the density of glycans within an array to be varied systematically. This printer combines microfluidics, a digital micromirror device, and a reactive surface to control the ratio of glycan and spacer within a feature. We use this tool to print features of variable-density mannosides onto a surface and study their binding to solutions with different concentrations of FITC-labeled ConA. Reduced feature dimensions provide a route to take multiple readings at each binding condition in order to obtain statistically robust data. Finally, we relate the changes in avidity to changes in mannoside spacing and suggest that ditopic binding is necessary for the protein to achieve sufficient avidity to remain surface bound. This work presents a versatile new way of studying quantitatively the relationship between surface density and avidity in situations where multivalency is central.

Materials and methods

Synthesis

The mannoside, pent-4-enyl-α-D-mannopyranoside (α-**Man**), was prepared as previously reported and characterized by ^{1}H NMR and mass spectrometry, and all spectroscopic data were consistent with the literature data.[20]

Surface preparation

The thiol-terminated Si/SiO$_x$ slides were prepared following previously reported literature procedures.[20,40] Si/SiO$_2$ wafers were purchased from Nova Electronic Materials (USA), (3-mercaptopropyl)trimethoxysilane was purchased from Gelest (USA), and all other chemicals were purchased from VWR and used as received. The wafers were cleaned by submerging them in piranha solution (3 : 1 H$_2$SO$_4$: H$_2$O$_2$) for 15 min and then taken out and rinsed with MilliQ H$_2$O. The substrates were dried under a stream of air. Once dried, the slides were placed in a 120 mL PhMe solution containing 4.5 mL of (3-mercaptopropyl)trimethoxysilane, which was heated to 37 °C in a H_2O bath for 4 h. The substrates were then rinsed with PhMe, a mixture of PhMe and EtOH, and then EtOH. Finally, the glass slides were cured in an oven at 105 °C for 18 h and stored in MeOH at 4 °C until used.

Printer design and microarray fabrication

A TERA Fab E-Series printer was modified to print the glycan microarrays. The system was equipped with an LED (405 nm) light source that reflects off a DMD that passes through an objective to focus the pattern onto the substrate, which was mounted on a piezoelectric stage. The patterns were designed on Adobe Illustrator and converted to BMP files for the DMD software. DMF solutions of the mannoside and allyl alcohol were made at concentrations of 200 mM and 100 mM of the photoinitiator diphenyl(2,4,6-trimethylbenzoyl)phosphine oxide (TPO). The solution was charged in a syringe and placed in a syringe pump (New Era Pump Systems Inc., USA) with PEEK tubing to flow the solution through the fluid cell. Eleven different 0.5 mL solutions were prepared for printing, where the concentration of alkene (glycan + spacer) remained at 200 mM, but the mole fraction, χ, of glycan varied from 1 to 0.01. The syringe pump flow rates were set to 5 µL min^{-1} during exposure time. A syringe with DMF was loaded into the syringe pump and set with a flow rate of 100 µL min^{-1} for 5 min in between each exposure, in order to wash away the remnant glycan solution in the fluid cell. After immobilization was completed for all prints, the substrate was rinsed in 200 proof EtOH and then sonicated in DMF for 10 min. The surface of the array was then passivated by immersing it in a 1% w/v BSA solution in PBS (10 mM PBS, pH = 7.4) for 30 min.

Lectin binding and analysis

A previously described microfluidic incubation chip[20] was used to introduce the lectin solutions to the glycan microarrays. FITC-labeled ConA was purchased from Vector Laboratories Inc. and was used as received. PBS solutions were prepared (PBS buffer, 10 mM; $MgCl_2$, 0.9 mM; $CaCl_2$, 0.5 mM; pH = 7.4; 0.01% Tween20) with lectin concentrations of 0.5–0.001 mg mL^{-1}. The ConA solutions were injected into the microarray using a custom built microfluidic chip with 250 µm-wide channels that can introduce 11 different lectin solutions to the microarray simultaneously. The surface was incubated for 16 h at room temperature with 8 different ConA solutions (0.5–0.001 mg mL^{-1}, 4.8 µM to 9.6 nM) and 3 channels containing only PBS buffer as controls. Following the incubation, the array was washed by immersing the chip into a buffer solution (PBS buffer, 10 mM; $MgCl_2$, 0.9 mM; $CaCl_2$, 0.5 mM; pH = 7.4; 0.01% Tween20) for 10 min, and this washing was repeated with fresh solution 3 times. Binding was analyzed with an Olympus BX60 fluorescence microscope (540–585 nm long pass filter). The images were analyzed with ImageJ[41] in order to measure the fluorescence of each feature, with fluorescence reported as normalized fluorescence (NF = feature fluorescence/background fluorescence) from the average measurement of 3 different features.

Results and discussion

Here, we investigate the multivalent binding of ConA with mannoside-patterned glycan microarrays (Fig. 1). ConA is a plant lectin that has two states: a homotetramer that exists at pH > 7, and a homodimer when the pH < 6, and both states

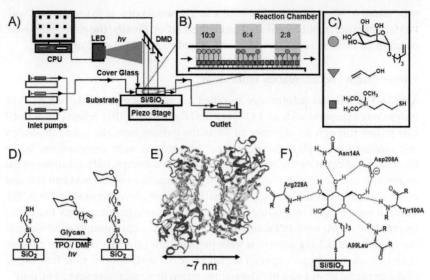

Fig. 1 (A) The printing platform combines a digital micromirror device, piezoelectric stage, and microfluidics to perform photochemical surface reactions. (B) The surface density is manipulated by varying the mole fraction, χ, of α-Man (green circles) to allyl alcohol spacer (blue triangles). (C) The three molecules involved in the surface functionalization. (D) Thiol–ene photochemical surface chemistry is used for the immobilization of α-Man and allyl alcohol. (E) An image from the RCSB PDB (http://rcsb.org) of PDB ID 3QLQ[42] shows the ribbon structure of ConA in its tetrameric form; each monomer binds two metal ions and one glycan (mannoside). The spacing between each glycan binding site is approximately 7 nm. (F) The bonding interactions presumed to occur between an immobilized mannoside and ConA.

exists in equilibria when the pH is 6–7. The monomeric subunit of ConA consists of 237 amino acids with a mass of 25.5 kDa and has three binding sites. Of the three sites, two are for metal ions, typically Ca^{2+} and Mn^{2+}, and the third for a glycan. ConA first binds to metal ions, which opens up the binding site for the glycan. ConA preferentially binds mannosides, and also binds glucosides, but with a lower affinity.[42] In order to study the effect of multivalency on ConA–mannoside avidity, a glycan microarray was prepared where the glycan density was varied systematically in each feature. This array was prepared using a new photochemical printer described recently by our group.[20] Briefly, the printer integrates a TERA-Fab E Series printer, which consists of a DMD (1024×768 mirrors) and an LED (405 nm, 32 mW cm^{-2}), which are coupled with a piezo-electric stage that supports the substrate (Fig. 1A). A CPU coordinates the spatiotemporal delivery of light onto the substrate and the movement of the stage. On the TERA-Fab E Series stage, we mounted a fluid cell,[40] in which photo-chemical reactions occur on the functionalized substrate and in solution, and microfluidics deliver and remove reagents from the fluid cell (Fig. 1B–C). We demonstrated the capabilities of this photochemical printer by preparing multi-plexed grafted-from brush polymer arrays[43] and multiplexed glycan microarrays.[20]

Here, a thiol–ene photochemical click reaction was used to immobilize different ratios of the alkene-labeled mannoside, pent-4-enyl-α-D-mannopyr-anoside (α-Man), and the spacer, allyl alcohol, in different ratios to form features on the substrate (Fig. 1D). The thiol–ene reaction is a popular reaction for the surface immobilization of biological probes because it is biorthogonal, proceeds rapidly and in high yield, and has few byproducts.[44–46] Previously, ourselves and others have shown that it can be used to immobilize dyes,[40] glycans[20] and glycopolymers,[22] confirming its compatibility for preparing microarrays, and we have studied the reaction kinetics to determine the reaction time required to proceed to completion. Here, we perform the thiol–ene reaction in the printer described above to create glycan arrays by mounting a Si/SiO$_2$ wafer that is functionalized with 3-(mercaptopropyl)trimethoxysilane into the fluid cell. Microfluidics introduced solutions containing different ratios of the alkene-functionalized α-**Man** and the spacer allyl alcohol (total concentration α-**Man** + allyl alcohol = 200 mM) in DMF with the photosensitizer diphenyl(2,4,6-trimethylbenzoyl)phosphine oxide (TPO, 100 mM). The ratio of α-**Man** : allyl alcohol was varied from 1 to 0.01. Features of 45 × 45 μm were prepared by irradiating the surface in the presence of the printing solution for 20 min at 8 mW cm^{-2} intensity, and 32 features from each α-**Man** : allyl alcohol ratio were printed. Between solutions, the substrate was washed with DMF for 5 min at a flow rate of 100 μL min^{-1} in order to minimize contamination. Once the printing was completed, the substrate was washed with EtOH, sonicated in DMF for 10 min, and then the unreacted areas of the surface were passivated by immersing it into a 1% (w/v) bovine serum albumin (BSA) solution. BSA was selected as a passivating agent because it has been previously used by us[20,22,30] and others[16,47] to prevent non-specific adsorption of GBPs to surfaces.

The binding of ConA to the variable density glycan microarray was measured using fluorescence microscopy. These experiments provided fluorescence values for 8 different ConA concentrations bound to glycans patterned at 11 different α-**Man** : allyl alcohol ratios. The lectin concentrations ranged from 4.8 μM to 9.6 nM. The fluorescence data for the binding of a 240 nM solution of FITC-ConA

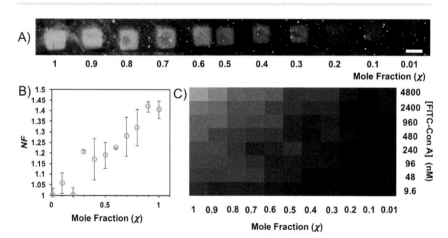

Fig. 2   (A) The fluorescence images of the printed features of α-**Man** following exposure to ConA with a concentration of 240 nM show a decrease in NF as the mole fraction, χ, decreases. The scale bar is 45 μm. (B) Data showing the decrease in NF as the χ of glycan decreases in each feature. The error bars are one standard deviation from the mean. (C) A heat map representing binding at all glycan mole fractions, χ, and ConA concentrations.

onto the different features are presented in Fig. 2A. The binding data were assembled into a heat map to represent the changes in fluorescence with the changes in both glycan density and lectin concentration (Fig. 2B), which illustrates that the fluorescence intensity decreases with decreasing ConA concentration and with decreasing mannoside χ in the printed feature. Finally, plotting the fluorescence data shows (Fig. 2C) a non-linear decrease with decreasing α-**Man** mole fraction (χ = [glycan]/([glycan] + [spacer])), with a drop-off to background levels at χ = 0.2. This nonlinear decrease in binding with decreasing χ is consistent with other studies of the multivalent binding of ConA in microarrays.[27,33] No fluorescence above the background level was observed in the control channels that contained only buffer solution.

The α-**Man**–ConA fluorescence data was used to determine the avidity (K_d) at each of the different ConA concentrations and at each χ by applying a Langmuir isotherm model (eqn (1)), which is commonly used to quantify binding in microarrays.[27,33,48]

$$F = \frac{[L]F_{max}}{[L] + K_d} \tag{1}$$

In this model, [L] is the concentration of the lectin, F is the observed fluorescence counts, and F_{max} is the maximum fluorescence observed when ConA binds to α-**Man** (F_{max} was observed at [ConA] = 4.8 μM and χ = 1). The binding data for all 88 different conditions (8 ConA concentrations × 11 χ) are presented in Table 1, and the K_d values range from 2700 to 43 nM. The following trends are seen when calculating K_d for each binding event. The observed K_d is dependent upon both χ and ConA concentration. For all ConA concentrations, the fluorescence measurements for $\chi \leq 0.2$ were identical to background levels, and no features were observed, indicating that no specific binding was occurring or that

Table 1 Dissociation constants, K_d, determined by fluorescence data and eqn (1) for each binding event between the ConA and immobilized mannose at different mole fractions, $\chi^{a,b}$

χ	[ConA] nM							
	4800	2400	960	480	240	96	48	9.6
	K_d (nM)							
1	270 ± 50	180 ± 42	240 ± 54	18 ± 2	110 ± 78	52 ± 5	43 ± 15	NA
0.9	270 ± 37	4 ± 1	350 ± 66	41 ± 15	100 ± 49	49 ± 11	28 ± 5	NA
0.8	710 ± 150	490 ± 110	400 ± 77	88 ± 26	130 ± 6	56 ± 14	29 ± 6	NA
0.7	910 ± 256	650 ± 190	420 ± 91	91 ± 39	140 ± 34	71 ± 8	37 ± 3	NA
0.6	380 ± 68	550 ± 200	460 ± 170	140 ± 57	160 ± 44	74 ± 31	39 ± 8	NA
0.5	2100 ± 426	1100 ± 290	570 ± 120	170 ± 100	170 ± 33	64 ± 14	51 ± 11	NA
0.4	1800 ± 370	1100 ± 100	660 ± 140	190 ± 61	180 ± 7	74 ± 6	66 ± 3	NA
0.3	2200 ± 490	1100 ± 340	550 ± 78	190 ± 85	200 ± 37	76 ± 10	70 ± 5	NA
0.2	NB	NB	NB	NB	NB	NB	NB	NB
0.1	NB	NB	NB	NB	NB	NB	NB	NB
0.01	NB	NB	NB	NB	NB	NB	NB	NB

[a] NB = no binding, NA = features undetectable. [b] Errors are reported as one standard deviation from the mean, taken from three different features in the array.

the fluorescence signal was too low to measure with our analytical methods. At the lowest ConA concentration (9.6 nM), the difference between $\chi = 1$ and $\chi = 0.01$ in terms of K_d is only 5 nM, which became statistically similar to the background level, although patterns were still observable at $\chi = 0.3$; we conclude that binding cannot be measured accurately with our fluorescence method at this concentration. We observed that, as glycan concentration decreased, K_d increased. For example, at 4.8 µM ConA (highest [ConA]), K_d increases 10-fold from 270 nM ($\chi = 1$) to 2200 nM ($\chi = 0.3$). Similarly, in the study done by Wong et al.,[33] as the printing concentrations of monomannose decreased, K_d increased, concluding that the spacing was too large for ConA to bind multivalently, which is in good agreement with our results. Alternatively, if χ is held constant, K_d increases with increasing [ConA]. For example, at $\chi = 0.8$, K_d ranges from 710 nM ([ConA] = 4.8 µM) to 28 nM ([ConA] = 48 nM). At high [ConA], the K_d values are higher than previously reported; however, as the concentration of [ConA] decreases, the K_d values begin to resemble those reported by Gildersleeve[27] (69 nM) and Wong[33] (80 nM). There is an aberrant data point when [ConA] is 2400 nM at $\chi = 0.9$, which does not seem to follow the binding trends. We believe that the unexpectedly low fluorescence read-out in this area was the result of the improper immobilization of α-Man during the printing of the features, which could have been caused by poor thiol monolayer formation or contamination in that area during printing.

The changes in avidity can be explained by considering the differing ability of ConA to participate in multivalent binding as glycan density is modulated. Here, we modify a simple model for surface density developed by Oyelaran that is based on average glycan spacing in order to explain the observed avidity trends. In this model, neoglycoprotein-coated surfaces are assumed to be two-dimensional arrays of ligands, and these ligands are evenly spaced across the surface.[28] In order to estimate the average spacing, we must first determine the density of molecules on the surface; here, we assume it to be $\sim 10^{13}$ molecules cm^{-2} with a range of 0.1–0.6 molecules nm^{-2}, which is based on previous work from our group on immobilizing electroactive probes onto functionalized substrates.[30] At this grafting density, there are between 2×10^8 and 1×10^9 bound molecules per 45×45 µm feature. The estimated average distance is found by dividing the feature area (2.025×10^9 nm^2) by the number of molecules in that feature, and taking the square root to find the distance. Using the lowest density estimate, we find that at $\chi = 1$, the estimated average distance between mannosides is 1 nm. As χ decreases, the spacing between mannosides increases, such that at $\chi = 0.2$, the calculated average distance between mannosides is 7.1 nm, which is slightly larger than the distance between the binding sites in ConA (7 nm), rendering multivalent binding nearly impossible, which is in good agreement with our data. Although this model is only a rough approximation, it does illuminate why binding decreases so dramatically at this critical threshold of $\chi = 0.2$. Assuming that at least two ConA–α-Man contacts are required for ConA to remain bound to the surface, at these low values of χ, this is not possible, and as such, the protein is easily washed off the surface (Fig. 3A). The calculations were also repeated under the assumption that there are 0.6 molecules nm^{-2}. Both calculations were plotted (Fig. 3B) to show the range in potential spacing between the immobilized glycans at different mole fractions and how that related to changes in the value of K_d.

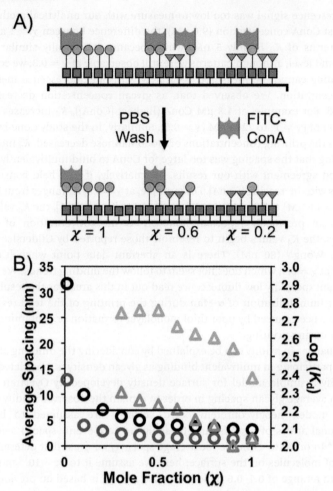

Fig. 3 (A) Model demonstrating how reducing χ decreases ConA recognition. As the number of glycans on the surface decreases, the average spacing between glycans becomes larger than the distance between the ConA binding sites. This resulting monovalent binding is not sufficiently strong enough to survive washing. (B) The estimated average spacing between mannosides as a function of χ is represented by the circles. The black circles assume a molecular packing of 0.1 molecules nm^{-2}, and the red circles represent changes in the spacing when assuming a molecular packing of 0.6 molecules nm^{-2}. The triangles are the $\log(K_d)$ values from Table 1 for [ConA] of 960 nM (green) and 240 nM (blue). The black line corresponds to an average spacing of 7 nm, which is the distance between the glycan-binding pockets in tetrameric ConA.

Conclusions

Here, we used a new photochemical printer combined with microfluidics to fabricate glycan microarrays with systematically varying glycan density. By varying the ratio of **α-Man** : allyl alcohol in the printing solution used during the photochemical immobilization, the average spacing between glycans was controlled. The association between the glycans in this microarray and FITC-

ConA was studied in a microfluidic chip, where solutions of varying [ConA] were exposed to the features printed at varying χ. Fluorescence microscopy analysis of the resulting array provided K_d values for 8 different ConA concentrations and 11 different mole fractions (χ). In addition to the increase in K_d with decreasing χ, we observed an abrupt decrease in fluorescence at $\chi = 0.2$. We explain this phenomenon using a model that considers the average spacing between glycans, and conclude that at $\chi = 0.2$, the glycan spacing is too great for the ConA to bind the surface multivalently. We believe that this versatile new printing and analysis strategy could help glycobiologists investigate quantitatively how multivalency affects association, and thereby understand how information is trafficked on the cell surface and other biointerfaces.

Conflicts of interest

The authors declare no competing financial interests.

Acknowledgements

This work was supported by funding from the Air Force Office of Scientific Research (FA9550-17-1-0356), the Department of Defense (MURI 15RT0675), a SEED grant from the City University of New York Advanced Science Research Center, and a SEED grant from the Professional Staff Congress of the City University of New York. A Research Centers in Minority Institutions Program grant from the National Institute of Health Disparities (MD007599) of the National Institutes of Health (NIH), which supports the infrastructure at Hunter College is also acknowledged.

References

1 Y. van Kooyk and G. A. Rabinovich, *Nat. Immunol.*, 2008, **9**, 593.
2 W. Van Breedam, S. Pöhlmann, H. W. Favoreel, R. J. de Groot and H. J. Nauwynck, *FEMS Microbiol. Rev.*, 2014, **38**, 598–632.
3 A. E. Smith and A. Helenius, *Science*, 2004, **304**, 237–242.
4 P. G. Wang and C. R. Bertozzi, *Glycochemistry: principles, synthesis, and applications*, Marcel Dekker, Inc., New York, 2001.
5 C. Müller, G. Despras and T. K. Lindhorst, *Chem. Soc. Rev.*, 2016, **45**, 3275–3302.
6 J. L. Jiménez Blanco, C. Ortiz Mellet and J. M. García Fernández, *Chem. Soc. Rev.*, 2013, **42**, 4518–4531.
7 C. Fasting, C. A. Schalley, M. Weber, O. Seitz, S. Hecht, B. Koksch, J. Dernedde, C. Graf, E.-W. Knapp and R. Haag, *Angew. Chem., Int. Ed.*, 2012, **51**, 10472–10498.
8 N. Jayaraman, *Chem. Soc. Rev.*, 2009, **38**, 3463–3483.
9 O. Renaudet and R. Roy, *Chem. Soc. Rev.*, 2013, **42**, 4515–4517.
10 J. J. Lundquist and E. J. Toone, *Chem. Rev.*, 2002, **102**, 555–578.
11 S. M. Dimick, S. C. Powell, S. A. McMahon, D. N. Moothoo, J. H. Naismith and E. J. Toone, *J. Am. Chem. Soc.*, 1999, **121**, 10286–10296.
12 S. Cecioni, A. Imberty and S. Vidal, *Chem. Rev.*, 2015, **115**, 525–561.

13 A. Bernardi, J. Jiménez-Barbero, A. Casnati, C. De Castro, T. Darbre, F. Fieschi, J. Finne, H. Funken, K.-E. Jaeger, M. Lahmann, T. K. Lindhorst, M. Marradi, P. Messner, A. Molinaro, P. V. Murphy, C. Nativi, S. Oscarson, S. Penadés, F. Peri, R. J. Pieters, O. Renaudet, J.-L. Reymond, B. Richichi, J. Rojo, F. Sansone, C. Schäffer, W. B. Turnbull, T. Velasco-Torrijos, S. Vidal, S. Vincent, T. Wennekes, H. Zuilhof and A. Imberty, *Chem. Soc. Rev.*, 2013, **42**, 4709–4727.

14 J. Heimburg-Molinaro, X. Song, D. F. Smith and R. D. Cummings, *Curr. Protoc. Protein Sci.*, 2011, **64**, 12.10.11–12.10.29.

15 X. Song, J. Heimburg-Molinaro, R. D. Cummings and D. F. Smith, *Curr. Opin. Chem. Biol.*, 2014, **18**, 70–77.

16 Y. Chevolot, C. Bouillon, S. Vidal, F. Morvan, A. Meyer, J.-P. Cloarec, A. Jochum, J.-P. Praly, J.-J. Vasseur and E. Souteyrand, *Angew. Chem., Int. Ed.*, 2007, **46**, 2398–2402.

17 A. R. de Boer, C. H. Hokke, A. M. Deelder and M. Wuhrer, *Anal. Chem.*, 2007, **79**, 8107–8113.

18 X. Song, B. Xia, Y. Lasanajak, D. F. Smith and R. D. Cummings, *Glycoconjugate J.*, 2008, **25**, 15–25.

19 F. Wojcik, S. Lel, A. G. O'Brien, P. H. Seeberger and L. Hartmann, *Beilstein J. Org. Chem.*, 2013, **9**, 2395–2403.

20 D. J. Valles, Y. Naeem, C. Carbonell, A. M. Wong, D. R. Mootoo and A. B. Braunschweig, *ACS Biomater. Sci. Eng.*, 2019, **5**(6), 3131–3138.

21 C. Wendeln, A. Heile, H. F. Arlinghaus and B. J. Ravoo, *Langmuir*, 2010, **26**, 4933–4940.

22 S. Bian, S. B. Zieba, W. Morris, X. Han, D. C. Richter, K. A. Brown, C. A. Mirkin and A. B. Braunschweig, *Chem. Sci.*, 2014, **5**, 2023–2030.

23 K. Neumann, A. Conde-González, M. Owens, A. Venturato, Y. Zhang, J. Geng and M. Bradley, *Macromolecules*, 2017, **50**, 6026–6031.

24 K.-S. Ko, F. A. Jaipuri and N. L. Pohl, *J. Am. Chem. Soc.*, 2005, **127**, 13162–13163.

25 S. Fukui, T. Feizi, C. Galustian, A. M. Lawson and W. Chai, *Nat. Biotechnol.*, 2002, **20**, 1011.

26 T. Feizi and W. Chai, *Nat. Rev. Mol. Cell Biol.*, 2004, **5**, 582–588.

27 Y. Zhang, Q. Li, L. G. Rodriguez and J. C. Gildersleeve, *J. Am. Chem. Soc.*, 2010, **132**, 9653–9662.

28 O. Oyelaran, Q. Li, D. Farnsworth and J. C. Gildersleeve, *J. Proteome Res.*, 2009, **8**, 3529–3538.

29 Y. Zhang, C. Campbell, Q. Li and J. C. Gildersleeve, *Mol. BioSyst.*, 2010, **6**, 1583–1591.

30 S. Bian, J. He, K. B. Schesing and A. B. Braunschweig, *Small*, 2012, **8**, 2000–2005.

31 C. Wendeln, S. Rinnen, C. Schulz, H. F. Arlinghaus and B. J. Ravoo, *Langmuir*, 2010, **26**, 15966–15971.

32 K. Godula, D. Rabuka, K. T. Nam and C. R. Bertozzi, *Angew. Chem., Int. Ed.*, 2009, **48**, 4973–4976.

33 P.-H. Liang, S.-K. Wang and C.-H. Wong, *J. Am. Chem. Soc.*, 2007, **129**, 11177–11184.

34 K. Godula, D. Rabuka, K. T. Nam and C. R. Bertozzi, *Angew. Chem., Int. Ed.*, 2009, **48**, 4973–4976.

35 K. Godula and C. R. Bertozzi, *J. Am. Chem. Soc.*, 2012, **134**, 15732–15742.

36 M. Mammen, S.-K. Choi and G. M. Whitesides, *Angew. Chem., Int. Ed.*, 1998, **37**, 2754–2794.

37 T. K. Dam, R. Roy, S. K. Das, S. Oscarson and C. F. Brewer, *J. Biol. Chem.*, 2000, **275**, 14223–14230.

38 T. K. Dam, T. A. Gerken and C. F. Brewer, *Biochemistry*, 2009, **48**, 3822–3827.

39 B. T. Houseman and M. Mrksich, *Chem. Biol.*, 2002, **9**, 443–454.

40 C. Carbonell, D. J. Valles, A. M. Wong, M. W. Tsui, M. Niang and A. B. Braunschweig, *Chem*, 2018, **4**, 857–867.

41 C. A. Schneider, W. S. Rasband and K. W. Eliceiri, *Nat. Methods*, 2012, **9**, 671–675.

42 B. Trastoy, D. A. Bonsor, M. E. Pérez-Ojeda, M. L. Jimeno, A. Méndez-Ardoy, J. M. García Fernández, E. J. Sundberg and J. L. Chiara, *Adv. Funct. Mater.*, 2012, **22**, 3191–3201.

43 C. Carbonell, D. J. Valles, A. M. Wong, M. Touve, N. C. Gianneschi and A. B. Braunschweig, *Nat. Commun.*, 2019.

44 C. E. Hoyle and C. N. Bowman, *Angew. Chem., Int. Ed.*, 2010, **49**, 1540–1573.

45 P. Jonkheijm, D. Weinrich, M. Köhn, H. Engelkamp, P. C. M. Christianen, J. Kuhlmann, J. C. Maan, D. Nüsse, H. Schroeder, R. Wacker, R. Breinbauer, C. M. Niemeyer and H. Waldmann, *Angew. Chem.*, 2008, **120**, 4493–4496.

46 A. B. Lowe, *Polym. Chem.*, 2014, **5**, 4820–4870.

47 E. A. Smith, W. D. Thomas, L. L. Kiessling and R. M. Corn, *J. Am. Chem. Soc.*, 2003, **125**, 6140–6148.

48 A. Kuno, N. Uchiyama, S. Koseki-Kuno, Y. Ebe, S. Takashima, M. Yamada and J. Hirabayashi, *Nat. Methods*, 2005, **2**, 851.

PAPER

Factors contributing to variability of glycan microarray binding profiles†

J. Sebastian Temme, [iD] Christopher T. Campbell and Jeffrey C. Gildersleeve [iD] *

Received 18th February 2019, Accepted 1st April 2019

DOI: 10.1039/c9fd00021f

Protein–carbohydrate interactions play significant roles in a wide variety of biological systems. Glycan microarrays are commonly utilized to interrogate the selectivity, sensitivity, and breadth of these complex protein–carbohydrate interactions. During the past two decades, numerous distinct glycan microarray platforms have been developed, each assembled from a variety of slide-surface chemistries, glycan-attachment chemistries, glycan presentations, linkers, and glycan densities. Comparative analyses of glycan microarray data have shown that while many protein–carbohydrate interactions behave predictably across microarrays, there are instances when various array formats produce different results. For optimal construction and use of this technology, it is important to understand sources of variances across array platforms. In this study, we performed a systematic comparison of microarray data from 8 lectins across a range of concentrations on the CFG and neoglycoprotein array platforms. While there was good general agreement on the binding specificity of the lectins on the two arrays, there were some cases of large discrepancies. Differences in glycan density and linker composition contributed significantly to variability. The results provide insights for interpreting microarray data and designing future glycan microarrays.

Introduction

Glycan-binding proteins are involved in numerous biological processes and are used extensively in basic and clinical research. For example, carbohydrate-binding antibodies and certain lectins are abundant in serum and are an important component of our immune defense system. Carbohydrate-binding monoclonal antibodies and purified plant lectins are used extensively to evaluate carbohydrate expression for both research and diagnostic purposes. In addition, several are in clinical trials for treating diseases. Antibody 3F8 is in clinical trials for the treatment of neuroblastoma and Ch14.18, marketed as

Chemical Biology Laboratory, Center for Cancer Research, National Cancer Institute, Frederick, MD, 21702, USA. E-mail: gildersj@mail.nih.gov

† Electronic supplementary information (ESI) available. See DOI: 10.1039/c9fd00021f

Unituxin by United Therapeutics, has been FDA approved for children with high-risk neuroblastoma.[1,2]

Information about binding affinity and specificity are critical for understanding the biological roles of glycan-binding proteins, using antibodies and lectins as reagents, developing agonists/antagonists, and pursuing clinical applications. Glycan microarrays provide a high-throughput format to rapidly evaluate binding of glycan-binding proteins to numerous potential glycan ligands in parallel while using only minimal amounts of samples.[3-9] Many groups have developed distinct glycan array platforms that use different slide surface chemistry, linkers, immobilization strategies, and/or glycan presentation formats. Several recent studies have compared binding profiles on different glycan array platforms and glycan presentation formats.[10-13] While these studies found general agreement across platforms, there were notable differences. To improve this technology and better understand how to interpret glycan array results, it is important to elucidate the factors that contribute to variability between array platforms. In this study, we investigate factors that give rise to differences in binding profiles on glycan microarrays.

Materials and methods

Lectins

The following biotinylated lectins were used in this study: concanavalin A (ConA, Cat #BA-1104-5), peanut agglutinin (PNA, Cat #BA-2301-1), *Ricinus communis* agglutinin I (RCA-I, Cat #BA-2001-5), soybean agglutinin (SBA, Cat #280828-1), and wheat germ agglutinin (WGA, Cat #BA-2101-5) were obtained from EY Labs (San Mateo, CA). *Maackia amurensis* lectin I (MAL-I, Cat #B1315) and *Sambucus nigra* lectin (SNA, Cat #B-1305) were obtained from Vector Laboratories (Burlingame, CA). *Helix pomatia* agglutinin (HPA, Cat #L6512) was obtained from Sigma-Aldrich (Burlington, MA).

CFG array data

Data for 5 of the lectins (ConA, HPA, Mal-I, SNA, and WGA) were reported previously.[10] Data for the other 3 lectins (PNA, RCA, SBA) were collected at the same time using the same protocols and materials. The data were obtained from the Consortium for Functional Glycomics website (CFG; http://www.functionalglycomics.org/). Detailed experimental procedures for the synthesis of NHS-reactive glycans, fabrication of the microarray, assaying the lectins, and analyzing the data have been reported previously.[14]

The CFG array is composed of glycans attached to a linker with a terminal amine printed on NHS-activated glass slides containing a hydrogel surface (Nexterion Hydrogel, Schott). The assay details for these data were reported previously.[10] Briefly, slides were incubated with freshly diluted biotinylated-lectin in binding buffer (20 mM Tris–HCl, pH 7.4 150 mM NaCl, 2 mM $CaCl_2$, 2 mM $MgCl_2$ + 0.05% Tween-20 + 1% BSA) for 1 hour at RT. The slide was washed and then incubated with AlexaFluor®-488 labeled streptavidin for 1 hour in the dark. The microarray slides were extensively washed, dried by centrifugation, imaged using a ProScan Array Scanner (Perkin Elmer), and then processed using Imagene software.

Neoglycoprotein microarray data

Data for 5 of the lectins (ConA, HPA, Mal-I, SNA, and WGA) were reported previously.[10] Data for the other 3 lectins (PNA, RCA, SBA) were collected at the same time using the same batch of slides, assay conditions, and materials. Detailed experimental procedures for the fabrication of the microarray, and the assay and data analysis have been reported previously.[15,16]

Our neoglycoprotein microarray is composed of glycans and glycopeptides chemically conjugated to either bovine serum albumin (BSA, A3059, Sigma-Aldrich, St. Louis, MO) or human serum albumin (HSA, A8763, Sigma-Aldrich, St. Louis, MO) to produce neoglycoproteins. The average number of glycans conjugated per molecule of albumin was determined by MALDI-TOF MS. Full details of the preparation and characterization of the neoglycoproteins have been previously published.[17] Neoglycoproteins and other glycoproteins are printed onto epoxide-coated glass slides (SuperEpoxy2, ArrayIt, Sunnyvale, CA) to generate microarrays. The assay details for these data were reported previously.[10] Briefly, slides were blocked with BSA overnight at 4 °C. Next, they were incubated with various concentrations of each biotinylated lectin in binding buffer (20 mM Tris–HCl, 150 mM NaCl, 10 mM $CaCl_2$, 10 mM $MgCl_2$ pH = 7.4 containing 0.05% Tween 20 and 0.1% BSA) for 1 hour at room temperature. After 3 washes, slides were incubated with Cy3-labeled streptavidin for 1 hour in the dark with gentle shaking. The microarray slides were extensively washed, dried by centrifugation, imaged using a GenePix 4000A scanner (Molecular Devices Corporation, Sunnyvale, CA) and analyzed with GenePix Pro 7.0 software (Molecular Devices Corporation, Sunnyvale, CA).

Results

Overview of data used in the study

For this study, we used data from 8 different lectins profiled on two different glycan microarray platforms. Our microarray[18–20] was composed primarily of neoglycoproteins (NGPs),[21] along with about 15 natural glycoproteins, and contained a total of 332 array components. To produce the neoglycoproteins, glycans and glycopeptides were attached *via* various linkers to bovine serum albumin or human serum albumin.[17] In many cases, we conjugated glycans or glycopeptides at a high ratio and a low ratio to produce both high and low-density conjugates. The number after the name represents the average number of determinants (*i.e.* glycans or glycopeptides) attached per molecule of albumin (*e.g.* Man-α 05 is a BSA conjugate with an average of 5 molecules of mannose per molecule of BSA). Neoglycoproteins and natural glycoproteins were then printed on epoxide-coated glass microscope slides. The Consortium for Functional Glycomics (CFG) array version 5.0 was composed of 611 glycans that contain an amine terminal linker attached to the reducing end.[14] The glycans were printed on an NHS-ester activated glass slide surface, which allows for covalent attachment *via* amide bond formation. The slides contain a hydrogel coating on the surface. A schematic of the two slide surfaces is shown in Fig. 1. Both arrays contained a diverse collection of *N*-linked glycans, *O*-linked glycans, glycolipid glycans, and other glycans.

The 8 lectins included in this study were ConA, RCA, HPA, Mal-I, PNA, SBA, SNA, and WGA. The 8 lectins have varying specificities, which provides

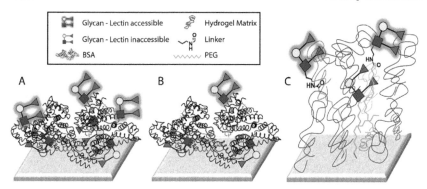

Fig. 1 Representations of the two glycan microarray surfaces. (A) Depiction of the neo-glycoprotein microarray with a high density of glycans. (B) Depiction of the neo-glycoprotein microarray with a low density of glycans. (C) Depiction of the CFG microarray with glycans coupled to an NHS-hydrogel coated slide. For both formats, some glycans might not be accessible for lectin binding.

comparisons across a diverse range of glycans. Data for 5 of the lectins were published as part of an international glycan microarray comparison.[10] Data for the other 3 lectins (PNA, SBA, RCA) were added for this study. Both array groups received lectin aliquots from the same parent lectin stocks, along with aliquots of BSA for use in blocking buffers. Each group incubated the lectins at room temperature in the same buffer with added calcium and magnesium. Incubation times were identical. All other parameters, such as washes, scanning, and processing were at the discretion of the array group. Finally, each of the lectins was profiled at multiple concentrations, which provides a greater dynamic range and accounts for saturation of binding. Collectively, the study includes a large dataset for analysis.

Analysis of consistency for each array platform

The first step for understanding differences between microarray binding profiles was identifying key differences. We wanted to focus on differences that were (1) large, (2) observed with consistency, and (3) substantially beyond what one would expect based on the inherent variability of the assay. To facilitate analysis of the data, we implemented a floor value of 100 RFU. For both arrays, there is a significant amount of noise for signals below 100, and we did not want to focus on large differences arising from very low signals (*e.g.* 1 *vs.* 100 RFU). In addition, we log transformed the data using base 2 to simplify comparisons and avoid focusing on differences that have a large RFU value but a small ratio (*e.g.* 50 000 *vs.* 30 000 RFU gives a difference = 20 000 but the ratio is less than 2-fold; alternatively, 5000 *vs.* 500 RFU has a difference of only 4500 but a ratio of 10-fold).

We started by evaluating the variability of each microarray platform. Ideally, we would compare replicate measurements for each lectin at the same concentration; however, these data were not available. Therefore, we compared datasets for lectins at the nearest possible lectin concentrations. The data for all lectins were grouped into one large analysis, and then Pearson correlation constants were determined. For the CFG array, this included comparisons of 100 μg mL^{-1}

vs. 50 µg mL^{-1}, 50 µg mL^{-1} *vs.* 10 µg mL^{-1}, 10 µg mL^{-1} *vs.* 1 µg mL^{-1}, 1 µg mL^{-1} *vs.* 0.5 µg mL^{-1}, and 0.5 µg mL^{-1} *vs.* 0.1 µg mL^{-1}. Across all 8 lectins, the Pearson correlation constants ranged from 0.80 to 0.98. For our NGP array, we included comparisons of 50 µg mL^{-1} *vs.* 10 µg mL^{-1}, 10 µg mL^{-1} *vs.* 1 µg mL^{-1}, 1 µg mL^{-1} *vs.* 0.3 µg mL^{-1}, and 0.3 µg mL^{-1} *vs.* 0.1 µg mL^{-1}. Across all 8 lectins, the Pearson correlation constants ranged from 0.85 to 0.96. Overall, both arrays demonstrated a very high degree of consistency.

Next, we identified instances where there was a large difference in signal from one lectin concentration to another for a given array component. Differences in the range of 2 to 10-fold were expected since the comparisons involved 2 to 10-fold differences in lectin concentration. Nevertheless, 20-fold differences only occurred about 1.2% of the time for the CFG array and 1.5% of the time for the NGP array. Differences of greater than 32-fold occurred about 0.5% of the time for both arrays. Differences of greater than 50-fold occurred 0.10% of the time for our array and 0.17% of the time for the CFG array. Differences in lectin concentration likely contribute to these totals. For example, the vast majority of cases where a difference of greater than 50-fold was observed were for comparisons of data points at 10 µg mL^{-1} *vs.* 1 µg mL^{-1}. Since a 10-fold difference in lectin concentration could easily give rise to a 10-fold difference in signal, a rate of 0.1–0.2% for 50-fold differences was considered very small. Also, one would expect even higher levels of consistency for replicate experiments carried out using the same lectin concentrations. Overall, these results demonstrate a very high degree of consistency for the two array platforms.

Identification of substantial differences between arrays

It is well appreciated that differences in glycan structure can give rise to large differences in lectin binding. Our goal for this study was to evaluate factors other than glycan structure. Therefore, we decided to limit the array-to-array comparison to the subset of components that have identical glycans on both array platforms. Differences in linker were allowed. This group included 177 comparisons for each lectin, totaling 1416 comparisons across all 8 lectins. The comparisons were carried out for identical or nearly identical lectin concentrations. These data included 6 comparisons: NGP at 50 µg mL^{-1} *vs.* CFG at 100 µg mL^{-1}, NGP at 50 µg mL^{-1} *vs.* CFG at 50 µg mL^{-1}, NGP at 10 µg mL^{-1} *vs.* CFG at 10 µg mL^{-1}, NGP at 1 µg mL^{-1} *vs.* CFG at 1 µg mL^{-1}, NGP at 0.3 µg mL^{-1} *vs.* CFG at 0.5 µg mL^{-1}, and NGP at 0.1 µg mL^{-1} *vs.* CFG at 0.1 µg mL^{-1}. Collectively, a total of 6018 array-to-array comparisons were considered (note: not all lectins were assayed at all concentrations; for example, only 3 lectins were assayed at 50 µg mL^{-1} on the CFG array; therefore, the total is not simply 6 × 1416).

Based on our analysis of within-array consistency above, we decided to focus on instances where signals for the same glycan and the same lectin varied by more than 50-fold from one array to the other. Differences of this size would occur very infrequently as a result of inherent variabilities in the assays. In addition, a high cutoff would allow us to avoid focusing on small variations resulting from differences in assay conditions or scanner settings/sensitivity. Finally, 50-fold or greater differences seemed substantial from a general molecular recognition perspective. In addition to the cutoff of 50-fold, we also wanted to focus on cases where the differences were observed consistently. Therefore, we only selected instances where there was a 50-fold or greater difference at one lectin concentration and a 10-fold or greater difference for the same glycan and lectin at one of the other lectin concentrations.

Using these criteria, we identified 101 instances (out of 6018 possible comparisons; ~1.7%) where there was a substantial difference between the arrays (see Fig. 2 and ESI Excel file,† sheet "CFG vs. NGP top 101 differences"). The percentage of instances is much higher than what would be expected based on the inherent variability of the assays. In addition, all 101 involved cases where the signal on our neoglycoprotein array was larger than the corresponding signal on the CFG array. Thus, we concluded that these instances are likely real differences between the array platforms and not a result of random variability.

The 101 instances involved some redundancy. When comparing the CFG data to the neoglycoprotein data, there were many cases where there were 2–3 densities of a glycan on the neoglycoprotein array. In these cases, the CFG glycan was compared to each density on the neoglycoprotein separately (e.g. Man-α-Sp8 on the CFG array was compared to Man-α at 5/BSA and 20/BSA on the neoglycoprotein array). In addition, there were cases where a particular glycan was attached to the surface via two different linkers. These situations were also counted as separate comparisons (e.g. GlcNAcβ-Sp0 and GlcNAcβ-Sp8 on the CFG array were both compared to GlcNAcβ-BSA on the neoglycoprotein array). If we only consider the glycan portion, the number of large discrepancies decreased from 101 to 62.

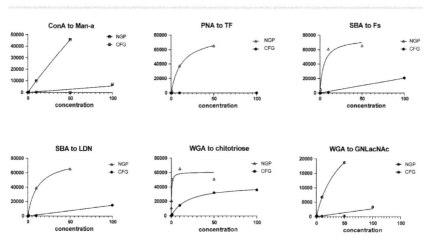

Fig. 2 Examples of discrepancies between the CFG and neoglycoprotein (NGP) array. Data are shown for 6 of the 101 instances of large discrepancies between the CFG and NGP array. Data are graphed over all measured concentrations of lectin, with lectin concentration on the x-axis and RFU on the y-axis. Abbreviations are as follows: Man-α (CFG = #3, mannose alpha linked to Sp8; NGP = #275, low density mannose alpha linked to BSA, ~5/BSA), TF [CFG = #141, Galβ1-3GalNAcα-Sp14 (threonine); NGP = #109, Ac-Ser-(Galβ1-3GalNAcα)Ser-Ser-Gly-Hex-BSA, ~16/BSA], Fs (Forssman disaccharide; CFG = #92, GalNAcα1-3GalNAcβ-Sp8; NGP = #108, GalNAcα1-3GalNAcβ-BSA), LDN (LacDiNAc; #100, GalNAcβ1-4GlcNAcβ-Sp8; NGP = #234, GalNAcβ1-4GlcNAcβ-BSA, ~16/BSA), chitotriose (CFG = #192, GlcNAcβ1-4GlcNAcβ1-4GlcNAcβ-Sp8; NGP = #129, GlcNAcβ1-4GlcNAcβ1-4GlcNAcβ-BSA, ~20/BSA), GNLacNAc (CFG = #183, GlcNAcβ1-3Galβ1-4GlcNAcβ-Sp0; NGP = #255, GlcNAcβ1-3Galβ1-4GlcNAcβ-BSA).

Factors contributing to differences between arrays

There are a variety of factors that could give rise to differences between arrays. Differences in linkers, glycan presentation, and slide surface are likely to have the biggest effects. From an initial qualitative assessment of the 101 cases, both the linker and glycan density appeared to be important factors. Both will be discussed in more detail below in separate sections. In addition, other factors could also influence the binding profiles.

Variability in the experimental conditions is one consideration. As mentioned above, many of the experimental parameters (*e.g.* batch of lectin, blocking agent, buffer, incubation temperature, incubation times) were held constant between the arrays. Some parameters were different. The washes were similar but not identical. We washed 3 times after the primary incubation and 7 times after the secondary incubation. The CFG washed 8 times after the primary incubation and 12 times after the secondary incubation; 4 of the final washes are with Milli-Q H_2O instead of buffer. In our experience, these differences in washes are not expected to have substantial effects. A second difference is the amount of BSA used in the incubation buffers. We used 0.1% while the CFG used 1%. This difference could potentially contribute to the differences in signals, but it was unclear how large the effect would be. Lastly, incubations on the neoglycoprotein microarray contained 10 mM $CaCl_2$ and 10 mM $MgCl_2$, whereas the CFG incubations included 2 mM $CaCl_2$ and 2 mM $MgCl_2$.

In addition to the experimental conditions described above, two other factors could lead to systematic differences: image acquisition/analysis and secondary reagent. For the secondary reagent, we used Cy3-labeled streptavidin and the CFG used AlexaFluor488-labeled streptavidin. Different fluorophores can potentially produce different signal intensities, but the relative signal strengths for various glycans would be the same. Additionally, differences in scanner settings/sensitivity would produce systematic variations in the two datasets. If the binding profiles were perfectly correlated but one scanner was more sensitive than the other, one would expect a perfect trendline but with a slope skewed from 1.0. We plotted our data *vs.* the CFG data for each of the identical or nearly identical lectin concentrations mentioned above and then determined the slopes of the trendlines. The average slope of the trendlines was 0.50 (ranging from 0.39–0.63). Since our data was plotted on the *x*-axis, a slope of 0.50 indicates that our signals were on average 2-fold higher than the CFG signals. Thus, we concluded that differences in the scanner sensitivity/settings and/or fluorophore on the secondary reagent produce a systematic variability between the arrays by a factor of about 2. Since 2-fold is relatively small, we did not re-scale or normalize the data to compensate for the difference.

The nature of the lectin could also affect binding profiles measured on arrays. When a lectin binds a glycan on an array surface, there can be a variety of additional interactions, such as interactions with the linker, adjacent molecules of lectin, and the slide surface. Various features of the lectin can influence these additional interactions, such as variations in binding pocket depth (shallow *versus* deep binding pockets), overall net charge, localized surface charge and/or hydrophobicity around the binding pocket, spacing of binding pockets, and steric accessibility of binding pockets. In our comparison, the 101 instances were not distributed evenly among the lectins. Forty of the 101 cases were signals derived from WGA, and 26 were signals derived from SBA. This over-representation was not simply due to having more abundant strong binding partners for WGA and SBA on the arrays. HPA had a similar number of strong binders, but only 9 of the 101 cases were from HPA data. The variation from one lectin to another highlights the importance of including many lectins/proteins in a cross-platform microarray comparison.

Effects of glycan density

Glycan density can have a large effect on recognition. Monovalent interactions between a glycan and a single lectin binding site are typically weak. Lectins and antibodies achieve tight binding through formation of multivalent complexes. The spacing of glycans on a surface can have a big effect on the ability to form a multivalent interaction. We have observed very large density effects for other lectins and monoclonal antibodies previously on our array.[22–25] In this study, 76 of the 101 instances of large differences between the arrays involved comparisons between the CFG array and a high-density glycan on the neoglycoprotein array, indicating that density is a key factor.

To isolate the effects of density, we focused our analysis on just data from our neoglycoprotein microarray. By comparing data from the same array experiments, all other parameters (*e.g.* incubation times, temperatures, buffers, slide surface, and linker) were identical. Our array has 73 components that are present at both high and low densities. Neoglycoproteins with <8 determinants per molecule of albumin (median = 5) were considered low density. High-density was defined as an average of >10 determinants per molecule of albumin (median = 16). Based on a simplified model of the array surface, 5/BSA would give an estimated spacing of roughly 80 Å from attachment point to attachment point, while 16/BSA would correspond to roughly 40 Å spacing (see ESI†). These estimates do not include the linker. We compared binding properties for these 73 components across all 8 lectins giving 584 data points at each of five lectin concentrations.

Our initial assessment involved determining Pearson correlation constants (R) for low-density *versus* high-density signals at each lectin concentration. The R values ranged from 0.84–0.90 indicating a high degree of correlation between low and high-density glycans. The correlations were slightly better if we compared low-density conjugates at a higher lectin concentration to high-density conjugates at a lower lectin concentration (R values ranged from 0.89–0.92). For example, with both datasets at 1 µg mL^{-1}, the R value for low *versus* high-density conjugates was 0.84. The correlation improved to 0.90 when comparing the low-density conjugates at 1 µg mL^{-1} to the high-density conjugates at 0.3 µg mL^{-1}.

While the overall correlation was high, there were a number of instances where there were large differences between a high and low-density pair. A summary of these differences is shown in Table 1. Ten-fold differences were observed 61 times at a lectin concentration of 50 µg mL^{-1} and 48 times at a lectin concentration of 10 µg mL^{-1}. For the vast majority of these cases (>96%), the signal for the high-density conjugate was larger than the low-density. There were 13–15 instances where there was a >50-fold difference in signal. All of the >50-fold differences favored the high-density conjugate and typically involved a strong positive signal at high-density and little or no signal at low-density (*e.g.* 180 RFU for low-density *vs.* 21 491 RFU at high-density). Thus, differences in glycan density can frequently

Table 1   Summary of large differences between high and low-density glycans

	10-Fold differences	20-Fold differences	50-Fold differences
50 µg mL^{-1}	61	37	15
10 µg mL^{-1}	48	27	13
1 µg mL^{-1}	28	15	1
0.3 µg mL^{-1}	8	3	0
0.1 µg mL^{-1}	7	4	0

give rise to 10-fold differences in signal and occasionally result in greater than 50-fold differences. We considered the possibility that larger high-density to low-density ratios for a given glycan pair (20/BSA *vs.* 4/BSA has a ratio of 5, whereas 10/BSA *vs.* 5/BSA has a ratio of 2) could produce larger differences in signals; however, instances of 10-fold differences were not correlated with the high-*vs.*-low ratio. Instances of large density effects were not distributed evenly among the 8 lectins. They were most frequently observed with SBA (31% of large differences were SBA binding).

   We next compared the CFG data to our low and high-density conjugates separately by determining Pearson correlation constants for each of the identical or nearly identical lectin concentrations (Table 2). For this analysis, we only included array components where the glycan was identical on both arrays. The CFG data correlated much better to our low-density conjugates. For example, at a concentration of 10 µg mL^{-1} the Pearson R value was 0.77 for the CFG/low-density comparison but only 0.55 for the CFG/high-density comparison. The analysis suggests that the surface density of glycans on the CFG array is similar to our low-density spots.

Effects of the linker

Another feature that could potentially have a major impact on differences between array binding profiles is the linker connecting the glycan to the array surface. In fact, 71 of the 101 instances of large differences discussed above involved comparisons with an identical glycan but a different linker. To isolate the effects of linkers, we focused our analysis on data from the CFG microarray. Since the data are all from the same microarrays and experiments, all other parameters (*e.g.* incubation times, temperatures, buffers, slide surface) were identical. To initiate this analysis, we identified glycans bearing two or more

Table 2   Comparison of R values for CFG data *vs.* high or low-density neoglycoproteins

NGP concentration	CFG concentration	R value for high-density	R value for low-density
50 µg mL^{-1}	100 µg mL^{-1}	0.58	0.80
50 µg mL^{-1}	50 µg mL^{-1}	0.44	0.64
10 µg mL^{-1}	10 µg mL^{-1}	0.55	0.77
1 µg mL^{-1}	1 µg mL^{-1}	0.75	0.87
0.3 µg mL^{-1}	0.5 µg mL^{-1}	0.81	0.88
0.1 µg mL^{-1}	0.1 µg mL^{-1}	0.65	0.85

linkers on the CFG array. From this group of 125 array components, there were 57 pairwise comparisons from glycans bearing two linkers. In addition, there were two cases where a glycan was present with 3 different linkers giving rise to 6 pairwise comparisons, and one case where a glycan was present with 4 linkers yielding 6 pairwise comparisons. This combined group of 69 pairwise comparisons was evaluated across 8 lectins at 3–6 concentrations ranging from 0.1 μg mL^{-1} to 100 μg mL^{-1}, producing a large dataset containing 2484 comparisons. As with previous datasets, we implemented a floor value of 100 RFU and log transformed the data using base 2.

We began by evaluating all pairwise comparisons across the entire dataset. The 2484 comparisons were highly correlated, with a Pearson correlation constant R value of 0.90. Since signals can vary more when there is higher background (*e.g.* higher lectin concentrations) or low signal strength (*e.g.* lower lectin concentrations), R values were determined for each lectin concentration separately (Table 3). The correlation constants ranged from 0.88 to 0.97, indicating the lectin concentration does not have a significant effect. In addition to evaluating overall correlations between linkers, we also identified specific cases where there were large differences in signals (Table 4). Ten-fold differences were observed 42 times for this dataset (a rate of 1.69%) and 50-fold differences were observed 10 times (rate = 0.4%). The majority (79%) of >10-fold differences occurred in the top 3 lectin concentration groups (100, 50, and 10 μg mL^{-1}), presumably because there are more signals and the signals are larger. Taken together, the results show that binding to glycans on different linkers are generally well correlated, but differences in the linker can give rise to 10-fold differences in signal and occasionally produce greater than 50-fold differences.

We next evaluated how differences in the linker affected each lectin. We examined the 69 pairwise comparisons against each individual lectin and found a range of R values of 0.63–0.97 (Table 1). PNA and SBA gave the lowest R values of 0.66 and 0.63, respectively. R values obtained for each lectin with concentration dependence (Table 3) showed moderate to good agreement across the dataset, indicating that lectin concentration is not a major source of large differences in this dataset. In addition to overall correlations, we also tabulated specific cases of large differences. For most lectins, the rate of a 10-fold difference was below 2%, but the rate for RCA was 6.3%. The higher rate for RCA was more pronounced at lower concentrations. RCA accounted for 8 of 9 total occurrences in the lowest

Table 3   Pearson correlation constants (R) for glycans with different linkers

	All lectins	ConA	WGA	SNA	MAL-I	HPA	PNA	RCA	SBA
All conc.	0.90	0.95	0.77	0.93	0.92	0.92	0.66	0.85	0.63
100 μg mL^{-1}	0.88	0.95	0.72	0.96	0.90	0.88	0.70	NA	0.67
50 μg mL^{-1}	0.93	0.95	0.82	0.93	NA	NA	NA	NA	NA
10 μg mL^{-1}	0.88	0.94	0.86	0.77	0.93	0.91	0.27	0.88	0.48
1 μg mL^{-1}	0.91	0.99	0.52	0.99	0.97	0.99	ND	0.80	ND
0.5 μg mL^{-1}	0.97	0.93	ND	1.00	NA	NA	NA	NA	NA
0.1 μg mL^{-1}	0.90	0.83	ND	0.99	NA	0.98	ND	0.76	ND

NA = no data available for analysis. ND = not determined, too few values were above the floor for the given lectin/conc.

Table 4 Pearson correlations constants for pairwise linker comparisons

Linker A	Linker A structure	Linker B	Linker B structure	R
Sp12		Sp13		0.99
Sp0		Sp9		0.98
Sp18		Sp8		0.98
Sp13		Sp24		0.97
Sp13		Sp21		0.97
Sp0		Sp23		0.96
Sp23		Sp8		0.96
Sp10		Sp12		0.95
Sp12		Sp21		0.95
Sp0		Sp8		0.93
Sp12		Sp24		0.91
Sp21		Sp24		0.90
Sp15		Sp8		0.83
Sp12		Sp25		0.83
Sp14		Sp8		0.60

three concentrations. Although the overall data for RCA were highly correlated ($R = 0.85$), there was considerable scatter about the trendline (see Fig. 3A *versus* Fig. 3B). Taken together, RCA appears to be more sensitive to effects of the linker than the other lectins in our study.

Next, we evaluated effects of specific linkers. We determined correlations for all pairwise comparison that involved the same linker pair (*e.g.* all pairwise comparisons of Sp0 to Sp8; Table 4). There are 20 linker-paired comparisons, but only 15 had sufficient data to determine an R value. The R values in this analysis ranged from 0.60–0.99 (Table 4). Examination of R values for both individual linkers and paired-linkers identified Sp14 as a statistically significant source of difference on the CFG array (compare Fig. 3C *vs.* Fig. 3D). The worst correlations were for Sp14 *versus* Sp8 with an R value of 0.60. In addition to general correlations, we also examined specific cases of large differences. The rate of 10-fold differences for individual linkers ranged from 0.00–5.56% (Table 5). Amino acids Sp14 and Sp15 stand out as linkers contributing to sources of large differences (2.98–5.56%). Additionally, the linker Sp21 had one of the highest rates of large

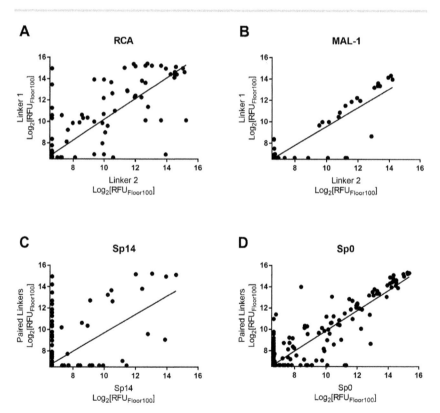

Fig. 3 Effects of linkers. Data shown for all linker pairwise comparisons with RCA (A) and MAL-1 (B). RCA linker pairwise comparisons are broadly distributed when compared to linker pairwise comparisons with MAL-1. Data are graphed over all measured concentrations of lectin from each of the 69 pairwise comparisons. Data shown for all pairwise comparisons of the poorly correlated linker Sp14 ((C) $R = 0.60$) as compared to well correlated linker Sp0 ((D) $R = 0.93$). Data are graphed over all measured concentrations of all lectins.

Table 5 Large differences for pairwise comparisons across lectins

Lectin	10-Fold differences (rate)	50-Fold differences (rate)
ConA	4 (0.97%)	1 (0.24%)
WGA	7 (1.69%)	3 (0.72%)
SNA	5 (1.21%)	None
MAL-I	4 (1.93%)	None
HPA	4 (1.45%)	None
PNA	3 (1.09%)	2 (0.72%)
RCA	13 (6.28%)	3 (1.45%)
SBA	2 (0.72%)	1 (0.36%)
All lectins	42 (1.69%)	10 (0.40%)

difference at 5.56%. Interestingly, all the large differences associated with Sp21 were within the RCA dataset and in all cases Sp21 gave a larger signal than the matched peptide or amino acid linkers (Sp24, Sp12, and Sp13). Sp24 and Sp25 also had large differences of 4.8% and 4.2%, respectively. Not surprisingly, the specific comparisons of Sp14 *vs.* Sp8 were a prominent source of large differences with a rate of 3.2% (Table S2†). Comparisons of Sp15 *vs.* Sp8 also had a high rate (5.6%). Sp15 (serine), and Sp14 (threonine) both have a free carboxylic acid on the linker which could have negative interactions with the lectins. Overall, the results suggest that the peptide portion of glycopeptides substantially influence recognition of the glycan portion on the array.

Estimation of other effects

From the results above, it was apparent that glycan density and linker structure can have major effects on glycan microarray data. It was not known how much other factors, such as slide surface and wash conditions, affect the data. To estimate the effects of other factors, we focused on comparing data obtained on the two arrays for the subgroup of array components with identical glycans and the fewest differences in linker composition and glycan density. Therefore, we limited this part of the analysis to 42 glycans we had obtained from the CFG. For these array components, the glycan portion and the core portion of the linker ($O–CH_2–CH_2–N$) were identical; however, the conjugation chemistry used to attach the linker to the surface or to albumin was different. Therefore, the stereochemistry of the glycosidic linkage and the first few atoms of the linker are identical. Moreover, the composition of the glycan and any impurities would be the same. We also limited this part of the study to our conjugates with low density, to best match the estimated density on the CFG array surface.

Of the 101 instances of large differences between our array and the CFG, only 10 came from the subgroup containing the 42 CFG glycans at low density. Thus, the vast majority of large discrepancies between arrays likely derive from differences in linkers and glycan density. Furthermore, 9 of the 10 involved binding by WGA, suggesting that certain lectins are much more sensitive to factors such as the slide surface, wash conditions, $CaCl_2/MgCl_2$ concentrations, and/or amount of BSA in the incubation buffer.

Discussion

Glycan microarrays have emerged as the primary tool used to evaluate binding properties of lectins and carbohydrate-binding antibodies.[3–9] In addition, they are frequently being used to study immune responses and identify biomarkers for various diseases.[26,27] The experiments are rapid and provide considerable information about relationships between structure and recognition. There are a variety of microarray platforms that have been developed, such as glycans on hydrogel modified surfaces,[14] neoglycolipid microarrays,[5] neoglycoprotein microarrays,[20] dendrimer microarrays,[28–35] glycopolymer microarrays.[36–38] microbial/pathogen glycan microarrays,[39–42] natural/shotgun microarrays,[43–47] liposome microarrays,[48] surfactant vesicle arrays,[49] nanoarrays,[50] and bead arrays.[51–54] These platforms use a variety of different chemistries to immobilize glycans, including non-covalent adsorption, amide bond formation, azide plus alkyne cycloaddition click chemistry, photoactivation with carbene/nitrene insertion, Michael additions, and epoxide opening. While different platforms provide good general agreement about binding properties of lectins, there can be key differences. Efforts to understand factors that give rise to these differences are important for interpreting data and optimal use of this technology. In addition, these efforts could lead to better microarray construction and improved assays.

There are several challenges when trying to understand factors that contribute to differences in microarray profiles. First, differences can arise from experimental issues, such as different sources of protein or different buffers. Second, the content on different arrays is often similar but not identical. For example, glycans on different arrays may vary by a single monosaccharide residue at the reducing end, such as Man9 attached to one GlcNAc residue *versus* Man9 attached to two GlcNAc residues. In addition, some molecules on different arrays will have identical glycans but different linkers. Features of presentation, such as glycan density, can also vary. Thus, it is difficult to determine if differences in binding profiles are due to the array platform, linker, glycan structure, and/or glycan density. Third, there is often insufficient data to observe general trends. For example, a given lectin or antibody may only bind a handful of glycans that are present on two different array platforms. Finally, there is some level of variability inherent in any assay. Thus, our first goal was to design a study that would address these challenges.

We selected two arrays for our study: our neoglycoprotein microarray and the CFG microarray. The CFG array is one of the largest and most widely used arrays in the world.[14] Since its inception, there have been thousands of array experiments run on the CFG array, and the data are publicly accessible *via* their website, providing a tremendous resource for the community. Our neoglycoprotein microarray provides a distinct platform with a unique presentation of glycans.[18–20] Both arrays contain a diverse collection of glycans with representation from many different glycan families, such as *N*-linked and *O*-linked, glycolipid glycans, non-human glycans, and Lewis/blood group antigens.

We utilized several experimental design strategies to facilitate our investigation. First, we started with data and protocols from a recent international comparison of glycan microarray platforms.[10] In that study, each array group received lectin aliquots from the same parent lectin stocks, along with aliquots of BSA for use in blocking buffers. Lectins were incubated for the same length of time at the same temperature in the same buffer. Salt concentrations were the same, and both groups used buffers supplemented with $MgCl_2$ and $CaCl_2$, albeit at somewhat difference concentrations. Therefore, most of the experimental conditions were identical or very similar. Second, we used a relatively large dataset in order to observe general trends. The data included 8 different lectins profiled on 2 different microarrays at 5–6 different concentrations producing over 36 000 data points. The lectins had differing specificities, allowing us to assess binding to different glycan structural families. In addition, we could evaluate effects from proteins with different architectures and amino acid compositions. To improve the overlap in glycan content between the array platforms, we obtained 42 glycans from the CFG. For these array components, the glycan portion and the core atoms of the linker were identical on both arrays.

While there was good agreement in the general binding properties of the lectins on different arrays, some significant differences were also observed. We focused on instances where the differences were large (>50-fold), consistent, and well beyond what one might expect from inherent variability of the assays. This group included 101 instances of sizeable differences from one array to the other. We then analyzed effects of various factors that could potentially contribute to the differences.

One key element of recognition by lectins is glycan density.[23–25,55–69] We isolated the effects of glycan density by focusing analyses on data derived from our neo-glycoprotein array where the same glycan was present at two different densities. By using this approach, all other factors (*e.g.* slide surface, linker composition, experiment conditions) were held constant. Ultimately, we found that differences in glycan density can frequently give rise to 10-fold differences in signal and sometimes produce greater than 50-fold differences. Therefore, glycan density can have a major effect on glycan microarray binding profiles, even within the same array platform.

We also compared the CFG data to our high and low-density components separately. The CFG data compared much better to our low-density conjugates. Therefore, we postulate that glycans on the surface of the CFG array are presented at low density, or at a density that is comparable to our low density. Of the 101 instances of big differences between array platforms, 76 involved comparing a CFG signal for a particular glycan to the corresponding high-density component on our array. Thus, variations in glycan density could account for a large proportion of the differences in binding profiles between our array and the CFG array.

At present, the optimal density or spacing of glycans on an array surface is not known. Our estimated spacings for glycans on our surface were ~40 Å for high density and ~80 Å for low density. These distances are on par with glycan spacings one might find in natural systems, such as glycans on a protein or cell surface. In addition, the binding sites of many lectins and antibodies are in the range of 30–100 Å apart. Thus, distances of 40–80 Å would allow formation of multivalent complexes for many natural glycan-binding proteins. That being said, the optimal density depends on the particular protein(s) being studied and likely varies for different studies. Therefore, variations in glycan density can be useful when constructing glycan microarrays.

In addition to glycan density, the composition of the linker is also an important factor to consider.[12,13,70–72] In our study, differences in linker could give rise to

10-fold variations in signal and occasionally 50-fold differences. Linkers with similar structure tended to produce very highly correlated signals. Linkers composed of amino acids had weaker correlations with other linkers and gave rise to higher rates of large discrepancies. The effects could be due to charge. Sp14 and Sp15 contain a free carboxylic acid in close proximity with the reducing end of the glycan. Interactions with this charged moiety could have a substantial effect on recognition. A prior study on binding to the CFG array also noted strong effects from linker Sp14 and suggested it was due to distinct conformational preferences.[13] In that study, Sp14 was found to give more false negative results, and molecular modeling indicated that this linker orients glycans down towards the surface making them less accessible for recognition.[13] Our results and prior results indicate that presentation of a glycan on a linear, flexible linker is quite different than presentation on a peptide. Thus, incorporation of glycopeptides on glycan microarrays can provide useful and distinct chemical diversity.

The optimal linker likely varies for different array platforms. Effects of linkers may differ depending on the surface composition and chemistry. For example, the CFG hydrogel arrays display glycans at the end of a long PEG chain, which may provide considerable flexibility and distance from the glass surface. Platforms where the linker attachment site is closer to the surface may have more pronounced effects. For example, short linkers that place the glycan in close proximity to a surface may hinder access by lectins and antibodies. Rigid linkers that limit conformational flexibility may also impede certain glycan-binding proteins.

When choosing a linker, the application is a key consideration. For establishing binding specificities or discovering new lectin, one may want a linker that is suitable for many glycan-binding proteins. For example, a long, flexible linker can allow glycans to move and adapt to accommodate different spacings, orientations, and steric environments of binding sites. In other cases, one may want glycan–linker pairs that can selectively capture lectins or glycan-binding antibodies. For example, when profiling lectins or antibodies from human serum, one needs to capture specific subpopulations of antibodies or specific lectins in the presence of a complex mixture of other glycan-binding proteins. Both the glycan and linker can contribute to selectivity. Arrays with variations in linkers could allow one to rapidly screen different glycan–linker pairings and identify ones that provide appropriate selectivity.

Overall, this study provides insights for interpreting glycan microarray data and designing better arrays. Both glycan density and the linker can have a substantial effect on recognition, a fact that is important to consider when analyzing both positive and negative signals. An absence of signal could easily arise if the glycan density is not well matched to the binding sites of the glycan-binding protein or the linker impedes recognition. When interpreting positive signals, glycans presented on a long flexible linker may provide different binding data to natural glycans. In nature, glycan determinants are often attached to a carrier glycan chain. For example, the blood group A determinant can be presented at the non-reducing termini of an N-linked glycan. Like other "linkers", the carrier chain may hinder or prevent binding. Alternatively, a glycan determinant may be displayed on a cell membrane *via* connection to a lipid. The lipid linker and close proximity to the membrane may provide a very different local environment than a long flexible linker on an array surface. In addition to interpretation, our results provide insights for designing future arrays. For example, array constructions with variations in glycan density and linker composition can provide a unique opportunity to enhance diversity in a way that complements variations in glycan composition. Lastly, the results indicate that different glycan microarray platforms are useful for the community. The ability of different microarray formats to provide distinct presentation of glycans enhances our understanding of glycan recognition and affords complementary, rather than redundant, information. In the future, it may be advantageous to design and assemble microarrays that combine desirable features of various microarray platforms and presentation formats.

Conflicts of interest

There are no conflicts to declare.

Acknowledgements

We thank the Consortium for Functional Glycomics (GM62116; The Scripps Research Institute), X. Huang (Michigan State University), Lai-Xi Wang (University of Maryland), and J. Barchi (National Cancer Institute) for generously contributing glycans for the array. This work was supported by the Intramural Research Program of the National Cancer Institute, NIH.

References

1 A. L. Yu, A. L. Gilman, M. F. Ozkaynak, W. B. London, S. G. Kreissman, H. X. Chen, M. Smith, B. Anderson, J. G. Villablanca, K. K. Matthay, H. Shimada, S. A. Grupp, R. Seeger, C. P. Reynolds, A. Buxton, R. A. Reisfeld, S. D. Gillies, S. L. Cohn, J. M. Maris and P. M. Sondel, Anti-GD2 antibody with GM-CSF, interleukin-2, and isotretinoin for neuroblastoma, *N. Engl. J. Med.*, 2010, **363**, 1324–1334.

2 N. K. V. Cheung, I. Y. Cheung, B. H. Kushner, I. Ostrovnaya, E. Chamberlain, K. Kramer and S. Modak, Murine anti-GD2 monoclonal antibody 3F8 combined with granulocyte-macrophage colony-stimulating factor and 13-cis-retinoic acid in high-risk patients with stage 4 neuroblastoma in first remission, *J. Clin. Oncol.*, 2012, **30**, 3264–3270.

3 A. Geissner and P. H. Seeberger, Glycan arrays: From Basic biochemical research to bioanalytical and biomedical applications, *Annu. Rev. Anal. Chem.*, 2016, **9**, 223–247.

4 X. Song, J. Heimburg-Molinaro, D. F. Smith and R. D. Cummings, Glycan microarrays of fluorescently-tagged natural glycans, *Glycoconjugate J.*, 2015, **32**, 465–473.

5 A. S. Palma, T. Feizi, R. A. Childs, W. Chai and Y. Liu, The neoglycolipid (NGL)-based oligosaccharide microarray system poised to decipher the meta-glycome, *Curr. Opin. Chem. Biol.*, 2014, **18**, 87–94.

6 S. Park, J. C. Gildersleeve, O. Blixt and I. Shin, Carbohydrate microarrays, *Chem. Soc. Rev.*, 2013, **42**, 4310–4326.

7 D. Wang, Carbohydrate antigen microarrays, in *Methods Mol. Biol.*, 2012, vol. 808, pp. 241–249.

8 C. D. Rillahan and J. C. Paulson, Glycan microarrays for decoding the glycome, *Annu. Rev. Biochem.*, 2011, **80**, 797–823.

9 J. Y. Hyun, J. Pai and I. Shin, The glycan microarray story from construction to applications, *Acc. Chem. Res.*, 2017, **50**, 1069–1078.

10 L. Wang, R. Cummings, D. Smith, M. Huflejt, C. Campbell, J. C. Gildersleeve, J. Q. Gerlach, M. Kilcoyne, L. Joshi, S. Serna, N. C. Reichardt, N. P. Pera, R. Pieters, W. Eng and L. K. Mahal, Cross-platform comparison of glycan microarray formats, *Glycobiology*, 2014, **24**, 507–517.

11 V. Padler-Karavani, X. Song, H. Yu, N. Hurtado-Ziola, S. Huang, S. Muthana, H. A. Chokhawala, J. Cheng, A. Verhagen, M. A. Langereis, R. Kleene, M. Schachner, R. J. De Groot, Y. Lasanajak, H. Matsuda, R. Schwab, X. Chen, D. F. Smith, R. D. Cummings and A. Varki, Cross-comparison of protein recognition of sialic acid diversity on two novel sialoglycan microarrays, *J. Biol. Chem.*, 2012, **287**, 22593–22608.

12 M. Kilcoyne, J. Q. Gerlach, M. Kane and L. Joshi, Surface chemistry and linker effects on lectin–carbohydrate recognition for glycan microarrays, *Anal. Methods*, 2012, **4**, 2721–2728.

13 O. C. Grant, H. M. Smith, D. Firsova, E. Fadda and R. J. Woods, Presentation, presentation, presentation! Molecular-level insight into linker effects on glycan array screening data, *Glycobiology*, 2014, **24**, 17–25.

14 O. Blixt, S. Head, T. Mondala, C. Scanlan, M. E. Huflejt, R. Alvarez, M. C. Bryan, F. Fazio, D. Calarese, J. Stevens, N. Razi, D. J. Stevens, J. J. Skehel, I. van Die, D. R. Burton, I. A. Wilson, R. Cummings, N. Bovin, C. H. Wong and J. C. Paulson, Printed covalent glycan array for ligand profiling of diverse glycan binding proteins, *Proc. Natl. Acad. Sci. U. S. A.*, 2004, **101**, 17033–17038.

15 C. T. Campbell, Y. Zhang and J. C. Gildersleeve, Construction and use of glycan microarrays, *Curr. Protoc. Chem. Biol.*, 2010, **2**, 37–53.

16 L. Xia and J. C. Gildersleeve, The glycan array platform as a tool to identify carbohydrate antigens, *Methods Mol. Biol.*, 2015, **1331**, 27–40.

17 Y. Zhang and J. C. Gildersleeve, General procedure for the synthesis of neoglycoproteins and immobilization on epoxide-modified glass slides, in *Carbohydrate Microarrays: Methods and Protocols*, ed. Y. Chevolot, Humana Press, 2012, vol. 808, pp. 155–165.

18 J. C. Manimala, T. A. Roach, Z. Li and J. C. Gildersleeve, High-throughput carbohydrate microarray profiling of 27 antibodies demonstrates widespread specificity problems, *Glycobiology*, 2007, **17**, 17C–23C.

19 J. C. Manimala, T. A. Roach, Z. T. Li and J. C. Gildersleeve, High-throughput carbohydrate microarray analysis of 24 lectins, *Angew. Chem., Int. Ed.*, 2006, **45**, 3607–3610.

20 J. Manimala, Z. Li, A. Jain, S. VedBrat and J. C. Gildersleeve, Carbohydrate array analysis of anti-Tn antibodies and lectins reveals unexpected specificities: Implications for diagnostic and vaccine development, *ChemBioChem*, 2005, **6**, 2229–2241.

21 C. P. Stowell and Y. C. Lee, Neoglycoproteins: the preparation and application of synthetic glycoproteins, *Adv. Carbohydr. Chem. Biochem.*, 1980, **37**, 225–281.

22 J. C. Gildersleeve and W. S. Wright, Diverse molecular recognition properties of blood group A binding monoclonal antibodies, *Glycobiology*, 2016, **26**, 443–448.

23 Y. Zhang, Q. Li, L. G. Rodriguez and J. C. Gildersleeve, An array-based method to identify multivalent inhibitors, *J. Am. Chem. Soc.*, 2010, **132**, 9653–9662.

24 Y. Zhang, C. T. Campbell, Q. Li and J. C. Gildersleeve, Multidimensional glycan arrays for enhanced antibody profiling, *Mol. BioSyst.*, 2010, **6**, 1583–1591.

25 O. O. Oyelaran, Q. Li, D. F. Farnsworth and J. C. Gildersleeve, Microarrays with varying carbohydrate density reveal distinct subpopulations of serum antibodies, *J. Proteome Res.*, 2009, **8**, 3529–3538.

26 S. Muthana and J. C. Gildersleeve, Glycan microarrays: Powerful tools for biomarker discovery, *Dis. Markers*, 2014, **14**, 29–41.

27 C. M. Arthur, R. D. Cummings and S. R. Stowell, Using glycan microarrays to understand immunity, *Curr. Opin. Chem. Biol.*, 2014, **18**, 55–61.

28 S. N. Narla, H. Nie, Y. Li and X. L. Sun, Multi-dimensional glycan microarrays with glyco-macroligands, *Glycoconjugate J.*, 2015, **32**, 483–495.

29 X. Li, J. Gao, D. Liu and Z. Wang, Studying the interaction of carbohydrate–protein on the dendrimer-modified solid support by microarray-based plasmon resonance light scattering assay, *Analyst*, 2011, **136**, 4301–4307.

30 M. Ciobanu, K. T. Huang, J. P. Daguer, S. Barluenga, O. Chaloin, E. Schaeffer, C. G. Mueller, D. A. Mitchell and N. Winssinger, Selection of a synthetic glycan oligomer from a library of DNA-templated fragments against DC-SIGN and inhibition of HIV gp120 binding to dendritic cells, *Chem. Commun.*, 2011, **47**, 9321–9323.

31 N. Parera Pera, H. M. Branderhorst, R. Kooij, C. Maierhofer, M. Van Der Kaaden, R. M. J. Liskamp, V. Wittmann, R. Ruijtenbeek and R. J. Pieters, Rapid screening of lectins for multivalency effects with a glycodendrimer microarray, *ChemBioChem*, 2010, **11**, 1896–1904.

32 X. Zhou, C. Turchi and D. Wang, Carbohydrate cluster microarrays fabricated on three-dimensional dendrimeric platforms for functional glycomics exploration, *J. Proteome Res.*, 2009, **8**, 5031–5040.

33 T. Fukuda, S. Onogi and Y. Miura, Dendritic sugar-microarrays by click chemistry, *Thin Solid Films*, 2009, **518**, 880–888.

34 S.-K. Wang, P.-H. Liang, R. D. Astronomo, T.-L. Hsu, S.-L. Hsieh, D. R. Burton and C.-H. Wong, Targeting the carbohydrates on HIV-1: Interaction of oligomannose dendrons with human monoclonal antibody 2G12 and DC-SIGN, *Proc. Natl. Acad. Sci. U. S. A.*, 2008, **105**, 3690–3695.

35 H. M. Branderhorst, R. Ruijtenbeek, R. M. J. Liskamp and R. J. Pieters, Multivalent carbohydrate recognition on a glycodendrimer-functionalized flow-through chip, *ChemBioChem*, 2008, **9**, 1836–1844.

36 K. Godula, D. Rabuka, Ki T. Nam and C. R. Bertozzi, Synthesis and microcontact printing of dual end-functionalized mucin-like glycopolymers for microarray applications, *Angew. Chem., Int. Ed.*, 2009, **48**, 4973–4976.

37 K. Godula and C. R. Bertozzi, Synthesis of glycopolymers for microarray applications *via* ligation of reducing sugars to a poly(acryloyl hydrazide) scaffold, *J. Am. Chem. Soc.*, 2010, **132**, 9963–9965.

38 M. L. Huang, M. Cohen, C. J. Fisher, R. T. Schooley, P. Gagneux and K. Godula, Determination of receptor specificities for whole influenza viruses using multivalent glycan arrays, *Chem. Commun.*, 2015, **51**, 5326–5329.

39 D. Wang, S. Liu, B. J. Trummer, C. Deng and A. Wang, Carbohydrate microarrays for the recognition of cross-reactive molecular markers of microbes and host cells, *Nat. Biotechnol.*, 2002, **20**, 275–281.

40 S. R. Stowell, C. M. Arthur, R. McBride, O. Berger, N. Razi, J. Heimburg-Molinaro, L. C. Rodrigues, J. P. Gourdine, A. J. Noll, S. Von Gunten, D. F. Smith, Y. A. Knirel, J. C. Paulson and R. D. Cummings, Microbial glycan microarrays define key features of host–microbial interactions, *Nat. Chem. Biol.*, 2014, **10**, 470–476.

41 A. Geissner, A. Reinhardt, C. Rademacher, T. Johannssen, J. Monteiro, B. Lepenies, M. Thépaut, F. Fieschi, J. Mrázková, M. Wimmerova, F. Schuhmacher, S. Götze, D. Grünstein, X. Guo, H. S. Hahm, J. Kandasamy, D. Leonori, C. E. Martin, S. G. Parameswarappa, S. Pasari, M. K. Schlegel, H. Tanaka, G. Xiao, Y. Yang, C. L. Pereira, C. Anish and P. H. Seeberger, Microbe-focused glycan array screening platform, *Proc. Natl. Acad. Sci. U. S. A.*, 2019, **116**, 1958–1967.

42 Y. A. Knirel, H. J. Gabius, O. Blixt, E. M. Rapoport, N. R. Khasbiullina, N. V. Shilova and N. V. Bovin, Human tandem-repeat-type galectins bind bacterial non-βGal polysaccharides, *Glycoconjugate J.*, 2014, **31**, 7–12.

43 A. R. De Boer, C. H. Hokke, A. M. Deelder and M. Wuhrer, Serum antibody screening by surface plasmon resonance using a natural glycan microarray, *Glycoconjugate J.*, 2008, **25**, 75.

44 A. R. De Boer, C. H. Hokke, A. M. Deelder and M. Wuhrer, General microarray technique for immobilization and screening of natural glycans, *Anal. Chem.*, 2007, **79**, 8107–8113.

45 X. Song, J. Heimburg-Molinaro, R. D. Cummings and D. F. Smith, Chemistry of natural glycan microarrays, *Curr. Opin. Chem. Biol.*, 2014, **18**, 70–77.

46 E. Lonardi, A. M. Deelder, M. Wuhrer and C. I. A. Balog, Microarray technology using glycans extracted from natural sources for serum antibody fluorescent detection, in *Methods Mol. Biol.*, 2012, vol. 808, pp. 285–302.

47 X. Song, Y. Lasanajak, B. Xia, J. Heimburg-Molinaro, J. M. Rhea, H. Ju, C. Zhao, R. J. Molinaro, R. D. Cummings and D. F. Smith, Shotgun glycomics: A microarray strategy for functional glycomics, *Nat. Methods*, 2011, **8**, 85–90.

48 Y. Ma, I. Sobkiv, V. Gruzdys, H. Zhang and X. L. Sun, Liposomal glyco-microarray for studying glycolipid–protein interactions, *Anal. Bioanal. Chem.*, 2012, **404**, 51–58.

49 M. A. Pond and R. A. Zangmeister, Carbohydrate-functionalized surfactant vesicles for controlling the density of glycan arrays, *Talanta*, 2012, **91**, 134–139.

50 S. Bian, J. He, K. B. Schesing and A. B. Braunschweig, Polymer pen lithography (PPL)-induced site-specific click chemistry for the formation of functional glycan arrays, *Small*, 2012, **8**, 2000–2005.

51 E. W. Adams, J. Ueberfeld, D. M. Ratner, B. R. O'Keefe, D. R. Walt and P. H. Seeberger, Encoded fiber-optic microsphere arrays for probing protein–carbohydrate interactions, *Angew. Chem., Int. Ed.*, 2003, **42**, 5317–5320.

52 R. Liang, L. Yan, J. Loebach, M. Ge, Y. Uozumi, K. Sekanina, N. Horan, J. Gildersleeve, C. Thompson, A. Smith, K. Biswas, W. C. Still and D. Kahne, Parallel synthesis and screening of a solid phase carbohydrate library, *Science*, 1996, **274**, 1520–1522.

53 S. Purohit, T. Li, W. Guan, X. Song, J. Song, Y. Tian, L. Li, A. Sharma, B. Dun, D. Mysona, S. Ghamande, B. Rungruang, R. D. Cummings, P. G. Wang and J. X. She, Multiplex glycan bead array for high throughput and high content analyses of glycan binding proteins, *Nat. Commun.*, 2018, **9**(1), 258.

54 K. Yamamoto, S. Ito, F. Yasukawa, Y. Konami and N. Matsumoto, Measurement of the carbohydrate-binding specificity of lectins by a multiplexed bead-based flow cytometric assay, *Anal. Biochem.*, 2005, **336**, 28–38.

55 E. A. Smith, W. D. Thomas, L. L. Kiessling and R. M. Corn, Surface plasmon resonance imaging studies of protein–carbohydrate interactions, *J. Am. Chem. Soc.*, 2003, **125**, 6140–6148.

56 B. T. Houseman and M. Mrksich, Carbohydrate arrays for the evaluation of protein binding and enzymatic modification, *Chem. Biol.*, 2002, **9**, 443–454.

57 M. M. Ngundi, C. R. Taitt, S. A. McMurry, D. Kahne and F. S. Ligler, Detection of bacterial toxins with monosaccharide arrays, *Biosens. Bioelectron.*, 2006, **21**, 1195–1201.

58 Y. Chevolot, C. Bouillon, S. Vidal, F. Morvan, A. Meyer, J. P. Cloarec, A. Jochum, J. P. Praly, J. J. Vasseur and E. Souteyrand, DNA-based carbohydrate biochips: A platform for surface glyco-engineering, *Angew. Chem., Int. Ed.*, 2007, **46**, 2398–2402.

59 P. H. Liang, S. K. Wang and C. H. Wong, Quantitative analysis of carbohydrate–protein interactions using glycan microarrays: Determination of surface and solution dissociation constants, *J. Am. Chem. Soc.*, 2007, **129**, 11177–11184.

60 E. Mercey, R. Sadir, E. Maillart, A. Roget, F. Baleux, H. Lortat-Jacob and T. Livache, Polypyrrole oligosaccharide array and surface plasmon resonance imaging for the measurement of glycosaminoglycan binding interactions, *Anal. Chem.*, 2008, **80**, 3476.

61 X. Song, B. Xia, Y. Lasanajak, D. F. Smith and R. D. Cummings, Quantifiable fluorescent glycan microarrays, *Glycoconjugate J.*, 2008, **25**, 15–25.

62 C. F. Grant, V. Kanda, H. Yu, D. R. Bundle and M. T. McDermott, Optimization of immobilized bacterial disaccharides for surface plasmon resonance imaging measurements of antibody binding, *Langmuir*, 2008, **24**, 14125–14132.

63 X. Tian, J. Pai and I. Shin, Analysis of density-dependent binding of glycans by lectins using carbohydrate microarrays, *Chem.–Asian J.*, 2012, **7**, 2052–2060.

64 K. Godula and C. R. Bertozzi, Density variant glycan microarray for evaluating cross-linking of mucin-like glycoconjugates by lectins, *J. Am. Chem. Soc.*, 2012, **134**, 15732–15742.

65 K. H. Mortel, R. V. Weatherman and L. L. Kiessling, Recognition specificity of neoglycopolymers prepared by ring-opening metathesis polymerization, *J. Am. Chem. Soc.*, 1996, **118**, 2297–2298.

66 C. W. Cairo, J. E. Gestwicki, M. Kanai and L. L. Kiessling, Control of multivalent interactions by binding epitope density, *J. Am. Chem. Soc.*, 2002, **124**, 1615–1619.

67 N. Horan, L. Yan, H. Isobe, G. M. Whitesides and D. Kahne, Nonstatistical binding of a protein to clustered carbohydrates, *Proc. Natl. Acad. Sci. U. S. A.*, 1999, **96**, 11782–11786.

68 M. Mammen, S. K. Choi and G. M. Whitesides, Polyvalent interactions in biological systems: Implications for design and use of multivalent ligands and inhibitors, *Angew. Chem., Int. Ed.*, 1998, **37**, 2754–2794.

69 S. Bashir, S. Leviatan Ben Arye, E. M. Reuven, H. Yu, C. Costa, M. Galinanes, T. Bottio, X. Chen and V. Padler-Karavani, Presentation mode of glycans affect recognition of human serum anti-Neu5Gc IgG antibodies, *Bioconjugate Chem.*, 2019, **30**, 161–168.

70 A. Chandrasekaran, A. Srinivasan, R. Raman, K. Viswanathan, S. Raguram, T. M. Tumpey, V. Sasisekharan and R. Sasisekharan, Glycan topology determines human adaptation of avian H5N1 virus hemagglutinin, *Nat. Biotechnol.*, 2008, **26**, 107–113.

71 D. M. Lewallen, D. Siler and S. S. Iyer, Factors affecting protein–glycan specificity: Effect of spacers and incubation time, *ChemBioChem*, 2009, **10**, 1486–1489.

72 T. K. Dam, S. Oscarson, R. Roy, S. K. Das, D. Pagé, F. Macaluso and C. F. Brewer, Thermodynamic, kinetic, and electron microscopy studies of concanavalin A and *Dioclea grandiflora* lectin cross-linked with synthetic divalent carbohydrates, *J. Biol. Chem.*, 2005, **280**, 8640–8646.

PAPER

A 'catch-and-release' receptor for the cholera toxin†

Clare S. Mahon,*[ab] Gemma C. Wildsmith,[a] Diksha Haksar,[c] Eyleen de Poel,[d] Jeffrey M. Beekman,[d] Roland J. Pieters,[c] Michael E. Webb[a] and W. Bruce Turnbull*[ab]

Received 11th February 2019, Accepted 20th March 2019

DOI: 10.1039/c9fd00017h

Stimuli-responsive receptors for the recognition unit of the cholera toxin (CTB) have been prepared by attaching multiple copies of its natural carbohydrate ligand, the GM1 oligosaccharide, to a thermoresponsive polymer scaffold. Below their lower critical solution temperature (LCST), polymers complex CTB with nanomolar affinity. When heated above their LCST, polymers undergo a reversible coil to globule transition which renders a proportion of the carbohydrate recognition motifs inaccessible to CTB. This thermally-modulated decrease in the avidity of the material for the protein has been used to reversibly capture CTB from solution, enabling its convenient isolation from a complex mixture.

Introduction

Interactions between proteins and other biological macromolecules or surfaces are crucial to the mediation of many physiological processes in healthy and diseased states,[1] and the development of synthetic materials which can perturb such interactions presents exciting opportunities for the production of new therapeutics and diagnostics.[2,3] Proteins often associate with their binding partners across large interface areas through multivalency,[4] harnessing the effects of multiple weak interactions to produce strong adhesive forces. The construction of receptors or inhibitors using polymer-based systems allows for convenient access to the large molecular architectures required to bridge recognition sites on the surfaces of proteins, eliminating the need for the demanding synthesis of

[a]School of Chemistry and Astbury Centre for Structural Molecular Biology, University of Leeds, Woodhouse Lane, Leeds, LS2 9JT, UK. E-mail: clare.mahon@york.ac.uk; w.b.turnbull@leeds.ac.uk

[b]Department of Chemistry, University of York, Heslington, York, YO10 5DD, UK

[c]Department of Chemical Biology & Drug Discovery, Utrecht Institute for Pharmaceutical Sciences, Utrecht University, Utrecht, The Netherlands

[d]Department of Pediatric Pulmonology, Wilhelmina Children's Hospital and Regenerative Medicine Center Utrecht, University Medical Centre Utrecht, Utrecht, The Netherlands

† Electronic supplementary information (ESI) available. See DOI: 10.1039/c9fd00017h

complex architectures such as dendrimers.[5] The choice of a synthetic polymer scaffold for the construction of receptors also enables the inclusion of secondary functionality, such as compositional adaptivity,[6,7] or the ability to respond to environmental stimuli such as temperature[8] or pH,[9] to produce an adaptive material.

The inclusion of thermoresponsive functionality within macromolecular receptors for proteins has yielded highly-functional materials which allow for tunable recognition of their targets. Gibson and coworkers[10] have produced temperature-responsive nanoparticle-based receptors by immobilising polymers with α-terminal aminogalactosyl units onto gold nanoparticles, along with a thermoresponsive polymer designed to act as a 'gate.' Below the lower critical solution temperature (LCST) of the 'gate' polymer, its steric bulk prevents the carbohydrate motifs from accessing the binding sites of soybean agglutinin (SBA), a complementary lectin. Heating the material induces collapse of the of the 'gate' polymer, exposing carbohydrate units on the surface of the nanoparticles which are then able to complex with SBA. A similar strategy has been used to control the surface presentation of receptor targeting ligands on polymer-decorated nanoparticles, and thereby control their endocytosis. Alexander and coworkers[11] have decorated gold nanoparticles with transferrin, an iron transporting protein, along with a 'shielding' thermoresponsive polymer. Below the LSCT of the polymer, nanoparticles display poor cellular uptake. An increase in temperature can be used to trigger collapse of the 'shielding' polymer, exposing the transferrin ligands and inducing cellular uptake. An impressive 'catch-and-release' system for a protein has been developed by Shea and coworkers,[12] where multicomponent polymer nanoparticles were demonstrated only to complex lysozyme above their LCST, enabling their binding to the protein and subsequent release upon cooling. Surface-bound polymer brushes have also been employed for the selective adsorption of proteins, by tuning the ionic strength of the surrounding medium to modulate electrostatic interactions between proteins and surfaces.[13]

Here, we have constructed a high-affinity multivalent receptor for a bacterial protein on a thermally responsive polymer scaffold, and demonstrated that an increase in temperature above the LCST of the polymer may modulate this recognition – a feature we have used to enable the isolation of a single protein from a complex mixture.

Our thermoresponsive receptor binds to the carbohydrate recognition domain of the cholera toxin,[14] a protein produced by *Vibrio cholerae* in the human intestine which is responsible for the acute diarrhoea associated with cholera. The cholera toxin is comprised of a single toxic A subunit and non-toxic B_5 pentamer of identical units (CTB), each possessing a recognition site for the GM1 ganglioside (GM1), which is displayed on intestinal epithelial surfaces. The pentasaccharide unit of the ganglioside, the GM1 oligosaccharide (GM1os), has been demonstrated by isothermal titration calorimetry to display a remarkably high affinity for CTB, with a reported K_d of ∼40 nM.[15] Already, the incorporation of multiple copies of GM1os on macromolecular scaffolds[16,17] such as dendrimers,[18] synthetic polymers[19] or proteins[20] has been shown to greatly enhance the inhibitory potency of recognition units, making CTB a particularly attractive target for the development of 'smart' multivalent receptors and inhibitors.

Results and discussion

Synthesis of thermoresponsive receptor

We chose to construct our receptor on a poly(N-isopropylacrylamide) (poly(-NIPAm)) scaffold, a class of polymer well known for its ability to reversibly desolvate upon increases in solution temperature.[21] NIPAm was copolymerised with *tert*-butyl 2-methacryloylhydrazine-1-carboxylate (M1) (Scheme 1), which was prepared *via* hydrazinolysis of methacrylic anhydride,[22] and subsequent BOC-protection of hydrazide units. RAFT polymerisation afforded a copolymer with a degree of polymerisation of approximately 99, displaying approximately 21 protected acylhydrazide functionalities as estimated using ^{1}H NMR spectroscopy. Subsequent deprotection of P1 yielded the multiply acylhydrazide-functionalised P2, which enabled the direct functionalisation of polymers with reducing sugars through the formation of β-glycoside linkages (Scheme 1). This strategy has been used for the preparation of surface-immobilised carbohydrate microarrays,[23] and avoids potentially complex synthetic modifications of the oligosaccharide in addition to presenting a general route to the production of receptors for other carbohydrate binding proteins. The GM1 ganglioside was treated with a bacterial endoglycoceramidase (EGCaseII, *Rhodococcus* sp. M-777) to yield GM1os,[24,25] which was combined with P2 under mildly acidic conditions to yield a polymer functionalised with multiple copies of GM1os (P2-GM1os). ^{1}H NMR spectroscopic analysis (ESI Fig. S2†) confirmed the oligosaccharide to be displayed on the polymer

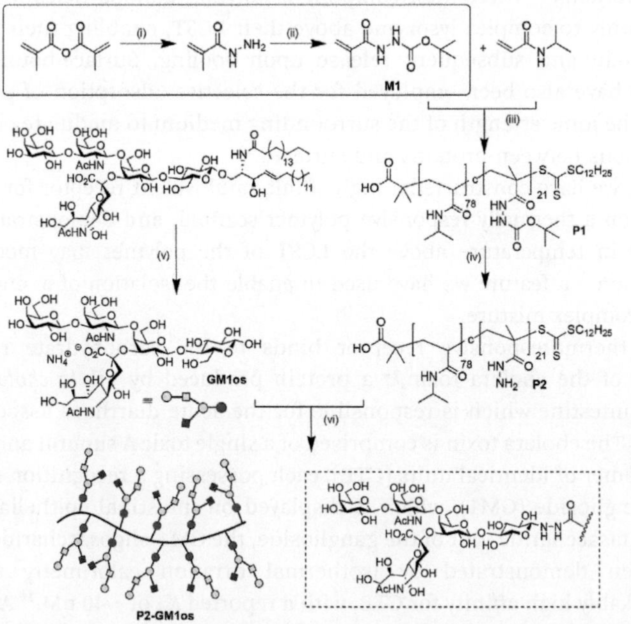

Scheme 1 Preparation of P2-GM1os. (i) N$_2$H$_4$·H$_2$O, CHCl$_3$, rt, 1 h. (ii) Di-*tert*-butyl dicarbonate, THF, rt, 24 h. (iii) S-Dodecyl-S′-(α,α′-dimethyl-α″-acetic acid)trithiocarbonate, 2,2′-azobis(2-methylpropionitrile), DMSO, 70 °C, 18 h. (iv) 50% v/v CH$_3$COOH, CH$_2$Cl$_2$, rt, 1 h. (v) EGCaseII, 25 mM NH$_4$OAc, pH 5.0, 0.2% v/v Triton X-100. (vi) 100 mM NH$_4$OAc, pH 4.5, rt, 18 h.

primarily as its β-glycoside. Dynamic light scattering (DLS) analysis (Fig. 2(c)) of **P2-GM1os** revealed a monomodal distribution with number average hydrodynamic diameter (D_h) of approximately 16 nm, a significant increase in hydrodynamic volume compared to **P2**, which displayed a D_h of approximately 5 nm.

Assessment of inhibitory potency of P2-GM1os for CTB

As other multivalent receptors for CTB have previously proven effective in inhibiting adhesion of the cholera toxin to GM1 functionalised surfaces,[26,27] we decided to investigate this effect with our system. Initially, the inhibitory potency of **P2-GM1os** towards CTB was assessed using an enzyme-linked lectin assay (ELLA, Fig. 1),[20] where the ability of CTB to bind to surface-immobilised GM1 was assessed over a range of **P2-GM1os** concentrations (Table 1) at 25 °C. **P2-GM1os** demonstrated an IC_{50} value of 3.78 nM, a significant enhancement in inhibitory potency compared to GM1os (Table 1, 505 nM), demonstrating that **P2-GM1os** can effectively disrupt interactions between CTB and surfaces displaying GM1os. We next investigated the ability of **P2-GM1os** to prevent the entry of the cholera toxin to human cells, using an intestinal organoid assay.[28] In this assay, intestinal organoids are exposed to the cholera toxin over a range of inhibitor concentrations, and inhibition is monitored by assessing the ability of the cholera toxin to enter the cells and to induce fluid secretion leading to an increase in organoid surface area. **P2-GM1os** was found to be a potent inhibitor of cholera toxin induced swelling, with an IC_{50} value of 5.68 nM, demonstrating an impressive improvement in inhibitory potency compared to GM1os (Table 1, 8.7 µM). These results demonstrate that the complexation of CTB by **P2-GM1os** can be harnessed to prevent the internalisation of the cholera toxin by intestinal cells. Synthetic multivalent inhibitors of cellular adhesion such as **P2-GM1os** may offer therapeutic potential for diseases caused by bacterial toxins, offering an attractive antibiotic-free route to the control of symptoms.[29]

Thermoresponsive behaviour

The thermoresponsive properties of **P2** and **P2-GM1os** were investigated by turbidimetry,[21] (Fig. 2(a) and (b)) providing an estimate of the LCST of each

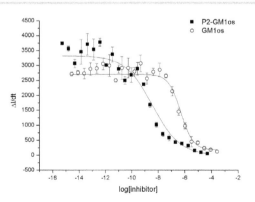

Fig. 1 Enzyme-linked lectin assay (ELLA) indicates the inhibitory potency of **P2-GM1os** compared to GM1os at 293 K. Error bars indicate the standard error of three measurements.

Table 1 Inhibitory potencies from enzyme-linked lectin assay (ELLA) and intestinal organoid assay. Curves were fitted using log[inhibitor] values, leading to the asymmetric distribution of fitting errors about the mean. These errors have therefore been omitted for simplicity[a]

Assay	Inhibitor	Valency	$\log[IC_{50}]$	IC_{50}/nM	Relative potency (per GM1os)
ELLA	GM1os	1	-6.30 ± 0.08	505	1
ELLA	P2-GM1os	21	-8.42 ± 0.14	3.78	6.38
Intestinal organoid assay	GM1os	1	-5.06 ± 0.11	8700	1
Intestinal organoid assay	P2-GM1os	21	-8.30 ± 0.16	5.68	72.9

[a] Potency is expressed relative to monovalent GM1os in the corresponding assay.

polymer. **P2** was observed to undergo a sharp phase transition with a cloud point of 44 °C at 2.0 mg mL^{-1} in phosphate-buffered saline (137 mM NaCl, 2.7 mM KCl, 10 mM Na$_2$HPO$_4$, 1.8 mM KH$_2$PO$_4$, pH 7.4 – PBS). The analogous transition for

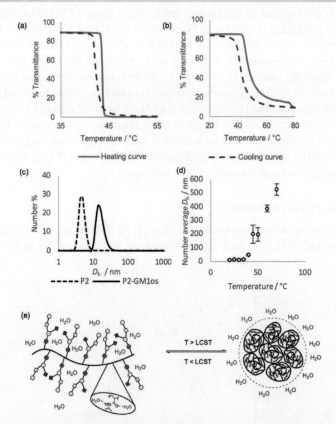

Fig. 2 Temperature–turbidity curves for **P2** (a) and **P2-GM1os** (b) at 2.0 mg mL^{-1} in PBS, pH 7.4 (137 mM NaCl, 2.7 mM KCl, 10 mM Na$_2$HPO$_4$, 1.8 mM KH$_2$PO$_4$). (c) Number average hydrodynamic diameters (D_h) for **P2** and **P2-GM1os** at 25 °C in PBS pH 7.4. (d) Number average D_h for **P2-GM1os** across the temperature range 20–80 °C. Error bars represent the standard distribution of at least five measurements. (e) The reversible desolvation of **P2-GM1os** at temperatures above its LCST.

P2-GM1os was observed to occur with a similar cloud point of 48 °C under the same conditions, but with the transition occurring over a broader temperature range, a feature we attribute to the numerous hydrogen bonding units incorporated onto **P2** by its functionalisation with GM1os. The cloud point determined for **P2-GM1os** by turbidimetry was in agreement with variable-temperature dynamic light scattering experiments conducted at 0.5 mg mL^{-1} in the same buffer (Fig. 2(d)), which showed the formation of large aggregates in solution around this temperature. We hypothesised that this triggered phase transition could be used to modulate the affinity of **P2-GM1os** for CTB, and decided to probe this hypothesis using isothermal titration calorimetry (ITC).

Isothermal titration calorimetry studies

Initially, recognition was investigated at 25 °C, below the LCST of the polymer, when chains adopt a hydrated coil conformation. Under these conditions, CTB

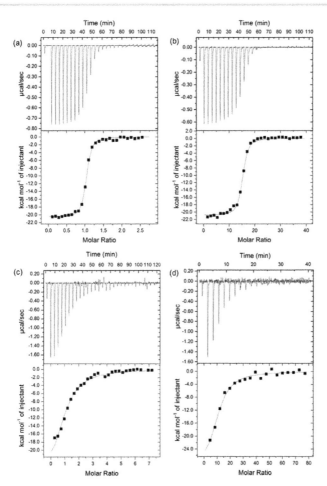

Fig. 3 Titrations of (a) 137 µM GM1os into 8.91 µM CTB at 25 C; (b) 110 µM CTB into 0.5 µM P2-GM1os at 25 C; (c) 376 µM GM1os into 9.32 µM CTB at 60 C; and (d) 378 µM CTB into 1.0 µM P2-GM1os at 60 C. CTB concentrations refer to the concentration of the B subunit.

Table 2 Thermodynamic parameters for GM1os and **P2-GM1os** binding to CTB at 25 °C and at 60 °C. Titrations were performed in PBS at pH 7.4 (137 mM NaCl, 2.7 mM KCl, 10 mM Na_2HPO_4, 1.8 mM KH_2PO_4)

Temp./°C	Ligand	$K_{d(app)}$/ nM (σ)	ΔG/kcal mol^{-1} (σ)	ΔH/kcal mol^{-1} (σ)	ΔS/cal K^{-1} mol^{-1} (σ)	n (σ)
25	GM1os	37.9 (3.7)	−10.1 (1.0)	−20.5 (0.1)	−34.9 (4.9)	1.0 (0.004)
25	**P2-GM1os**	31.4 (4.1)	−10.2 (1.3)	−20.8 (0.2)	−35.5 (6.5)	14.8 (0.07)
60	GM1os	4460 (430)	−8.15 (0.8)	−30.1 (0.9)	−66.0 (8.9)	1.0^a
60	**P2-GM1os**	3660 (810)	−8.29 (1.83)	−34.2 (4.3)	−77.7 (19.7)	9.9 (0.99)

^{a} Binding stoichiometry fixed during fitting.

was observed to recognise GM1os with a K_d of 37.9 nM (±3.7 nM) (Fig. 3(a), Table 2). **P2-GM1os** also displayed nanomolar affinity for CTB, with an apparent K_d of 31.4 nM (±4.1 nM) and a binding stoichiometry of 15 CTB subunits per polymer chain, demonstrating that multiple GM1os residues on each polymer can interact with CTB with a similar affinity to that observed for the monovalent interaction. The apparent lack of improvement in K_d, despite a significant enhancement in inhibitory potency, has been observed previously with other multivalent inhibitors of CTB.[20,30] We next repeated these experiments at 60 °C, above the LCST of the polymer, when polymer chains are expected to exist in a collapsed, globular conformation. At this temperature CTB, which has been determined to be thermally stable at temperatures of up to 74 °C,[31] was shown to recognise GM1os with a K_d of 4.46 μM (±0.43 μM) (Fig. 3(c), Table 2). This decrease in affinity at elevated temperature may be accounted for by the unfavourable entropic contribution to binding, and in particular the increase in flexibility of the loop around the carbohydrate recognition site of the protein.[32] Under these conditions, **P2-GM1os** binds CTB with an apparent K_d of 3.66 μM (±0.81 μM) (Fig. 3(d), Table 2), similar to the monovalent interaction. Notably, however, the binding stoichiometry of the interaction decreased to 9.9, suggesting fewer GM1os residues per polymer chain interact with CTB at elevated temperature. We propose that the thermally-triggered collapse of **P2-GM1os** renders a proportion of GM1os recognition units inaccessible to CTB on the interior of the globule, whilst the GM1os residues which remain on the surface still interact with CTB with a comparable affinity to the monovalent interaction. We hypothesised that this thermally-induced change in binding stoichiometry could be harnessed to modulate the overall avidity of the material for CTB. With this aim in mind, we immobilised **P2-GM1os** onto a solid support, to investigate if the polymer could be used to facilitate 'catch-and-release' isolation of the protein.

Reversible complexation of CTB

P2-GM1os was conveniently immobilised onto a solid support in a one-pot procedure. The trithiocarbonate end groups of were first converted to thiols by aminolysis, and immediately subjected to thiol–ene coupling[33] with commercially available maleimide functionalised agarose beads (Scheme 2). Upon incubation with a 10 μM solution of CTB in PBS, beads were shown to complex with CTB using gel electrophoresis (Fig. 4(a)). The CTB complexed with the beads was not

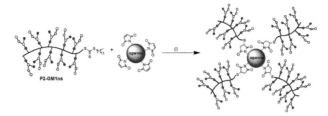

Scheme 2 Immobilisation of **P2-GM1os** onto maleimide-functionalised agarose beads. (i) Hexylamine, Et₃N, DMSO.

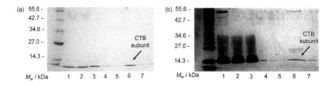

Fig. 4 The 'catch-and-release' behaviour of **P2-GM1os** functionalised beads applied to (a) a 10 µM solution of CTB in PBS pH 7.4; and (b) *Vibrio* sp. 60 growth medium, as demonstrated by gel electrophoresis. In each case lane 1 contains the analyte; lane 2 beads after exposure to analyte; lane 3 the supernatant after analyte loading; lanes 4 and 5 contain the bead washes with PBS at room temperature; lane 6 contains the eluate obtained after treatment with PBS at 60 °C and lane 7 the beads after elution. Gels were stained using Coomassie blue (a); or silver stain (b).

eluted during washes with PBS at ambient temperature, but when incubated with the same buffer at 60 °C, above the determined LCST of **P2-GM1os**, CTB was successfully recovered from the beads. To determine if this 'catch-and-release'[12] behaviour could translate to the isolation of a single protein from a highly complex mixture, we attempted to isolate CTB from a solution of *Vibrio* sp. 60 growth medium in which CTB is expressed, and exported from the cells to the surrounding medium. The cells were separated from the rest of the growth medium by centrifugation, and the supernatant incubated with **P2-GM1os** functionalised beads at room temperature. The beads were washed with PBS at room temperature to remove any weakly adhering material, before treatment with the same buffer at 60 °C triggered elution of CTB (Fig. 4(b)). This response demonstrates that the combined effects of high-affinity multivalent recognition and the stimuli-responsive nature of our polymer scaffold may be used to selectively isolate a single bacterial toxin from a complex biological medium.

Conclusions

In summary, we report a stimuli-responsive receptor which can complex CTB with nanomolar affinity, and release the protein on demand in response to a thermal stimulus. Our polymeric receptor has been shown to be effective in preventing the internalisation of the cholera toxin by intestinal cells, with inhibitory potency in the low nanomolar range. The temperature-mediated 'catch-and-release' behaviour has been used to isolate CTB from bacterial growth medium, a complex biological medium containing a multitude of other proteins. Current 'gold-standard' identification protocols for cholera[34] and other diarrheal diseases rely on bacterial culture, processes which require time and access to appropriate facilities. We anticipate that our system could contribute to the development of diagnostics for diseases such as cholera, by providing a route to conveniently isolate the toxin for identification.

Finally, our method presents a straightforward route to the synthesis of highly functional multivalent receptors for lectins, incorporating carbohydrate recognition motifs directly from reducing sugars. Our approach could be easily adapted to enable the production of receptors for the many other carbohydrate-binding proteins implicated in bacterial disease,[16] by judicious choice of carbohydrate recognition units.

Experimental

Materials and characterisation

All reagents were purchased from Sigma Aldrich, Alfa Aesar, Carbosynth or Flu-orochem and used as received unless otherwise stated. N-Isopropylacrylamide was recrystallised from hexane prior to use. Phosphate buffered saline (PBS) refers to a solution of 137 mM NaCl, 2.7 mM KCl, 10 mM Na_2HPO_4 and 1.8 mM KH_2PO_4 in MilliQ water. ^1H and ^{13}C NMR spectra were recorded on a Bruker Avance 500 spectrometer at 500 MHz and 125 MHz respectively, with the residual solvent signal as an internal standard. Mass spectra were collected using a Bruker MicroTOF instrument. Gel permeation chromatography (GPC) was conducted on a Varian ProStar instrument (Varian Inc.) equipped with a Varian 325 UV-Vis dual wavelength detector (254 nm), a Dawn Heleos II multi-angle laser light scattering detector (Wyatt Technology Corp.), a Viscotek 3580 differential RI detector, and a pair of PL gel 5 μm mixed D 300 × 7.5 mm columns with guard column (Polymer Laboratories Inc.) in series. Near monodisperse methyl methacrylate standards (Agilent Technologies) were used for calibration. Data collection was performed with Galaxie software (Varian Inc.) and chromatograms analyzed with Cirrus software (Varian Inc.) and Astra software (Wyatt Technology Corp.). Turbidimetric analysis was performed on a Cary 100 UV-Vis spectrometer, with the absorbance of 2.0 mg mL^{-1} solutions monitored at 550 nm. Cloud points are reported as the mid-point of a fit to a standard dose–response curve in Origin 8. Dynamic light scattering (DLS) instrumentation consisted of a Malvern Zetasizer μV with a 633 nm HeNe laser module. Measurements were made at an angle of 173° (back scattering) and Malvern Zetasizer software was used to analyse the data. Solutions were filtered through a 0.22 μm PTFE filter prior to analysis. At each temperature point the solution was incubated for 30 min before measurements were made.

Methacryloyl hydrazide[22] (1)

Methacrylic anhydride (9.6 mL, 65 mmol) in $CHCl_3$ (50 mL) was added dropwise to hydrazine hydrate (13.2 mL, 272 mmol) at 0 °C. The mixture was left to stir at room temperature for 1 h, then the organic fraction was removed. The aqueous fraction was extracted with $CHCl_3$ (3 × 25 mL) and the combined organic extracts were dried over $MgSO_4$ and evaporated to dryness, yielding a while solid. The title product was recrystallised from 10 : 1 toluene : CH_2Cl_2 and isolated as colourless, needle-like crystals (2.50 g, 38%). ^{1}H NMR (500 MHz, CDCl$_3$): δ 1.95 (s, 3H, CH_3), 3.98 (s, 2H, NH_2), 5.35 (s, 1H, CH_2), 5.71 (s, 1H, CH_2), 7.39 (s, 1H, NH). ^{13}C NMR (125 MHz, CDCl$_3$): δ 18.5, 120.5, 138.2, 169.5. Melting point 83.0–86.5 °C (84–86 °C (ref. 22)). ESI-HRMS: calculated for C$_4$O$_2$N$_2$H$_9$ [M + H$^+$]: 101.0709; found 101.0019.

tert-Butyl 2-methacryloylhydrazine-1-carboxylate[35] (M1)

Di-*tert*-butyl dicarbonate (5.12 g, 24 mmol) in THF (50 mL) was added dropwise to **1** (2.38 g, 24 mmol) in THF (50 mL). The reaction mixture was left to stir at room temperature for 24 h before it was evaporated to dryness. The white solid obtained was dissolved in CH$_2$Cl$_2$ (100 mL) and washed with sat. NaCl$_{(aq)}$ (100 mL). The organic fraction was dried over MgSO$_4$ and evaporated to dryness, yielding a white solid. **M1** was recrystallised from 10 : 1 toluene : CH$_2$Cl$_2$ and isolated as white crystals (2.65 g, 55%). ^{1}H NMR (500 MHz, CDCl$_3$): δ 1.47 (s, 9H, C(CH_3)$_3$), 1.98 (s, 3H, CH_3), 5.42 (s, 1H, CH_2), 5.81 (s, 1H, CH_2), 6.70 (s, 1H, NHNHCOC(CH$_3$)$_3$), 7.89 (s, 1H, NHNHCO C(CH$_3$)$_3$). ^{13}C NMR (125 MHz, CDCl$_3$): δ 18.5, 28.3, 82.1, 83.62, 133.3, 155.7, 167.8. Melting point: 115.5–117.6 °C. ESI-HRMS: calculated for C$_9$H$_{16}$N$_2$NaO$_3$ [M + Na$^+$]: 223.1053; found 223.1057.

BOC-protected acylhydrazide copolymer P1

S-1-Dodecyl-*S'*-(α,α-dimethyl-α''-acetic acid)trithiocarbonate[36] (DDMAT) (20 mg, 5.5 × 10^{-5} mol, 1.0 eq.), α,α'-azoisobutyronitrile (AIBN) (1.8 mg, 1.1 × 10^{-5} mol, 0.2 eq.), *N*-isopropylacrylamide (NIPAm) (0.993 g, 8.78 mmol, 160 eq.) and **M1** (0.241 g, 1.20 mmol, 20 eq.) were combined in DMSO (3 mL) and degassed *via* three freeze–pump–thaw cycles. The vessel was backfilled with N$_{2(g)}$ and allowed to warm to room temperature, then placed in a preheated oil bath at 70 °C for 18 h. The polymerisation was quenched by rapid cooling in N$_{2(l)}$, followed by exposure to air. The solution was added dropwise to a large excess of ice-cold diethyl ether, yielding a yellow–white solid which was redissolved in THF and the precipitation repeated twice before drying under high vacuum. **P1** was isolated as a yellow–white solid (0.350 g). ^{1}H NMR (500 MHz, CDCl$_3$): δ 1.15 (br, NHCH(CH_3)$_3$), 1.48 (br, (COOC(CH_3)$_3$), 1.6–1.9 (br, (CH_2, CH_3), polymer backbone), 2–2.4 (br, (CH), polymer backbone, 3.04 (br, NHCHCH$_3$), 4.01 (br, NHCHCH$_3$), 6.0–7.0 (br, NHNHCOOC(CH$_3$)$_3$).

Acylhydrazide copolymer P2

P1 (30 mg, 2.2 × 10^{-6} mol) was dissolved in CH$_2$Cl$_2$ (1 mL). Trifluoroacetic acid (1 mL) was added and the solution was left to stir at room temperature for 1 h. The solution was concentrated *in vacuo*, yielding a yellow glassy film which was redissolved in H$_2$O, dialysed against H$_2$O and lyophilised, yielding **P2** as a yellow–white solid (21 mg, 83% yield). ^{1}H NMR (500 MHz, CDCl$_3$): δ 1.13 (br, NHCH(CH_3)$_3$, polymer backbone), 1.6–1.9 (br, (CH_2, CH_3), polymer backbone), 2–2.4 (br, (CH), polymer backbone), 4.01 (br, NHCHCH$_3$).

GM1os functionalised copolymer P2-GM1os

GM1os (6.2 mg, 6.1 × 10^{-6} mol) and **P2** (3.0 mg, 2.6 × 10^{-6} mol) were combined in 100 mM NH$_4$OAc, pH 4.5, D$_2$O (500 µL). The solution was left to stir at room temperature for 24 h. The solution was then dialysed against H$_2$O and lyophilised, yielding the title product as a yellow–white solid (3.5 mg, 41% yield). ^{1}H NMR (500 MHz, D_2O): ESI Fig. S2.†

GM1os preparation[15]

GM1 ganglioside (100 mg, 6.37×10^{-5} mol) and BSA (2.0 mg) were dissolved in 25 mM NH_4OAc, pH 5.0, 0.2% v/v Triton X-100 (9 mL). The solution was split into 450 µL aliquots, to which EGCaseII (50 µL at 100 µM in 20 mM Na_2HPO_4, pH 7.0, 500 mM NaCl) was added. The solutions were incubated at 37 °C for 12 days, at which point TLC analysis confirmed complete reaction of the ganglioside. The solutions were combined and washed with Et_2O (2×10 mL), before the combined aqueous fraction was passed through a 0.45 µm syringe filter, then loaded onto a C18 solid phase extraction cartridge and eluted with H_2O. Fractions containing GM1os were combined and twice lyophilised from H_2O, yielding GM1os as a white solid (50.7 mg, 4.97×10^{-5} mol, 78% yield). The ^{1}H NMR spectrum recorded (ESI Fig. S13†) is in agreement with previous reports.[37]

Immobilisation of P2-GM1os onto agarose beads

Procedure adapted from Boyer et al.[33] Maleimide functionalised agarose beads (Cube Biotech.) (500 µL suspension) were washed with H_2O (3×1 mL) and lyophilised. **P2-GM1os** (3.2 mg, 1.0×10^{-7} mol, 1.0 eq.) was dissolved in DMSO (600 µL). The solution was sparged with $N_{2(g)}$ before addition of Et_3N (25 µL, 1.8×10^{-5} mol, 180 eq.) and hexylamine (64 µL, 4.5×10^{-4} mol, 4500 eq.). The suspension was left to stir at room temperature for 16 h, then beads were isolated by centrifugation (500g, 1 min). The supernatant was removed and the beads were washed with H_2O (3×1 mL), PBS pH 7.4 (3×1 mL) and resuspended in 2 mM β-mercaptoethanol in PBS (500 µL). The suspension was agitated gently at room temperature for 2 h, then beads were isolated by centrifugation (500g, 1 min), washed with PBS pH 7.4 (3×1 mL), H_2O (3×1 mL) and stored in 20% v/v $EtOH_{(aq)}$. UV-Vis spectra of the initial solution of **P2-GM1os** were compared to that of the supernatant after coupling (Fig. S4†), with the loss of the absorbance of the tri-thiocarbonate unit at 309 nm demonstrating reduction of polymer end groups.

Isothermal titration calorimetry (ITC)

CTB samples were dialysed into PBS using SpectraPor dialysis tubing (MWCO 2.5–3 kDa, SpectrumLabs). **P2-GM1os** samples were dialysed against MilliQ H_2O, lyophilised and dissolved in the same buffer solution. The concentration of GM1os solutions in MilliQ H_2O were determined using ^{1}H NMR spectroscopic analysis[15] using EtOH as an internal standard. These solutions were then lyophilised and reconstituted in the CTB dialysis buffer prior to the titration to ensure an exact buffer match during the experiment. Concentrations of CTB stated refer to the concentrations of the subunits.

The reference cell was filled with water and the analysis cell filled with **P2-GM1os** in the case of **P2-GM1os** – CTB titrations, or CTB in the case of GM1os – CTB titrations. Both cells were allowed to reach thermal equilibrium at 25 °C, or 60 °C, before titration. Titrations typically consisted of a single 2 µL injection, followed by 29×8 µL injections when using the VP-ITC instrument, and single 0.5 µL injection, followed by 19×2 µL injections when using the ITC-200 instrument. Separate titrations of the binding partner in the syringe into buffer were used to assess the heat of dilution.

Protocol for reversible complexation of CTB

P2-GM1os beads (stored in 20% v/v EtOH$_{(aq)}$) were isolated by centrifugation (500g, 1 min) and resuspended in 10 μM CTB in PBS, pH 7.4 (500 μL). The tube was agitated gently at room temperature for 1 h before the beads were isolated by centrifugation (500g, 1 min) and the supernatant removed. **P2-GM1os** beads were washed with PBS (2 × 500 μL) at room temperature, then incubated at 60 °C in preheated PBS for 15 min. Beads were isolated by centrifugation (500g, 30 s) and the elution repeated.

Protocol for purification of CTB from bacterial growth medium

Cells from a glycerol stock of *Vibrio* sp. 60 harbouring plasmid pATA13 (ref. 38) (kindly provided by Prof. Tim Hirst) were used to inoculate growth medium (100 mL, 25 g L^{-1} LB mix, 15 g L^{-1} NaCl, ampicillin 100 μg mL^{-1}). The culture was grown overnight at 30 °C with shaking at 200 rpm, then used to inoculate fresh growth medium (6 × 1 L, 25 g L^{-1} LB mix, 15 g L^{-1} NaCl, ampicillin 100 μg mL^{-1}). These cultures were incubated at 30 °C with shaking at 200 rpm until A$_{600}$ reached 0.6 before the protein expression was induced by addition of isopropyl β-D-1-thiogalactopyranoside to a concentration of 0.5 mM. Cultures were incubated (30 °C, 200 rpm) for a further 24 h, then cells were removed by centrifugation (7500g, 15 min). **P2-GM1os** beads (stored in 20% v/v EtOH$_{(aq)}$) were isolated by centrifugation (500g, 1 min) and resuspended in the supernatant (4 × 500 μL) with a 15 min incubation time and gentle agitation at room temperature. The beads were isolated by centrifugation (500g, 1 min) and the supernatant removed. **P2-GM1os** beads were washed with PBS (2 × 500 μL) at room temperature, the incubated at 60 °C in preheated PBS for 15 min. Beads were isolated by centrifugation (500g, 30 s) and the elution repeated.

Enzyme-linked lectin assay[20] (ELLA)

96-Well, high-binding, black, flat-bottomed plates were treated with 1.3 μM GM1 ganglioside in EtOH (100 μL per well), and the solution was left to evaporate under laminar flow overnight. Wells were washed with PBS (3 × 200 μL) before treatment with 1.0% w/v bovine serum albumin (BSA) in PBS (100 μL) for 30 min at 37 °C. Wells were then washed with PBS (3 × 200 μL).

Inhibitor solutions were prepared in 0.1% w/v BSA, 0.05% Tween 20, PBS, pH 7.4 (521 μM GM1os, 100 μM **P2-GM1os**). The concentration of GM1os was determined by ^{1}H NMR spectroscopic analysis using EtOH as an internal standard.[15] A 24-step three-fold dilution series of both inhibitor solutions was performed in triplicate. Each solution was combined with a CTB–horseradish peroxidase conjugate (CTB-HRP) at 5.0 ng mL^{-1} CTB-HRP, 0.01% w/v BSA, 0.05% Tween 20, PBS, pH 7.4. These inhibitor–toxin solutions were allowed to incubate at room temperature for 2 h, then were transferred to GM1 functionalised plates (200 μL per well). The plates were left at room temperature for 30 min, then emptied and wells were washed with 0.1% w/v BSA, 0.05% Tween 20, PBS, pH 7.4 (3 × 200 μL). 5 μM H$_2$O$_2$, 5 μM Amplex red, PBS, pH 7.4 (100 μL) was added and fluorescence was monitored over a 5 min period (λ_{ex} 531 nm, λ_{em} 595 nm) at 25 °C. The slope of each plot of fluorescence intensity against time was recorded and these values were plotted against log [inhibitor] and fitted using a standard dose–response model in Origin 8.

Human rectal biopsies

Rectal biopsies were collected from a healthy human subject after acquiring approval by the Ethics Committee of the University Medical Center Utrecht (UMCU) and after obtaining informed consent of the individual.

Generating and culturing organoids

The generation and biobanking of the intestinal organoids differed slightly from previously described protocols.[39] After washing with PBS, crypts were isolated from the biopsies, *via* 60–90 min incubation in 10 mM EDTA at 4 °C on a rocking platform. Crypts were collected, centrifuged at 130g for 5 min at 4 °C, and supernatant removed. The crypts pellet was resuspended in 50% Matrigel (Corning, diluted in culture medium), and droplets of crypts suspension were plated onto pre-warmed 24-well plates. After polymerization of the Matrigel (±15 min, 37 °C/5% CO_2), droplets were immersed in culture medium (advanced DMEM/F12 supplemented with HEPES, GlutaMAX, penicillin, streptomycin, N-2, B-27, mEGF (50 ng mL^{-1}, Life Technologies), *N*-acetylcysteine (1.25 mM), nicotinamide (10 mM), SB202190 (10 μM, Sigma), A83-01 (500 nM Tocris), Primocin (100 μg mL^{-1}, Invivogen), and 50% Wnt3a-, 20% Rspo-1-, and 10% Noggin-conditioned media). In 5–9 days, crypts grew out into full grown organoids, which were passaged *via* mechanical disruption of the organoids. Medium was refreshed every 2–3 days, and organoids were passaged at least two times before assays were performed.

Organoid swelling assay

CFTR function measurements in intestinal organoids was performed slightly differently than the previously described procedures.[26,39] Organoids cultured for 6–8 days were mechanically disrupted and seeded into flat-bottom 96-well plates in 50% Matrigel and incubated overnight at 37 °C, 5% CO_2. The next day, cholera toxin (Sigma C8052) was incubated together with cholera toxin inhibitors (titration) for 4 h at rt followed by staining of the organoids with 3 μM calcein AM (Invitrogen) for 15–30 min prior to organoid stimulation. Organoid swelling was monitored using the Zeiss LSM 800 confocal microscope (images were taken every 15 min during cholera/inhibitor stimulation and every 10 min during forskolin stimulation) while the organoids were kept at 37 °C/5% CO_2. The organoid area increase (2D) was analyzed using Zen Blue 2.0 analysis software, and the area under the curve (AUC) calculations (of the total area increase measured in 4 h) were conducted with GraphPad Prism 5.0.

CTB expression and purification

Cells from a glycerol stock of *Vibrio* sp. 60 harbouring plasmid pATA13 were used to inoculate growth medium (100 mL, 25 g L^{-1} LB mix, 15 g L^{-1} NaCl, 100 μg mL^{-1} ampicillin). The culture was grown overnight at 30 °C with shaking at 200 rpm, then used to inoculate fresh growth medium (6 × 1 L, 25 g L^{-1} LB mix, 15 g L^{-1} NaCl, ampicillin 100 μg mL^{-1}). These cultures were incubated at 30 °C with shaking at 200 rpm until A_{600} reached 0.6 before the protein expression was induced by addition of isopropyl β-D-1-thiogalactopyranoside to a concentration of 0.5 mM. Cultures were incubated (30 °C, 200 rpm) for a further 24 h, then cells were removed by centrifugation (7500g, 15 min). The combined supernatant was treated with ammonium sulphate (550 g L^{-1}) and left to stir at 5 °C overnight. Crude protein was isolated by centrifugation (17 000g, 25 min) and redissolved in 100 mM NaH$_2$PO$_4$, pH 7.0, 500 mM NaCl (60 mL). Insoluble material was removed by centrifugation (5000g, 10 min) before the solution was passed through a 0.22 μm filter then loaded onto a lactose–sepharose 6B column and eluted with 300 mM lactose, 100 mM NaH$_2$PO$_4$, pH 7.0, 500 mM NaCl. CTB was dialysed against PBS, pH 7.4, freeze-dried and stored at −80 °C.

Expression of EGCaseII[24]

A synthetic gene for endoglycosylceramidase from (*Rhodococcus* sp. M-777) bearing an *N*-terminal His-tag and codon-optimised for expression in *E. coli* in a pET28a backbone was purchased from Genscript (Piscataway, USA). Cells from glycerol stock (*E. coli* BL21 (DE3) gold) were used to inoculate TYP broth (100 mL, 16 g L^{-1} tryptone, 16 g L^{-1} yeast, 5 g L^{-1} NaCl, 2.5 g L^{-1} K$_2$HPO$_4$, 50 μg mL^{-1} kanamycin) and the culture was grown at 37 °C with shaking at 200 rpm overnight. The temperature was decreased to 20 °C before protein expression was induced by addition of isopropyl β-D-1-thiogalactopyranoside to a final concentration of 0.1 mM. The culture was incubated at 20 °C with shaking at 200 rpm for 7.5 h before cells were isolated by centrifugation (5000g, 10 min) and cell pellets stored at −20 °C overnight. The cell pellets were resuspended in BugBuster (5 mL) and left at room temperature for 30 min before the cell debris was removed by centrifugation (5000g, 15 min) and the supernatant was applied to a Ni-NTA column which had been pre-equilibrated with 20 mM Na$_2$HPO$_4$, pH 7.0, 0.5 M NaCl. The flow-through was passed through the column again, before the column was washed with 20 mM Na$_2$HPO$_4$, pH 7.0, 0.5 M NaCl (20 mL), then eluted with a stepwise imidazole gradient (10 mL aliquots: 50, 100, 200, 300, 400, 500 mM imidazole) in the same buffer. Elutions at 100 mM and 200 mM imidazole were combined and dialysed against 20 mM Na$_2$HPO$_4$, pH 7.0, 0.5 M NaCl. ECGaseII was concentrated to 100 μM, divided into 50 μL aliquots and stored at −80 °C.

Conflicts of interest

There are no conflicts to declare.

Acknowledgements

C. S. M. wishes to thank EPSRC and the University of Leeds for the award of a Doctoral Prize Fellowship which enabled this work. W. B. T. acknowledges the support of BBSRC (BB/M005666/1). The authors thank Dr David A. Fulton, Newcastle University, UK, for access to gel permeation chromatography equipment.

Notes and references

1 L.-G. Milroy, T. N. Grossmann, S. Hennig, L. Brunsveld and C. Ottmann, *Chem. Rev.*, 2014, **114**, 4695–4748.

2 H. Yin and A. D. Hamilton, *Angew. Chem., Int. Ed.*, 2005, **44**, 4130–4163.

3 V. Azzarito, K. Long, N. S. Murphy and A. J. Wilson, *Nat. Chem.*, 2013, **5**, 161–173.

4 J. D. Badjić, A. Nelson, S. J. Cantrill, W. B. Turnbull and J. F. Stoddart, *Acc. Chem. Res.*, 2005, **38**, 723–732.

5 M. V. Walter and M. Malkoch, *Chem. Soc. Rev.*, 2012, **41**, 4593–4609.

6 C. S. Mahon, C. J. McGurk, S. M. D. Watson, M. A. Fascione, C. Sakonsinsiri, W. B. Turnbull and D. A. Fulton, *Angew. Chem., Int. Ed.*, 2017, **56**, 12913–12918.

7 C. S. Mahon, M. A. Fascione, C. Sakonsinsiri, T. E. McAllister, W. B. Turnbull and D. A. Fulton, *Org. Biomol. Chem.*, 2015, **13**, 2756–2761.

8 Y. Gou, J. Geng, S.-J. Richards, J. Burns, C. Remzi Becer and D. M. Haddleton, *J. Polym. Sci., Part A: Polym. Chem.*, 2013, **51**, 2588–2597.

9 G. Kocak, C. Tuncer and V. Butun, *Polym. Chem.*, 2017, **8**, 144–176.

10 S. Won, S.-J. Richards, M. Walker and M. I. Gibson, *Nanoscale Horiz.*, 2017, **2**, 106–109.

11 E. J. Sayers, J. P. Magnusson, P. R. Moody, F. Mastrotto, C. Conte, C. Brazzale, P. Borri, P. Caliceti, P. Watson, G. Mantovani, J. Aylott, S. Salmaso, A. T. Jones and C. Alexander, *Bioconjugate Chem.*, 2018, **29**, 1030–1046.

12 K. Yoshimatsu, B. K. Lesel, Y. Yonamine, J. M. Beierle, Y. Hoshino and K. J. Shea, *Angew. Chem., Int. Ed.*, 2012, **51**, 2405–2408.

13 A. Bratek-Skicki, V. Cristaudo, J. Savocco, S. Nootens, P. Morsomme, A. Delcorte and C. Dupont-Gillain, *Biomacromolecules*, 2019, **20**, 778, DOI: 10.1021/acs.biomac.8b01353.

14 D. Vanden Broeck, C. Horvath and M. J. S. De Wolf, *Int. J. Biochem. Cell Biol.*, 2007, **39**, 1771–1775.

15 W. B. Turnbull, B. L. Precious and S. W. Homans, *J. Am. Chem. Soc.*, 2004, **126**, 1047–1054.

16 A. Bernardi, J. Jimenez-Barbero, A. Casnati, C. De Castro, T. Darbre, F. Fieschi, J. Finne, H. Funken, K.-E. Jaeger, M. Lahmann, T. K. Lindhorst, M. Marradi, P. Messner, A. Molinaro, P. V. Murphy, C. Nativi, S. Oscarson, S. Penades, F. Peri, R. J. Pieters, O. Renaudet, J.-L. Reymond, B. Richichi, J. Rojo, F. Sansone, C. Schaffer, W. B. Turnbull, T. Velasco-Torrijos, S. Vidal, S. Vincent, T. Wennekes, H. Zuilhof and A. Imberty, *Chem. Soc. Rev.*, 2013, **42**, 4709–4727.

17 S. Cecioni, A. Imberty and S. Vidal, *Chem. Rev.*, 2015, **115**, 525–561.

18 A. V. Pukin, H. M. Branderhorst, C. Sisu, C. A. G. M. Weijers, M. Gilbert, R. M. J. Liskamp, G. M. Visser, H. Zuilhof and R. J. Pieters, *ChemBioChem*, 2007, **8**, 1500–1503.

19 C. L. Schengrund and N. J. Ringler, *J. Biol. Chem.*, 1989, **264**, 13233–13237.

20 T. R. Branson, T. E. McAllister, J. Garcia-Hartjes, M. A. Fascione, J. F. Ross, S. L. Warriner, T. Wennekes, H. Zuilhof and W. B. Turnbull, *Angew. Chem., Int. Ed.*, 2014, **53**, 8323–8327.

21 E. S. Gil and S. M. Hudson, *Prog. Polym. Sci.*, 2004, **29**, 1173–1222.

22 Y. Chen, N. Ballard, O. D. Coleman, I. J. Hands-Portman and S. A. F. Bon, *J. Polym. Sci., Part A: Polym. Chem.*, 2014, **52**, 1745–1754.

23 M.-r. Lee and I. Shin, *Org. Lett.*, 2005, **7**, 4269–4272.

24 M. D. Vaughan, K. Johnson, S. DeFrees, X. Tang, R. A. J. Warren and S. G. Withers, *J. Am. Chem. Soc.*, 2006, **128**, 6300–6301.

25 M. Ito, *Trends Glycosci. Glycotechnol.*, 1990, **2**, 399–402.

 This journal is © The Royal Society of Chemistry 2019

26 V. Kumar and W. B. Turnbull, *Beilstein J. Org. Chem.*, 2018, **14**, 484–498.

27 H. Zuilhof, *Acc. Chem. Res.*, 2016, **49**, 274–285.

28 D. D. Zomer-van Ommen, A. V. Pukin, O. Fu, L. H. C. Quarles van Ufford, H. M. Janssens, J. M. Beekman and R. J. Pieters, *J. Med. Chem.*, 2016, **59**, 6968–6972.

29 M. Garland, S. Loscher and M. Bogyo, *Chem. Rev.*, 2017, **117**, 4422–4461.

30 C. Sisu, A. J. Baron, H. M. Branderhorst, S. D. Connell, C. A. G. M. Weijers, R. de Vries, E. D. Hayes, A. V. Pukin, M. Gilbert, R. J. Pieters, H. Zuilhof, G. M. Visser and W. B. Turnbull, *ChemBioChem*, 2009, **10**, 329–337.

31 B. Goins and E. Freire, *Biochemistry*, 1988, **27**, 2046–2052.

32 E. A. Merritt, T. K. Sixma, K. H. Kalk, B. A. M. van Zanten and W. G. J. Hol, *Mol. Microbiol.*, 1994, **13**, 745–753.

33 X. Huang, C. Boyer, T. P. Davis and V. Bulmus, *Polym. Chem.*, 2011, **2**, 1505–1512.

34 M. Alam, N. A. Hasan, M. Sultana, G. B. Nair, A. Sadique, A. S. G. Faruque, H. P. Endtz, R. B. Sack, A. Huq, R. R. Colwell, H. Izumiya, M. Morita, H. Watanabe and A. Cravioto, *J. Clin. Microbiol.*, 2010, **48**, 3918–3922.

35 A. K. Pearce, B. E. Rolfe, P. J. Russell, B. W. C. Tse, A. K. Whittaker, A. V. Fuchs and K. J. Thurecht, *Polym. Chem.*, 2014, **5**, 6932–6942.

36 J. T. Lai, D. Filla and R. Shea, *Macromolecules*, 2002, **35**, 6754–6756.

37 S. Sabesan, K. Bock and R. U. Lemieux, *Can. J. Chem.*, 1984, **62**, 1034–1045.

38 A. T. Aman, S. Fraser, E. A. Merritt, C. Rodigherio, M. Kenny, M. Ahn, W. G. J. Hol, N. A. Williams, W. I. Lencer and T. R. Hirst, *Proc. Natl. Acad. Sci. U. S. A.*, 2001, **98**, 8536–8541.

39 J. F. Dekkers, C. L. Wiegerinck, H. R. de Jonge, I. Bronsveld, H. M. Janssens, K. M. de Winter-de Groot, A. M. Brandsma, N. W. M. de Jong, M. J. C. Bijvelds, B. J. Scholte, E. E. S. Nieuwenhuis, S. van den Brink, H. Clevers, C. K. van der Ent, S. Middendorp and J. M. Beekman, *Nat. Med.*, 2013, **19**, 939–945.

Faraday Discussions

DISCUSSIONS

Preparation of multivalent glycan micro- and nano-arrays: general discussion

Adam Braunschweig, Joseph P. Byrne, Ryan Chiechi, Yuri Diaz Fernandez, Jeff Gildersleeve, Kamil Godula, Laura Hartmann, Clare Mahon, Yoshiko Miura, Alshakim Nelson, Stephan Schmidt, W. Bruce Turnbull, Daniel Valles, Jin Yu and Dejian Zhou

DOI: 10.1039/C9FD90062D

Adam Braunschweig opened the discussion of the paper by Jeffrey Gildersleeve: In your contribution to this Discussion, you have shown that the avidity between protein and glycan is sensitively dependent upon glycan–glycan spacing, linker composition, linker length, *etc*. As a result, different microarrays produce different results for specificity and avidity towards a particular lectin. Given this inherent complexity, is there a "true value" for avidity that reflects most accurately how this recognition may be occurring in biology?

Jeff Gildersleeve answered: This is an interesting consideration. A glycan determinant can be present in varying contexts in biology. The avidity will depend on the structure of the determinant as well as the presentation. For example, a lectin might have four binding sites. In some settings, the lectin might only be able to engage two binding sites. In other settings, the lectin might bind glycans using all four of its binding sties. These could both be biologically relevant, but the avidities could be quite different. So, there are likely many "true values" for avidity for each glycan determinant. When connecting apparent K_d values measured on a glycan microarray to biological systems, I think of it as potential – "in the right context, this lectin could bind with an apparent K_d value of X to this glycan". It has the potential to bind, but it will depend on other factors, such as the nature of the carrier chain and the spacing and orientation of the glycans.

Daniel Valles asked: Since there has been much discussion about the varying binding constants throughout different glycan arrays, do you think binding should be defined as a range rather than a single number?

Kamil Godula responded: Of course a single binding constant can be measured for the binding to a single glycan or a multivalent glycopolymer. However, it is not that clear how much practical information these measurements provide if one is considering the presence of a large number of similar glycoconjugates at the cell surface. Perhaps an apparent binding constant to the glycocalyx surface may be a more meaningful measure.

Jeff Gildersleeve then added: This is a really interesting perspective. There are so many factors that influence binding – monovalent *vs.* multivalent ligand, type and extent of multivalency, nature of the linker or carrier glycan chain, assay conditions, *etc.* As a result, the measured binding interaction can vary quite a lot. How one defines binding or discusses binding depends on the situation. For an individual protein binding to a specific ligand, I would probably refer to a single number. If one is talking about how well a lectin binds its ligands, a range would make more sense.

Clare Mahon asked: To what extent does the observed dissociation constant depend on the technique you use to measure it? Would you expect to determine similar dissociation constants for wholly solution-phase measurements compared to those determined on surfaces?

Jeff Gildersleeve answered: The method can have a significant effect. We refer to the values we measure on the array as "apparent K_d values". One might also refer to them as "surface K_d values". In most cases, they represent the apparent binding avidity of a multivalent binding interaction. Binding in solution to a multivalent conjugate can be quite different. We have discuss this at length in a paper.[1] Briefly, on a surface, a lectin can form a multivalent complex involving 1 lectin and 1 neoglycoprotein, or it can form a multivalent complex bridging 2 or more neoglycoprotein molecules on the surface. In solution, only the 1:1 complex forms. In many cases, lectins can form multivalent complexes with a surface but not with a neoglycoprotein in solution (*ie.* the spacing and orientation of the glycans on the protein may not be suitable to form a 1:1 complex). Binding in solution to a monovalent glycan will often give much weaker binding than interactions with multivalent glycans on a surface.

1 Y. Zhang, Q. Li, L. Rodriguez and J. C. Gildersleeve, An array-based method to identify multivalent inhibitors, *J. Am. Chem. Soc.*, 2010, **132**, 9653–9662.

Dejian Zhou opened the discussion of the paper by Daniel Valles: The use of the right linker to ensure that all sugars are accessible to lectin binding is important to make glycan microarrays more robust. Have you studied how linker length and flexibility affect the lectin binding? Also the sugar density reported in your paper appears quite low, *e.g.* at 0.1–0.6 molecule per nm^2, suggesting that the sugars are rather loosely packed on the surface. How did you determine the sugar density of such surfaces?

Daniel Valles answered: At the moment, we have not done a thorough investigation of different linker lengths. However, we understand that different linkers can lead to more robust systems, which is why we plan to examine this in the future. Unfortunately, the packing is just pure estimation based on the size of the monosaccharide. We decided to show a range from 0.1–0.6 nm^2 to help draw conclusions with our experimental data. The theoretical calculations show that if the molecules were packed at 0.1 nm^2, multivalency could not occur if the glycan density falls below 20%, which agrees with the experimental data we have shown in the paper.

Ryan Chiechi then enquired: Rather than having to discuss effective binding constants, would it make more sense to measure the binding of an internal standard and report a value relative to that? That way the internal standard experiences the same local environment as the surface-binding event that is under investigation, potentially facilitating comparisons across a variety of experimental platforms. This is a concept that is common in other areas such as colloidal self-assembly, where what matters is the relative strengths of interactions in competition with each other rather than scalar values on an absolute unit scale.

Jeff Gildersleeve replied: This is a really interesting idea and something that could be very helpful. We print a variety of control spots on our array, including IgG, IgM, and IgA. I think at least some other groups also print these controls. These controls could potentially serve as standards for assessing relative affinity between array platforms. I would like to note that the apparent K_d values measured on our array and the CFG array have similar values. Thus, in the majority of cases, the apparent affinities are quite similar; there are just certain specific instances where there are large differences in binding.

Stephan Schmidt asked: The presented results indicate a critical spacing of mannose units above which ConA is bound to the chip surface. Since this mannose spacing correlates with the minimum binding site distance of ConA ($\sim$7 nm), do you think that ConA binds in a chelate-like fashion on the chip? Furthermore, when further increasing the mannose density, there is a proportionality between mannose density and bound ConA. Does this suggest that this linear regime is governed by statistical multivalency, *i.e.* the mannose units contribute additively?

Daniel Valles answered: To the first part of your question, ConA is a tetramer constructed of four monomers, each of which has its own binding site. Due to the location of each binding pocket, we believe that only two sites are available to bind to a monolayer of mannose. That being said, our paper suggests that if both of those pockets cannot reach a mannose attached to the surface, then the overall binding will not be strong enough to remain attached to the surface. In this particular example, it may seem to be linear, but what we expected to see was step-like drops-off of fluorescence when spacing between the immobilized mannose starts to make them further away from one another. The best example of this is the difference between $\chi = 0.3$ and $\chi = 0.2$. There is a sudden drop off to the point where the fluorescence of the ConA binding becomes scarce to the point where it is immeasurable.

Joseph P. Byrne commented: This was a really interesting piece of research. I was fascinated at how your estimated density of sugars on the surface, and hence the average distance between mannose units, correlated well with the experimentally-known distance between binding sites for ConA to interact in a multivalent way. This seems like an encouraging result.

Do you think that this system that you and your colleagues have described could be used as a tool for estimating, or perhaps even quantifying, the distance between binding sites or carbohydrate-recognition domains, in new, less-studied, or even unknown lectins?

Adam Braunschweig replied: Thanks for this question. What we reported in our *Faraday Discussion* manuscript was that no binding between printed mannose and ConA in solution was observed when our estimate of the average spacing between mannosides reached ∼7 nm, which is the same as the distance between the binding sites of ConA. So, one could assume that for ConA to persist on the surface after washing, then the protein would need to be able to reach at least two mannosides, so if we consider this assumption valid, then, potentially this method could be used to estimate the distance between binding sites. Although the data are promising, there is, however, considerable uncertainty in estimating the distance between mannosides on the surface. You will see in the paper that we had two estimates for the distance between mannosides on the surface, with the high-density as well as a low-density estimate based upon previous results from our group. A consistent theme in this general discussion was that characterizing the nature of the surface with nanoscale precision remains a major challenge. Without more accurate and easy-to-use methods to characterize directly the average spacing of the mannosides, I would be reticent to draw any firm conclusions, but I do consider this work to be a step forward in the exact direction that your question implies.

Yuri Diaz Fernandez opened the discussion of the paper by Clare Mahon: Within your paper, the thermodynamic parameters for the monomer and the polymer-anchored receptors are very similar at the same temperature. Particularly, we would expect the entropy to be different considering the change in degrees of freedom, going from a monomer to the polymeric receptor. Could you explain why this intuitively expected difference is not observed in your system?

W. Bruce Turnbull then added: Regarding the entropy of binding of the polymers to the CTB lectin, it is important to note that we do not see any apparent enhancement in affinity by calorimetry for the glycopolymers relative to the monovalent GM1 oligosaccharide, even though we do see an enhancement in inhibitory potency. While others studying different systems have often observed affinity enhancements using ITC, we have never seen this for multivalent GM1 ligands binding to CTB regardless of whether the multivalent scaffold is a polymer, dendrimer or tailored glycoprotein. Yet in most of these cases we do see substantial increases in inhibitor potency up to a few thousand times relative to the monovalent ligand. I suspect that the differences we observe between the binding and inhibition experiments are a result of observing distinct processes happening on different timescales: the ITC reports rapid binding to the GM1 while a slower rearrangement to the most stable multivalent configuration during the inhibition assay would be effectively invisible calorimetrically because it would likely have slow kinetics and relatively small net enthalpy changes. The result is that it is challenging to interpret the measured entropies in terms of multivalent binding.

Clare Mahon replied to both: In both cases, the binding of GM1os to CTB is entropically unfavourable – the geometry of the binding site requires rigidification of the oligosaccharide conformation to enable complexation. You are correct to note that we observed a similar value for the entropy of complexation in both cases. This observation may suggest that major reorganisation of the polymer conformation is not required to enable complexation. Below their LCST we expect the polymers to exist in solution as hydrated random coils. In this conformation a significant proportion of GM1os residues are expected to be displayed on the surface of the coil, available for complexation without significant rearrangement of the polymer backbone.

Yuri Diaz Fernandez then asked: Have you observed any effect of the pH on the affinity of the receptor?

Clare Mahon replied: We have not performed any experiments to investigate the complexation of our polymer with CTB under conditions other than physiological pH. We don't expect, however, that changing the pH would alter the affinity of the interaction to a significant extent – provided, of course, that the protein itself was stable at the pH under study.

Dejian Zhou said: The GM1 conjugated thermal response polymer can catch and release CTB in a temperature-dependent manner. Since CTB–glycan binding became weaker at higher temperature, have you tested whether GM1-conjugated to a non-thermal responsive polymer can also do the same job?

Clare Mahon responded: Yes, we had looked at similar receptors constructed on a non-thermoresponsive dimethylacrylamide scaffold and we didn't see a change in the binding stoichiometry at elevated temperature – so the 'catch and release' behaviour is dependent on the thermoresponsive nature of the poly(-NIPAm) scaffold.

Yoshiko Miura asked: The catch and release of CTB was attained by PolyNIPAAm-GM1 conjugates. The system is based on the LCST of polyNIPAAm. Considering the LCST of the polymer, the entropy effect is important. Is it possible to measure thermodynamic parameters like entropy?

Clare Mahon responded: Yes, some techniques allow for the determination of entropic contributions too binding. In our case, we used isothermal titration calorimetry to determine all thermodynamic parameters of interaction. The recognition of GM1os by CTB is entropically unfavourable, as it requires rigidification of the oligosaccharide conformation. We found that the binding of GM1os-decorated polymers to CTB was also unfavourable both above and below the LCST of the polymer.

Yoshiko Miura opened a general discussion by addressing Jeffrey Gildersleeve: Glycan arrays have been examined by many molecules and methods.

When considering the interaction between glycan and protein, it is thought that not only the enthalpy benefits between glycan and protein, but also the mobility of molecules, the shape of clusters, the flexibility of molecules, and so on, are involved. Do you have any idea about the structure of the linker that links the glycans?

Jeff Gildersleeve replied: The linker certainly affects recognition in many cases. The portion of a glycan that interacts with a lectin or antibody is often referred to as the glycan determinant. In nature, glycan determinants are often attached to a carrier glycan chain or lipid. For synthetic systems, glycan determinants are often attached *via* a non-natural linker to a surface or multivalent scaffold. The carrier chain, lipid, or synthetic linker can all significantly influence recognition. Sometimes, they can provide some additional contacts or interactions that enhance binding. At other times, they can decrease or even prevent recognition – for example, they might block access of the glycan determinant to the binding pocket through steric interference. I am not able to predict these types of inter-actions, and they likely vary quite a bit from one protein to another. Therefore, I believe variations in linkers and carrier glycan chains are valuable elements of diversity to include on a glycan microarray to allow one to evaluate their effects empirically.

Alshakim Nelson then addressed Jeff Gildersleeve and Kamil Godula: The field has been trying to probe the nature of carbohydrate–ligand interactions (spacing, ligands, linkers, *etc.*), so has anyone looked at using artificial intelligence or machine learning to identify systems to probe these interactions?

Jeff Gildersleeve and **Kamil Godula** have not yet replied.

Jin Yu then enquired: How did you compare between two different array, since the signal-to-noise ratio is different on each slide. In common cases we used the value by subtracting the total signal to local background, but even the same epitope can have different background from different slides, therefore resulting in different observed signal intensities. How did CFG address to this issue?

A follow-up question is: despite your great work with the glycoprotein conju-gate microarray, if you could choose again, which method of microarray setup would you prefer; between an NHS coated covalent array, an NGL noncovalent array, a glycoprotein array, or another platform like the Luminex beads system?

Jeff Gildersleeve replied: For the comparison described in our paper, we did not adjust for differences in signal-to-noise ratios. There were several reasons. First, the CFG and our own data both used background-subtracted RFU signals, so this methodology was consistent on both arrays. Second, the noise levels were similar for the vast majority of the data (One way to evaluate noise is to look at the variations in signals for the lowest 50% of RFU values. For example, the lowest 50% of RFU values across the array might vary between +50 and -50 RFU.) There was one set of data where our array had a lot of noise – SNA at 50 μg mL^{-1}. Since there was a lot of noise and our noise was considerably higher than the CFG, we did not include this dataset in our comparison. Third, our analyses were focused on signals that were well above background levels for both arrays. Thus, we concluded that variations due to noise would have only very minor influence on the analysis.

As for your second question, I think each of the platforms has their advantages and disadvantages. The optimal system really depends on the application. One of the really neat features of the NGL platform is the ability of the neoglycolipids to move and adjust to match the spacing of different glycan binding proteins. Because of this feature, one does not have to have the right spacing on the surface – the glycans can adjust to accommodate different proteins. However, there are situations where one would like to distinguish between different proteins that have different spacing preferences. For example, some subpopulations of anti-bodies in human serum will only bind glycans at high density while others will bind at either high or low density. These different subpopulations can have different clinical significance. To detect or monitor these different subpopula-tions of antibodies, one needs defined/stable densities of glycans on the surface.

There are also advantages and disadvantages for a bead-based array versus a slide. In terms of sample throughput, the Luminex bead system is really powerful. For example, one can easily evaluate hundreds to thousands of human serum samples. The slide based system has advantages in terms of total number of glycans that can be evaluated and using much less material. For example, one could print 20,000 different glycans on a slide and test a lectin for binding to all of them in a single experiment. The Luminex system can accommodate up to about 500 different glycans in one experiment, but the people I know that use Luminex prefer to just use more like 100–200 different types of beads per experiment. There are several features I really like about the neoglycoprotein system. First, it is easy to translate results from the array to other assays and experiments. When using array binding information to design multivalent inhibitors or reagents, one needs a multivalent presentation that matches the presentation on the array surface. For many array platforms, finding this match can be challenging. For example, what multivalent scaffold should one use – a nanoparticle, liposome, glycopolymer, *etc.*? In addition, what type of linker should one use to attach the glycan to the scaffold? What density of glycan will best mimic the density on the array surface. These are not trivial questions. With the neoglycoprotein platform, we can use the same neoglycoprotein from our array in other experiments. For example, going from the array to an ELISA or Western blot is really simple. Also, one can attach a neoglycoprotein to a Luminex bead and the signals are nearly identical because many features of presentation, such as the linker and glycan density, are preserved. Second, we have flexibility for how we attached glycans to the neo-glycoprotein. We don't need a specific group, like a free amine, to get a glycan onto the surface. This allows us to use glycans from many different sources, including glycans with an azide linker, a free lactol, or a carboxylic acid linker. Third, we can print a lot of different entities using the same conditions – we print neoglycoproteins, natural glycoproteins, and DNA on our surfaces using the same conditions, settings, and slide surface. That being said, if I was to start over, I'm not sure I would use bovine serum albumin as the carrier protein for our neo-glycoproteins. This protein is a lot more complex than I realized when I first got started. I would also love a system where we have much better control over spacing and orientation of glycans on the surface. We can modulate average differences on the surface, but it would be really neat to control them more precisely on a molecular level.

W. Bruce Turnbull asked: Is there any correlation between lectin architecture and the differences in binding that you observe between the different formats of the glycan microarray? For example cases where binding sites are pointing in the same direction *vs.* in opposing directions?

Jeff Gildersleeve replied: While lectin architecture is likely to be an important contributor, we are not yet able to correlate lectin architecture with differences in binding. The ability to form a multivalent complex depends on many factors, including the spacing and orientation of the lectin binding sites, the density of the glycans on the surface, the linker length, and the linker flexibility. At this point, we don't have a good system to isolate the effects of architecture. In addition, we have very little information about how these lectins are interacting with glycans on the surface for any glycan microarray.

W. Bruce Turnbull then asked: Could the differences in binding you see between glycan microarrays with different architectures provide insights into the native ligands for a lectin? For example, glycolipid *vs.* glycoprotein?

Jeff Gildersleeve responded: Using the differences in binding to gain insight about the native ligands would be very useful, but it is challenging. A key barrier is that we know very little about how the lectins interact with glycans on a micro-array surface at a molecular level.

Dejian Zhou commented: The glycan microarrays based on bovine serum albumin (BSA) neoglycoprotein scaffolds appear to be effective in promoting lectin binding. However, given the sugars are conjugated to BSA *via* surface lysine residues which are not evenly distributed throughout the protein surface, it may be difficult to produce a uniform sugar coating with the same inter-sugar distance. In addition, have you checked whether such BSA layers are homogenous on the surface? Do they form a uniform, complete monolayer? Have you considered using other nanoparticle scaffolds (*e.g.* quantum dots or gold nanoparticles) which have uniform surface reactivity toward glycan ligands and therefore may be able to offer potentially better control over the sugar densities and inter-glycan spacing. Moreover, the unique optical properties of these nanoparticles, *e.g.* strong fluorescence for quantum dots[1] and efficient fluorescence quenching for gold nanoparticles,[2] can be harnessed for binding confirmation and quantification. For example, we have recently demonstrated that quantum dots displaying polyvalent specific glycan ligands (glycan = mannose or Man–α-1,2-Man) can act as multifunctional nanoprobes to dissect the exact binding modes of a pair of closely related, almost identical tetrameric lectins, DC-SIGN and DC-SIGNR, *via* a multimodal readout strategy combining fluorescence resonance energy transfer (FRET), transmission electron microscopy and hydrodynamic size analysis. Moreover, they can also be used to quantify the binding affinity with DC-SIGN/R *via* a sensitive, ratiometric FRET readout. We have further revealed that the apparent K_d measured by our FRET method matches well to their inhibition potency (IC_{50}) against DC-SIGN mediated pseudo-Ebola virus infection of target cells.[3,4] Given such potential advantages, are you interested in making glycan microarrays using such functional nanoparticle scaffolds? If so, I will be very happy to collaborate with you on this development.

1 I. L. Medintz, A. R. Clapp, H. Mattoussi, E. R. Goldman, B. Fisher and J. M. Mauro, *Nat. Mater.*, 2003, **2**, 630–638.
2 B. Dubertret, M. Calame and A. J. Libchaber, *Nat. Biotech.*, 2001, **19**, 365–370.
3 Y. Guo, C. Sakonsinsiri, I. Nehlmeier, M. A. Fascione, H. Zhang, W. Wang, S. Pöhlmann, W. B. Turnbull and D. Zhou, *Angew. Chem., Int. Ed.*, 2016, **55**, 4738–4742.
4 Y. Guo, I. Nehlmeier, E. Poole, C. Sakonsinsiri, N. Hondow, A. Brown, Q. Li, S. Li, J. Whitworth, Z. Li, A. Yu, R. Brydson, W. B. Turnbull, S. Pöhlmann and D. Zhou, *J. Am. Chem. Soc.*, 2017, **139**, 11833–11844.

Jeff Gildersleeve replied: These are excellent points. The neoglycoprotein format has a number of limitations. We have very minimal molecular level information about our surfaces, but our expectation is that there is heterogeneity to the spacing and orientation of the glycans. In addition, we do not know if we have a monolayer on the microarray surface. We use a print concentration that saturates the surface in its ability to capture neoglycoproteins. This provides consistency from one array to another and from spot to spot, but we have not been able to characterize the surface in great detail. From some prior work, we think the glass surface is pretty rough at the molecular level and that there are a variety of crevices of varying size.[1] We also believe there is a consistent coating of neo-glycoprotein over the surface area of the spot.[2]

We have considered many different multivalent formats for constructing arrays. So far, the neoglycoprotein format has been the most convenient. We would be very interested in methods to control spacing and orientation of glycans in a more precise way. The use of your nanoparticles could offer some unique properties and advantages for constructing glycan microarrays. We can talk more later *via* email about a potential collaboration, but it could be pretty interesting.

1 Y. Wang, J. C. Gildersleeve, A. Basu and M. B. Zimmt, Photo- and bio-physical studies of lectin-conjugated fluorescent nanoparticles: Reduced sensitivity in high density assays, *J. Phys. Chem. C.*, 2010, **114**, 14487–14494.
2 Y. Zhang, Q. Li, L. Rodriguez and J. C. Gildersleeve, An array-based method to identify multivalent inhibitors, *J. Am. Chem. Soc.*, 2010, **132**, 9653–9662.

Stephan Schmidt addressed Clare Mahon: Some work suggest that above the LCST carbohydrate functionalized PNIPAM shows higher affinity due to presentation of the carbohydrates on the outside of the hydrophobic globule. This is different for the presented GM1 functionalized PNIPAM, where the higher affinity was observed below LCST. Can you explain the different behaviors?

Clare Mahon replied: There are some studies in the literature where poly(-NIPAm)s have been used as steric 'shields' to prevent or frustrate complexation of a receptor below the LCST of the poly(NIPAm). At temperatures above the LCST these chains collapse, exposing the carbohydrate recognition motifs and consequently increasing the affinity of the receptor to the target lectin.

In our case, we have made linear statistical copolymers of NIPAm and a GM1os-modified monomer unit. When these polymers undergo coil-to-globule collapse we expect they will rearrange to some extent to display their hydrophilic GM1os residues on the outside of these globules. The surface of these globules must display net hydrophobic character, however, as we observe the formation of large aggregates in solution by DLS. The size of these aggregates dictates that they must incorporate many collapsed chains, so overall we have whole polymeric receptors buried within the aggregates. The GM1os residues on the surface of the aggregate can interact with CTB, but those in the centre cannot, so overall we see a decrease in the avidity of the interaction.

Laura Hartmann addressed Jeff Gildersleeve & Alshakim Nelson: On the one hand the discussion has shown the importance of model systems to address fundamental questions on carbohydrate–lectin interactions such as 2D arrays. On the other hand, it has been discussed that we should think more about the application and what requirements will have to be met outside of an array setting going towards biological systems. In order to potentially bridge these two areas of research, do you think it would be possible to print onto cells and create an array on cells?

Jeff Gildersleeve replied: This is an intriguing idea. We have never tried to print onto cells. Peng Wu is developing a cell-based glycan array platform, although the approach is to construct glycans on the surface of cells *via* enzymatic modification rather than printing. Glycolipids and neoglycolipids can insert into cell membranes under the right conditions, so it might be interesting to print them onto cells.

Alshakim Nelson responded: The advantage of a 2D array is that we can control patterns at the nanoscale with sub-10 nm resolution. We still do not have the tools to produce 3D arrays with the same resolution with spatial control. There is some inspiring work by Kristi Anseth and Cole DeForest (among others), who are developing methodologies that allow one to produce 3D patterns within hydrogels with spatiotemporal control. These developments may be the key to developing model systems that allow us to investigate multivalency at the cell surface.

Ryan Chiechi concluded by remarking: In defense of molecular nano-fabrication: Commercial photolithographic technology is not only constrained in the types of structures it can fabricate, but also the materials. Furthermore, the reported feature-size of photolithographic processes is the smallest measurable feature. A 7-nm process does not imply that one can write arbitrary nano-structures in arbitrary materials with 7-nm resolution. Molecular self-assembly can provide Angstrom-level resolution and capture molecule-scale interactions between nano- and micro-scale objects, which is particularly relevant *in vivo*.

Conflicts of interest

Peter Seeberger holds significant shares in GlycoUniverse, the company that produces the automated glycan synthesizer.

Faraday Discussions

PAPER

Engineering of spectator glycocalyx structures to evaluate molecular interactions at crowded cellular boundaries†

Daniel J. Honigfort, [ID] Michelle H. Zhang, Stephen Verespy‡, III and Kamil Godula [ID] *

Received 28th February 2019, Accepted 10th April 2019

DOI: 10.1039/c9fd00024k

In the mucosal epithelium, the cellular glycocalyx can project tens to hundreds of nanometers into the extracellular space, erecting a physical barrier that provides protective functions, mediates the exchange of nutrients and regulates cellular interactions. Little is understood about how the physical properties of the mucosal glycocalyx influence molecular recognition at the cellular boundary. Here, we report the synthesis of PEG-based glycopolymers with tunable glycan composition, which approximate the extended architecture of mucin glycoproteins, and tether them to the plasma membranes of red blood cells (RBC) to construct an artificial mucin brush-like glycocalyx. We evaluated the association of two lectins, ConA and SNA, with their endogenous glycan ligands on the surface of the remodelled cells. The extended glycocalyx provided protection against agglutination of RBCs by both lectins; however, the rate and magnitude of ConA binding were attenuated to a greater degree in the presence of the glycopolymer spectators compared to those measured for SNA. The different sensitivity of ConA and SNA to glycocalyx crowding likely arises from the distinct presentation of their mannoside and sialoside receptors, respectively, within the native RBC glycocalyx.

Introduction

The cellular glycocalyx is a dynamic carbohydrate-rich macromolecular system populating the boundary between a cell and its surroundings.[1] It is composed of glycolipids and glycoproteins, which can extend tens to hundreds of nanometers above the plasma membrane, providing cells with a physical protective barrier

Department of Chemistry and Biochemistry, University of California San Diego, 9500 Gilman Drive, La Jolla, CA 92093-0358, USA. E-mail: kgodula@ucsd.edu

† Electronic supplementary information (ESI) available: ¹H NMR and IR spectra for polymers **1**, **3**, **4**, and **5**. Extended data for Table 1 and Fig. 2 and 4. See DOI: 10.1039/c9fd00024k

‡ Current address: Encodia, Inc., 11125 Flintkote Ave., Suite B, San Diego, CA 92121, USA.

while also facilitating their interactions with their environment (Fig. 1). The glycan structures distributed throughout the glycocalyx are recognized by a wide array of glycan binding proteins (GBPs), including adhesion molecules found in the extracellular matrix and on neighboring cells, lectins and antibodies of the immune system, or signaling proteins such as growth factors and cytokines.[2] Clustering of multiple copies of glycans on glycoprotein scaffolds or in glycolipid patches provides a mechanism to overcome the characteristically weak binding of individual glycans to their protein partners and produces molecular patterns within the glycocalyx that elicit high-avidity recognition by GBPs.[3]

Macromolecular glycoconjugates isolated from natural sources[4] or generated synthetically[5] have provided important tools for investigating and formulating the paradigms that define our current understanding of how avidity in glycan–protein interactions is achieved.[6] Employed as soluble ligands or integrated into glycan arrays, surface plasmon resonance or biolayer interferometry platforms, synthetic materials (*e.g.*, glycoclusters, spherical glycodendrimers, or linear glycopolymers) have provided key insights into how ligand parameters, such as glycan valency and spacing or scaffold architecture, influence GBP binding and higher-order association (*e.g.*, oligomerization or crosslinking).[7,8] More recently, cell surface engineering approaches have begun to emerge that allow for the integration of these materials directly into the glycocalyx of cells,[9] to further evaluate how the compositional heterogeneity, dynamics and nanoscale organization of the glycocalyx influence biological interactions at the cell surface.

While most work in this area has focused primarily on introducing constellations of specific glycan structures within the glycocalyx and evaluating their recognition by cognate receptors, the effects of the bulk properties of the glycocalyx on cellular interactions and functions are also becoming increasingly pursued. For instance, the remodelling of plasma membranes of mammary epithelial cells with lipidated glycopolymer mimetics of mucin glycoproteins was found to exert biomechanical forces that promoted the assembly of integrin adhesion complexes and enhanced signaling associated with cell proliferation,

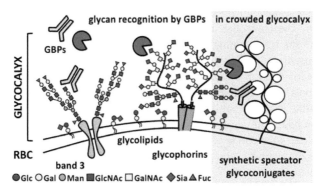

Fig. 1   Recognition of glycans by protein receptors occurs in the dynamic and compositionally heterogeneous environment of the cellular glycocalyx. These interactions can be influenced by the presence and physical properties of non-binding spectator glycoconjugates. Synthetic mimetics of membrane-associated glycoproteins, such as mucins, can be generated and installed into the native glycocalyx to evaluate how changes in glycocalyx crowding and physical properties impact ligand–receptor interactions.

thus identifying a glycocalyx-dependent mechanism for enhancing the survival and metastatic potential of circulating tumor cells.[10] Further studies using more minimalistic models of bulky glycocalyx components (*i.e.*, linear polyethylene glycol and hyperbranched polyglycerol polymers) covalently grafted to membrane proteins of red blood cells (RBCs) afforded shielding of ABO antigens and protection from complement-mediated cell lysis according to polymer size[11,12] and localization within the glycocalyx.[13]

Despite these advances, systematic surveys of the influence of the "spectator" components of the cellular glycocalyx, with respect to the composition, physical properties and organization of glycans, on molecular recognition events at the cell surface are still lacking. Here, we report the synthesis of well-defined linear poly(ethylene glycol) (PEG) backbones that approximate the architecture and properties of mucin glycoproteins and their use for the scaffolding of spectator glycocalyx structures with varied monosaccharide composition at the surfaces of RBCs. The resulting mucinous glycocalyx models exerted differential effects on lectin association with imbedded endogenous glycan receptors according to the structure of the spectator glycoconjugate and lectin identity.

Results and discussion

Synthesis of mucin–mimetic glycocalyx building blocks

We initiated our study by establishing a synthetic strategy to generate glycopolymer structures that exhibit key architectural features of mucin glycoproteins while allowing for the facile introduction of a range of glycan structures. Mucins are large, heavily glycosylated proteins populating the surfaces of epithelial cells, and are composed of characteristic variable number of tandem repeat regions with a high frequency of *O*-glycosylation sites.[14] The close packing of glycans along the polypeptide chain forces the protein into an extended conformation with a high persistence length.[15] This property of mucins can be recapitulated in synthetic linear glycopolymers by controlling the positioning of glycan appendages in sufficient density and close proximity to the polymer backbone.

The Bertozzi group has developed two mucin-mimetic polymer architectures based on glycosylated poly(methylvinyl ketone) (pMVK) and poly(serine) scaffolds generated by the controlled reversible addition–fragmentation chain transfer (RAFT)[16] and the ring-opening *N*-carboxyanhydride (NCA)[17] polymerization techniques, respectively. Both approaches afforded functional mucin-like polymeric structures with distinct advantages and drawbacks. While the pMVK scaffold allows for post-polymerization modification with a wide array of glycan structures, its utility is somewhat limited due to the hydrophobicity of the core polyhydrocarbon backbone, which at low glycan content may give rise to amphiphilic behaviour. The polypeptide-based approach obviates this problem; however, it necessitates the synthesis and purification of glycosylated NCA monomers, which can be challenging.

We decided to address these challenges by building mucin mimetics on hydrophilic PEG scaffolds suitably modified to allow for later stage introduction of glycans, as well as additional functional elements such as surface anchors and optical probes for characterization. We decided to target a set of glycopolymers displaying the monosaccharides glucose (Glc), galactose (Gal), fucose (Fuc), and glucuronic acid (GlcA) (Fig. 2A). The sugars were selected based on their common

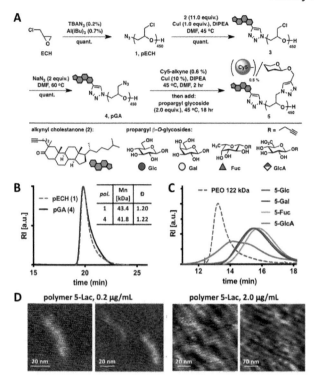

Fig. 2 Synthesis and characterization of mucin mimetics **5**. (A) Glycopolymers **5** were generated from poly(epichlorohydrin) **1** *via* chloride-to-azide sidechain substitution followed by the CuAAC conjugation of propargyl β-*O*-glycosides (Glc, Gal, Fuc, and GlcA). The polymers were furnished with a hydrophobic cholestanone moiety for anchoring into cell membranes and a Cy5 fluorescent tag for imaging and quantification. (B) SEC analysis indicated narrow molecular weight and chain-length distributions before and after the chloride-to-azide exchange. (C) Glycopolymers **5** exhibited increased retention in aqueous SEC compared to a PEO standard of similar molecular weight (122 kDa), a characteristic behavior of glycopolymers with extended molecular conformations. (D) AFM imaging of lactose-modified glycopolymers confirmed elongated, mucin-like morphology of the PEG-based glycopolymers.

presence in mammalian glycans and their distinct chemical and physical properties. For instance, Glc and Gal share the same molecular composition, but differ in the orientation of the C4 hydroxyl group, which may impact their packing along the polymer chain. In addition, the increased hydrophobicity of Fuc compared to Glc and Gal (log P values of -2.02, -3.24 and -3.38, respectively) and the negative charge of GlcA at physiological pH ($pK_a = 3.2$), may differentially affect the persistence length and physical properties of the resulting glycopolymers.

Using a monomer-activated anionic ring opening polymerization developed by Carlotti and colleagues,[18] we have synthesized an azide-terminated poly(epichlorohydrin) (pECH) polymer (**1**, Fig. 2A) with well-defined size ($M_n = 43\,400$ Da, DP ~ 450) and narrow chain-length distribution ($Đ = 1.20$). Although the use of a racemic epichlorohydrin monomer produces an atactic pECH polymer **1**, we reasoned that the stereochemical relationships of the pendant sidechains would not significantly impact the overall architecture and physical properties of the desired mucin mimetic glycopolymers. The polymeric precursor 1 was elaborated into fluorescent cell surface-targeting mucin mimetics in a three-step synthetic sequence (Fig. 2A). First, we introduced a hydrophobic anchor for plasma membrane targeting *via* the copper-catalysed alkyne–azide cycloaddition (CuAAC)[19] between the chain-end azide group in 1 and an alkynyl cholestanone derivative 2 (ref. 20) (11.0 equiv. per chain-end azide). Treatment of the chlor-omethyl sidechains in the cholestanone modified pECH polymer 3 with sodium azide[21] (2.0 equiv. per Cl) generated a reactive polymer intermediate 4 primed for a sequential CuAAC conjugation of alkynyl fluorophores (Cy5, ~0.5% sidechains) and propargyl glycosides to complete the desired glycopolymers 5 (Fig. 2A). ^{1}H NMR spectroscopy and size exclusion chromatography (SEC) analyses of poly-meric intermediate 4 confirmed quantitative azide-chloride replacement without any observable increase in polymer dispersity (Fig. 2B and S8†). The levels of fluorophore labelling in glycopolymers 5 were assessed based on UV-VIS absorption profiles and matched values predicted based on reaction stoichiom-etry. The efficiency of glycan conjugation to the polymer was difficult to establish accurately based on ^{1}H NMR spectroscopy alone, due to the overlap between glycan and polymer proton signals; however, IR spectroscopy analysis of glyco-polymers 5 revealed the disappearance of the characteristic azide stretching absorption frequency at $\nu = 2100$ cm^{-1} (Fig. S17†) after conjugation, indicating quantitative side-chain modification.

We anticipated that the close positioning of the pendant glycan residues with respect to the polymer backbone would force the glycopolymers into an extended conformation. This behaviour is supported by aqueous SEC analysis of polymers 5 exhibiting significantly increased retention on the stationary phase compared to a PEO standard of similar molecular weight (Fig. 2C, $M_{w,PEO} = 122$ kDa). Atomic force microscopy (AFM) of glycopolymers carrying larger lactose disaccharide residues to facilitate visualization provided direct evidence for extended polymer structures (Fig. 2D and S18†). We were able to clearly observe individual glyco-polymer molecules with length distributions measured at 41 ± 6 nm (Fig. 2D) confirming that the new glycopolymer architecture can mimic the nanoscale topology of native mucins.

Generation of spectator glycocalyx structures on RBCs

Molecular recognition events at the cell surface occur in the context of the gly-cocalyx, which often defines the properties of the surrounding microenviron-ment, such as polarity, pH, permittivity, and crowding, that may influence the energetics of binding as well as molecular diffusion and transport. However, these properties of the glycocalyx environment are rarely considered when evaluating molecular interactions at the cell surface, despite observations that, for instance, the concentrations of the negatively charged sialic acids in the glycocalyx of B cells can reach ~100 mM (ref. 22) or that transmembrane epithelial mucins produce dense glycoprotein brushes spanning lengths up to ~1500 nm.[23]

Some of these conditions can be modelled in a controlled fashion at the surfaces of cells by augmenting their endogenous glycocalyx with synthetic gly-comaterials. For instance, glycopolymers affixed to the outer leaflet of the plasma membrane *via* a lipid anchor exhibit dynamic behaviour, such as lateral membrane diffusion and spatial reorganization in response to cell adhesion forces,[9,10] and display molecular recognition motifs to promote cell–cell interaction and intra-cellular signalling.[24] We set out to construct a synthetic glycocalyx on the surfaces of RBCs to model the barrier functions of the mucinous epithelial glycocalyx using glycopolymers 5 (Fig. 3A). RBCs offer a practical choice for this purpose due to their relatively modest glycocalyx, which extends less than 10 nm above the plasma membrane, the lack of active endocytosis preventing the uptake and clearance of the exogenously added materials, and their amenability to a range of protein-binding assays.

The treatment of RBCs with Cy5-labelled glycopolymers 5 (c_{pol} = 115 nM to 10.0 μM), resulted in a dose-dependent membrane incorporation of the polymers based on flow cytometry analysis (Fig. 3A). Incubation of RBCs with 5-Glc lacking the cholestanone anchor gave no appreciable cell-surface incorporation of the glycopolymer (Fig. S20†), indicating that the introduction of the anchor into the glycopolymers *via* precursor 3 is successful and necessary for their plasma membrane insertion. As anticipated, the polyanionic GlcA-modified polymer exhibited lower membrane grafting efficiency (∼65%) at the highest polymer concentration (10.0 μM) compared to 5-Glc and 5-Gal polymers. Interestingly, we also observed somewhat decreased (∼12%) grafting for the polymer bearing

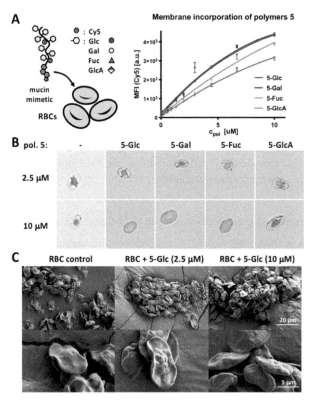

Fig. 3   RBC glycocalyx remodeling with glycopolymers 5. (A) Polymers 5 incorporate dose-dependently into RBC membranes according to their pendant glycans. Increasing glycopolymer density in RBC membranes induces cell rounding, swelling, and lysis at polymer concentrations above 2.5 μM *via* brightfield microscopy (B) or SEM (C).

fucose (**5-Fuc**), which is a more hydrophobic C6-deoxy analogue of L-galactose (Fig. 3B). Visual inspection of the remodelled RBCs in a 96 well plate indicated a decreased ability of cells treated with the non-ionic Glc, Gal and Fuc glycopolymers to settle to the bottom of the wells starting at polymer concentrations of ~1.8, 3.2, and 5.6 μM, respectively (Fig. S22†). We saw no changes in sedimentation properties for RBCs treated with the polyanionic **5-GlcA** polymer over the entire range of concentrations (115 nM to 10.0 μM). These observations are consistent with prior findings showing that increasing the size and surface density of non-ionic polymers, such as methoxypolyethylene glycol (mPEG), covalently grafted to proteins on RBCs resulted in the shielding of cell surface charge and prevented cell sedimentation and plasma protein-induced cell stacking (Rouleaux formation).[11]

Increasing the surface density of the mucin-mimetic glycopolymers on RBCs may introduce membrane deformations due to entropic pressures resulting from mushroom to brush transitions in the compressing polymer brush.[25] Indeed, we observed visible rounding of RBCs at polymer concentrations above 2.5 μM with signs of membrane disintegration in cells treated with 10.0 μM polymers using optical microscopy (Fig. 3B and S21†). To exclude contributions from altered membrane morphologies to protein binding in the engineered glycocalyx structures, we imaged the remodelled cells using scanning electron microscopy (SEM, Fig. 3C). We observed no significant effects on cell shape or membrane structures for polymer treatments at concentrations up to 2.5 μM and confirmed the breakdown of RBCs at 10.0 μM polymer concentration (Fig. 3C).

Our analyses indicate that RBCs can be effectively remodelled with mucin mimetics **5** to introduce synthetic glycocalyx structures to the cell surface. The efficiency of polymer incorporation and the resulting physical properties of the remodelled RBCs were strongly influenced by the structure of the monosaccharide component in **5**, and a too high polymer density at the plasma membrane ($c_{pol} > 2.5$ μM) led to observable rounding of cells and an eventual cell lysis. Based on these findings, we identified an optimal polymer concentration of 2.5 μM for RBC remodelling, which provides maximal polymer incorporation without adverse effects on membrane morphology.

Effects of spectator glycocalyx structures on lectin interactions with RBCs

The influence of the compositional heterogeneity and organization of the glycocalyx microenvironment on ligand–receptor interactions at the cell surface has recently been drawing attention, as glycans are increasingly considered for therapeutic targeting.[26] For instance, glycan array screens have identified avidity enhancements for the binding of anti-glycan antibodies in the presence of neighboring non-binding glycan epitopes.[27,28] Similarly, ABH blood group antigens presented on the extracellular domains of band 3 proteins on RBCs have been found to modulate the binding of proteins, such as the *Sambucus nigra* agglutinin (SNA) or the human Siglec-2, to sialic acid receptors primarily carried by glycophorins (Fig. 1).[29] The localization of the A and B antigens either to the periphery or the center of sialylated glycoprotein clusters in the RBC membranes, respectively, resulted in a differential stabilization of sialoglycan patches and changes in their binding avidity toward SNA and Siglec-2. The phenomenon of glycan clustering is not limited to RBCs and has been found to contribute to a range of biological processes, including host–pathogen interactions, tumor antigen recognition and T-cell activation *via* a mucinous glycosynapse.[30] The mucin-rich glycocalyx covering epithelial cells is believed to provide a semi-permeable physical barrier limiting the diffusion of proteins and particulates to the cell surface and protecting membrane protein structures from chemical and enzymatic degradation.[14] Mucins contribute directly to the organization of gly-cocalyx structures and drive the clustering of cell adhesion molecules by exerting forces against the surrounding extracellular environment.[10] We set out to evaluate how the presence of a mucin-like glycocalyx brush at the cell surface would influence the interactions of lectins with endogenous glycan receptors. We selected two lectins, Concanavalin A (ConA) and SNA, for their respective speci-ficity in the targeting of two distinct glycan classes prominently represented in the RBC glycocalyx, the oligomannose core of *N*-linked glycans found primarily on band 3, and to a smaller extent, on glycophorins, and the α2,6-linked sialic acids displayed prominently at terminal positions of *O*-linked glycans on glycophorins and glycosphingolipids.

First, we investigated the ability of glycopolymers **5** to protect against RBC agglutination with both lectins. Using agglutination assays with non-remodeled RBCs, we identified the minimal lectin concentrations (c_{agg}) required to induce cell crosslinking and aggregation (Fig. S23 and S24†). SNA showed ~17-fold greater agglutination capacity $(c_{agg,SNA} = 33.0$ ng mL$^{-1})$ compared to ConA $(c_{agg,ConA} = 0.57$ µg mL$^{-1})$. This may, presumably, be due to the greater abundance and accessibility of sialic acid modifications within the glycocalyx compared to the *N*-linked core mannose structures targeted by ConA.

We next examined the ability of polymers **5** to shield RBCs from agglutination by SNA and ConA by determining the changes in the minimal agglutination concentration, (c_{agg}), required for each lectin to agglutinate RBCs remodeled with polymers **5** $(c_{pol} = 2.5$ µM (Table 1, Fig. S23 and S24†)). All glycopolymers **5** provided protection against agglutination by both lectins in the following order: Glc < Gal ~ Fuc < GlcA (Table 1). It should be noted that the polyanionic glyco-polymer, **5-GlcA**, afforded the most shielding effect (3- to 4-fold increase in c_{agg}), despite its lower density in the glycocalyx (~65% surface grafting efficiency of the non-ionic polymers, **5-Glc** and **5-Gal**, Fig. S19†). Each polymer within the set showed a similar level of shielding against agglutination by both lectins, as measured by the relative change in c_{agg} with respect to non-remodeled cells. This suggests that the bulk properties of the spectator glycocalyx rather than lectin

Table 1 Agglutination of RBCs remodeled with glycopolymers **5** (2.5 µM) by ConA (0–1.8 µg mL^{-1}) and SNA (0–105 ng mL^{-1}) lectins

	ConA						SNA								
c_{ConA} [µg mL^{-1}]	0–0.43	0.57	0.76	1.00	1.30	1.80	c_{SNA} [ng mL^{-1}] 0–25	33	45	60	80	105	140	187	250
Glc							Glc								
Gal							Gal								
Fuc							Fuc								
GlcA							GlcA								

identity provides the dominant contribution toward protection against agglutination in these experiments. Thus, all polymers established a physical barrier against crosslinking with additional repulsive forces arising from increased negative charge density at the surface of cells presenting the glucuronic acid polymer, **5-GlcA**.

These observations led us to investigate whether the spectator glycoconjugates might exert an influence on lectin association within the glycocalyx at sub-agglutination concentrations. Using a flow-cytometry assay, we measured the binding of ConA and SNA to the remodeled RBCs (Fig. 4). At ConA and SNA concentrations of 8 µg mL^{-1} and 1 µg mL^{-1}, respectively, saturation binding was attained for both lectins in ~600 s without any apparent cell aggregation. We observed 7–13% and 6–9% decreases in the total amount of bound ConA and SNA, respectively, at saturation in the presence of polymers **5** compared to untreated control cells (Fig. 4). Distinct from the results obtained from hemagglutination experiment, the structure of the glycan appendages in glycopolymers **5** had no significant impact on the magnitude of inhibition of lectin binding; however, all polymers appeared to inhibit ConA binding to the RBC surface to a somewhat greater extent compared to SNA.

Analysis of the linear regions of the binding curves revealed a ~6–10% decrease in initial rates for the association of ConA in the presence of the polymers, compared to the non-remodelled RBCs (Fig. 4 and S26†). In contrast, the initial rates for SNA association in the presence of all polymers **5** were indistinguishable from those observed in control cells (Fig. 4 and S26†), with differences beginning to appear at later times, closer to the point of saturation binding.

Collectively, these observations indicate the distinct influence of the spectator glycocalyx on the association of the two lectins (Fig. 5). The ConA lectin targets the

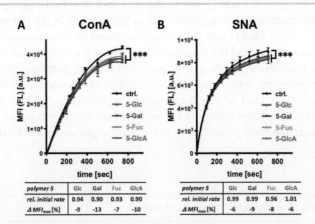

Fig. 4 Association of ConA and SNA lectins with glycocalyx-remodeled RBCs. (A) Binding of fluorescein (FL)-labeled lectins to remodeled cells were assessed via flow cytometry. The presence of spectator glycopolymers **5** at the surface of RBCs attenuates both the initial rate as well as saturation binding of ConA to cell surface glycans. (B) Glycopolymers **5** have no effect on the initial rates of SNA binding but attenuate lectin association near saturation. Relative initial rates were calculated from the linear regions of lectin binding curves and normalized to control cells without polymer treatment. ΔMFI$_{max}$ corresponds to the change in median fluorescence intensity of cells at saturation lectin binding (ANOVA, Tukey's multiple comparisons test; ***$p < 0.001$).

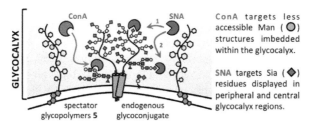

Fig. 5 Glycocalyx crowding with spectator glycoconjugates differentially affects the association of ConA and SNA with endogenous glycans depending on their distribution throughout the glycocalyx. The glycopolymer spectators inhibit both the initial rates and saturation binding of ConA to the less accessible mannose structures. Glycocalyx crowding influences the association of SNA with sialic acid residues only at later time points near saturation binding, once the more available peripheral glycans have become occupied.

less accessible mannoside residues of *N*-linked glycans placed in close proximity to the polypeptide chains of glycoproteins and generally positioned deeper within the glycocalyx. As such, the lectin diffusion into the glycocalyx and binding to its glycans is likely to be more sensitive to the presence of crowding spectator glycoconjugates. This is reflected in the initial rates of ConA binding to RBCs in the presence of glycopolymers **5**, and the greater decrease in the total amount of ConA bound at saturation. In contrast, the SNA lectin targets the outermost sialic acid residues in both *N*- and *O*-linked glycans, which are prominently displayed at the periphery of the glycocalyx and, thus, more accessible. As such, the spectator glycoconjugates would not be expected to interfere with SNA binding until all available peripheral glycans are occupied and the diffusion of the lectin deeper into the glycocalyx is required for additional binding. The observed rapid onset of SNA binding with initial rates indistinguishable between polymer-remodeled RBCs and untreated controls followed by spectator-induced inhibition of binding at a later timepoint supports this model. As the binding of the lectins reaches saturation, the effects of the spectator glycocalyx would be expected to become exacerbated due to the increasing crowding from the newly introduced proteins.

Conclusions

In this paper, we have described a new synthetic route to generate glycopolymer mimetics of mucin glycoproteins with well-defined architectures and tuneable glycosylation. Using a non-covalent cell membrane engineering approach, we have delivered these materials to the surfaces of red blood cells to augment the physical properties of their glycocalyx and evaluated the effect of increased molecular crowding on lectin recognition of endogenous glycan receptors. While the polymers attenuated the binding of both lectins near saturation, only the initial rates of ConA association were affected due to increased glycocalyx crowding in the presence of the synthetic glycoconjugate spectators. The unequal sensitivity of ConA and SNA can be rationalized based on the differential accessibility of their respective glycan targets within the glycocalyx.

Methods

Materials

All chemicals, unless stated otherwise, were purchased from Sigma Aldrich and used as received. Reaction progress was checked by analytical thin-layer chromatography (TLC, Merck silica gel 60 F-254 plates) monitored either with UV illumination, or by staining with iodine, ninhydrin, or CAM stain. Solvent compositions are reported on a volume/volume (v/v) basis unless otherwise noted. Alkynyl cholestanone **2**,[20] and Glc,[31] Gal,[31] Fuc,[32] GlcA,[33] and Lac[34] propargyl glycosides were prepared according to published procedures. Turkey red blood cells as a 10% solution were obtained from Lampire Biological Laboratories (cat #724908). *Sambucus nigra* agglutinin (SNA) and Concanavalin A (ConA) lectin were purchased from Vector Labs. Fluorescein (FL)-NHS for lectin labeling was purchased from Thermo Scientific and Cyanine 5 (Cy5)-alkyne for labeling of polymers was obtained from Sigma Aldrich (fluorophore structures are shown in ESI on page S4†). Lectins were labeled according to manufacturer protocols and the extent of labeling and lectin concentration was determined by UV-VIS spectroscopy and BCA assay, respectively. A detailed list of analytical instruments and general procedures used for the purification and structural characterization of synthetic materials and for polymer and cell imaging (optical microscopy, AFM, and SEM) can be found in the ESI.†

Synthesis of azide-terminated poly(epichlorohydrin), pECH (1)

Epichlorohydrin was polymerized according to the procedure developed by Carlotti.[18] Briefly, a flame-dried Schlenk flask (10 mL) equipped with a magnetic stirrer and fitted with a PTFE stopcock was charged with tetrabutylammonium azide (TBAN$_3$, 20 mg, 0.037 mmol, 0.002 equiv.) under argon. A solution of freshly distilled epichlorohydrin (1.29 mL, 16.5 mmol) was prepared in anhydrous toluene (4 mL). A solution of triisobutylaluminum in toluene (1.07 M, 104 µL, 0.111 mmol, 0.007 equiv.) was added *via* a syringe under argon at −30 °C. The reaction was stirred for 4 hours and then stopped by the addition of ethanol. The resulting pECH polymer **1** was precipitated into hexanes and dried under vacuum to yield a clear viscous oil (1500 mg, 99% yield). The polymer was analyzed by SEC (0.2% LiBr in DMF): $M_w = 52\,000$, $M_n = 43\,400$, $Đ = 1.20$.

Synthesis of cholestanone-terminated poly(epichlorohydrin) (3)

In a flame-dried Schlenk flask (10 mL), pECH polymer **1** (15 mg, 0.3 µmol) was dissolved in degassed anhydrous DMSO (200 µL). Alkynyl cholestanone **2** (1.7 mg, 3.8 µmol, 11.0 equiv.) was added, followed by CuI (~0.05 mg, 0.3 µmol, 1.0 equiv.) and one drop diisopropylethyl amine (DIPEA, ~5 µL). The reaction was stirred at 40 °C for 12 h, at which time it was quenched by the addition of water to precipitate the polymer. The resultant polymer was triturated with hexanes to remove unreacted **2** and dried on vacuum to yield a clear viscous oil in (16 mg, 100% yield). The polymer was analyzed by SEC (0.2% LiBr in DMF): $M_w = 52\,000$, $M_n = 43\,000$, $Đ = 1.20$.

Synthesis of cholestanone-terminated poly(glycidyl azide), pGA (4)

The chloride to azide exchange in pECH polymer **3** was accomplished according to a previously published procedure.[21] Briefly, in a flame-dried Schlenk flask (10 mL), polymer **3** (15 mg, 0.16 mmol) was dissolved in dry DMF (300 μL). To the solution was added NaN_3 (21 mg, 0.32 mmol, 2.0 equiv.), and the reaction was stirred at 60 °C for 3 days under argon to allow complete conversion. The polymer solution was filtered and precipitated in ethanol to yield a clear viscous oil (16 mg, 100% yield). The polymer was analyzed by SEC (0.2% LiBr in DMF): $M_w = 51\ 000$, $M_n = 41\ 800$, $Đ = 1.22$.

Synthesis of Cy5-labeled glycopolymers 5

In a flame-dried Schlenk flask (10 mL), polymer **4** (9.00 mg, 0.09 mmol) was dissolved in degassed dry DMSO (250 μL). To the solution was added Cy5-alkyne (0.50 mg, 0.50 μmol) in DMSO (50 μL), followed by CuI (2.00 mg, 9.00 μmol) and DIPEA (16 μL, 0.09 mmol). The reaction was stirred in the dark under Ar at 40 °C for 2 h. After this time, the reaction mixture was aliquoted (50 μL) into separate vials containing β-propargyl glucoside, galactoside, fucoside, and glucuronoside (0.03 mmol, 2.00 equiv. per azide side-chain) in degassed anhydrous DMSO (50 μL). The reactions were stirred in the dark at 40 °C overnight. After this time, the reactions were diluted with DI water and treated with Cuprisorb beads (SeaChem labs) for 18 h to sequester copper. The resulting copper-free solutions were filtered through Celite to remove the resin and lyophilized. The dry residues were triturated 3× with methanol with monitoring by TLC to remove excess unreacted glycosides. The resulting Cy5-labeled glycopolymers **5** were dissolved in D_2O and lyophilized to give a blue solid in a quantitative yield for each polymer (note: the blue color of the glycopolymers arises from the presence of the Cy5 label and not residual copper contamination. Glycopolymers lacking the Cy5 label were isolated as white solids). The polymers **5** were characterized using ^{1}H NMR ($CDCl_3$, 300 MHz) and UV-Vis ($\lambda_{max} = 633$ nm) spectroscopy and analyzed by aqueous SEC. Absorbance readings at known concentrations of glycopolymers **5** indicated the presence of ~2 Cy5 molecules per polymer chain (0.5% sidechain occupancy). The resulting glycopolymers were analyzed by aqueous SEC (0.2 M $NaNO_3$ in 0.01 M Na_2HPO_4, pH = 7.0). **5-Glc:** $M_{n,calc.} = 143$ kDa, $M_{n,SEC} = 20\ 578$, $Đ = 1.369$; **5-Gal:** $M_{n,calc.} = 143$ kDa, $M_{n,SEC} = 21\ 273$, $Đ = 1.374$; **5-Fuc:** $M_{n,calc.} = 136$ kDa, $M_{n,SEC} = 18\ 562$, $Đ = 1.306$; **5-GlcA:** $M_{n,calc.} = 149$ kDa, $M_{n,SEC} = 31\ 725$, $Đ = 2.099$.

Remodeling of RBC glycocalyx with glycopolymers 5

RBCs (4% w/v in PBS) were incubated with Cy5-labeled glycopolymers **5** at increasing concentrations ($c_{pol} = 0.1$–10.0 μM) for 1 h at 37 °C. The cells were washed 1× with PBS, then were probed for the presence of Cy5 fluorescence using flow cytometry. The data were analyzed on Cytobank online software. Cells were gated to exclude debris, and the median fluorescence intensities (MFI) of the cells are reported. Means and standard deviations for each condition were calculated from three independent biological replicates.

Determination of cell morphology by electron microscopy

RBCs treated with glycopolymers **5** at concentrations of 2.5 μM and 10.0 μM, as well as untreated cells, were prepared fixed in glutaraldehyde solution in PBS (2.5%) at 4 °C overnight. The cells were gradually transferred into EtOH by washes with DI water containing gradually increasing concentrations of EtOH (0–100%, 10% increments). The samples were dried using a Tousimis AutoSamdri 815A critical point dryer and sputter coated with iridium for 8 seconds using Emitech K575X Iridium Sputter coater. SEM imaging was done with ETD detector at HV 4.00 kV with 0.1 nA current.

Agglutination of glycocalyx-remodeled RBCs in the presence of ConA and SNA lectins

To round-bottom, 96-well plates containing RBCs (25 µL, 1% in PBS) treated with glycopolymers 5 (2.5 µM) or alkynyl cholestanone 2 (2.5 µM) or untreated cells, were added fluorescein-labeled ConA and SNA lectins (25 µL) at increasing concentrations (c_{ConA} = 0–1.8 µg mL^{-1} and c_{SNA} = 0–0.25 µg mL^{-1}). The cells were mixed gently but thoroughly using a pipette tip and, then, allowed to agglutinate for 45 min. After this time, the plates were scanned on an EPSON Perfection V700 Photo scanner (Digital ICE Technologies), and the lowest lectin concentrations required to induce RBC agglutination were determined. The settling of RBCs to the bottom of a well to form a solid dot shape indicated a lack of agglutination. Each condition was evaluated in three independent biological replicates.

Determination of lectin association with glycocalyx-remodeled RBCs by flow cytometry

In 96-well, round-bottom plates, to RBCs (0.33% in PBS) treated with glycopolymers 5 (2.5 µM) or alkynyl cholestanone 2 (2.5 µM) or to untreated cells, were added fluorescein-labeled ConA and SNA lectins at sub-agglutination concentrations (c_{ConA} = 8 µg mL^{-1} and c_{SNA} = 1 µg mL^{-1}). The cells were vortexed vigorously for ~10 s and then analyzed by flow cytometry (Canto II) for the presence of a fluorescein signal at discrete time points until saturation lectin binding was observed. The data were analyzed on Cytobank software. Cells were gated to exclude debris, and median fluorescence intensities (MFI) of the cells are reported. Means and standard deviations were calculated from two independent biological experiments, and p-values corresponding to each condition *vs.* an untreated RBC control were calculated using 2-way ANNOVA tests with PRISM software. The linear regions of the lectin binding curves were determined (t = 0–200 s) and fitted using a linear regression in PRISM software. The slopes designating the initial rates of lectin association and the R^2 values for the linear fits were extracted for each condition and their significance with respect to untreated RBC controls was assessed based on p-values calculated using 1-way ANNOVA tests.

Conflicts of interest

There are no conflicts to declare.

Acknowledgements

We thank the UCSD Microscopy Core facility (*via* p30 grant NS047101 from NINDS) for assistance with fluorescence microscopy imaging, and the Glycobiology Research and Training Center for access to tissue culture facilities and analytical instrumentation. SEM and AFM work was performed at the San Diego Nanotechnology Infrastructure (SDNI) of UCSD, a member of the National

Nanotechnology Coordinated Infra-structure (NNCI), which is supported by the National Science Foundation (Grant ECCS-1542148). We also wish to thank Dr Meghan Altman for help for her expertise in and valuable comments on RBC assay development. This work was supported by the NIH Director's New Innovator Award (NICHD: 1DP2HD087954-01). K. G. is supported by the Alfred P. Sloan Foundation (FG-2017-9094) and the Research Corporation for Science Advancement *via* the Cottrell Scholar Award (Grant #24119).

References

1 S. Weinbaum, J. M. Tarbell and E. R. Damiano, The structure and function of the endothelial glycocalyx layer, *Annu. Rev. Biomed. Eng.*, 2007, **9**, 121–167.

2 A. Varki and P. Gagneux, Biological functions of glycans, in *Essentials of Glycobiology*, ed. A. Varki, R. D. Cummings, J. D. Esko, *et al.*, 3rd edn, Cold Spring Harbor Laboratory Press, Cold Spring Harbor, NY, ch. 7, 2015–2017.

3 R. D. Cummings, R. L. Schnaar, J. D. Esko, K. Drickamer and M. E. Taylor, Principles of glycan recognition, in *Essentials of Glycobiology*, ed. A. Varki, R. D. Cummings, J. D. Esko, *et al.*, Cold Spring Harbor Laboratory Press, Cold Spring Harbor, NY, 3rd edn, ch. 29, 2015–2017.

4 C. F. Brewer, M. C. Miceli and L. G. Baum, Clusters, bundles, arrays and lattices: novel mechanisms for lectin–saccharide-mediated cellular interactions, *Curr. Opin. Struct. Biol.*, 2002, **12**, 616–623.

5 M. L. Huang and K. Godula, Nanoscale materials for probing the biological functions of the glycocalyx, *Glycobiology*, 2016, **26**, 797–803.

6 S. Bhatia, L. C. Camacho and R. Haag, Pathogen inhibition by multivalent ligand architectures, *J. Am. Chem. Soc.*, 2016, **138**, 8654–8666; M. Mammen, S.-K. Choi and G. M. Whitesides, Polyvalent interactions in biological systems: Implications for design and use of multivalent ligands and inhibitors, *Angew. Chem., Int. Ed.*, 1998, **37**, 2754–2794; J. E. Gestwicki, C. W. Cairo, L. E. Strong, K. A. Oetjen and L. L. Kiessling, Influencing receptor–ligand binding mechanisms with multivalent ligand architecture, *J. Am. Chem. Soc.*, 2002, **124**, 14922–14933.

7 L. L. Kiessling, J. E. Gestwicki and L. E. Strong, Synthetic multivalent ligands as probes of signal transduction, *Angew. Chem., Int. Ed.*, 2006, **45**, 2348–2368.

8 S. C. Purcell and K. Godula, Synthetic glycoscapes: addressing the structural and functional complexity of the glycocalyx, *Interface Focus*, 2019, **9**, 20180080.

9 D. Rabuka, M. B. Forstner, J. T. Groves and C. R. Bertozzi, Noncovalent cell surface engineering: Incorporation of bioactive synthetic glycopolymers into cellular membranes, *J. Am. Chem. Soc.*, 2008, **130**, 5947–5953.

10 M. J. Paszek, *et al.*, The cancer glycocalyx mechanically primes integrin-mediated growth and survival, *Nature*, 2014, **511**, 319–325.

11 A. J. Bradley, K. L. Murad, K. L. Regan and M. D. Scott, Biophysical consequences of linker chemistry and polymer size on stealth erythrocytes: Size does matter, *Biochim. Biophys. Acta, Biomembr.*, 2002, **1561**, 147–158.

12 R. Chapanian, *et al.*, Influence of polymer architecture on antigens camouflage, CD47 protection and complement mediated lysis of surface grafted red blood cells, *Biomaterials*, 2012, **33**, 7871–7883.

13 E. M. J. Siren, R. Chapanian, I. Constantinescu, D. E. Brooks and J. N. Kizhakkedathu, Oncotically driven control over glycocalyx dimension

for cell surface engineering and protein binding in the longitudinal direction, *Sci. Rep.*, 2018, **8**, 7581.

14 C. L. Hattrup and S. J. Gendler, Structure and function of the cell surface (tethered) mucins, *Annu. Rev. Physiol.*, 2008, **70**, 431–457.

15 R. Bansil and B. S. Turner, Mucin structure, aggregation, physiological functions and biomedical applications, *Curr. Opin. Colloid Interface Sci.*, 2006, **11**, 164–170.

16 K. Godula, D. Rabuka, K. T. Nam and C. R. Bertozzi, Synthesis and microcontact printing of dual end-functionalized mucin-like glycopolymers for microarray applications, *Angew. Chem., Int. Ed.*, 2009, **48**, 4973–4976.

17 J. R. Kramer, B. Onoa, C. Bustamante and C. R. Bertozzi, Chemically tunable mucin chimeras assembled on living cells, *Proc. Natl. Acad. Sci. U. S. A.*, 2015, **112**, 12574–12579.

18 M. Gervais, A. Labbé, S. Carlotti and A. Deffieux, Direct synthesis of α-azido,ω-hydroxypolyethers by monomer-activated anionic polymerization, *Macromolecules*, 2009, **42**, 2395–2400.

19 C. W. Tornøe, C. Christensen and M. Meldal, Peptidotriazoles on solid phase [1,2,3]-triazoles by regiospecific copper(ɪ)-catalyzed 1,3-dipolar cycloadditions of terminal alkynes to azides, *J. Org. Chem.*, 2002, **67**, 3057–3064.

20 C. Alarcón-Manjarrez, R. Arcos-Ramos, M. F. Álamo and M. A. Iglesias-Arteaga, Synthesis, NMR and crystal characterization of dimeric terephthalates derived from epimeric 4,5-seco-cholest-3-yn-5-ols, *Steroids*, 2016, **109**, 66–72.

21 J. Meyer, H. Keul and M. Möller, Poly(glycidyl amine) and copolymers with glycidol and glycidyl amine repeating units: Synthesis and characterization, *Macromolecules*, 2011, **44**, 4082–4091.

22 B. E. Collins, *et al.*, Masking of CD22 by *cis* ligands does not prevent redistribution of CD22 to sites of cell contact, *Proc. Natl. Acad. Sci. U. S. A.*, 2004, **101**, 6104–6109.

23 M. Kesimer, *et al.*, Molecular organization of the mucins and glycocalyx underlying mucus transport over mucosal surfaces of the airways, *Mucosal Immunol.*, 2013, **6**, 379–392.

24 J. E. Hudak, S. M. Canham and C. R. Bertozzi, Glycocalyx engineering reveals a Siglec-based mechanism for NK cell immunoevasion, *Nat. Chem. Biol.*, 2014, **10**, 69–75.

25 J. G. Gandhi, D. L. Koch and M. J. Paszek, Equilibrium modeling of the mechanics and structure of the cancer glycocalyx, *Biophys. J.*, 2019, **116**, 694–708.

26 J. E. Hudak and C. R. Bertozzi, Glycotherapy: New advances inspire a reemergence of glycans in medicine, *Chem. Biol.*, 2014, **21**, 16–37.

27 C.-H. Liang, S.-K. Wang, C.-W. Lin, C.-C. Wang, C.-H. Wong and C.-Y. Wu, Effects of neighboring glycans on antibody–carbohydrate interaction, *Angew. Chem., Int. Ed.*, 2011, **50**, 1608–1612.

28 V. S. Shivatare, S. S. Shivatare, C.-C. D. Lee, C. H. Liang, K.-S. Liao, Y.-Y. Cheng, G. Saidachary, C.-Y. Wu, N.-H. Lin, P. D. Kwong, D. R. Burton, C.-Y. Wu and C.-H. Wong, Unprecedented role of hybrid *N*-glycans as ligands for HIV-1 broadly neutralizing antibodies, *J. Am. Chem. Soc.*, 2018, **140**, 5202–5210.

29 M. Cohen, N. Hurtado-Ziola and A. Varki, ABO blood group glycans modulate sialic acid recognition on erythrocytes, *Blood*, 2009, **114**, 3668–3676.

30 M. Cohen and A. Varki, Modulation of glycan recognition by clustered saccharide patches, in *International Review of Cell and Molecular Biology*, Academic Press, Burlington, 2014, vol. 308, pp. 75–125.

31 A. L. M. Morotti, K. L. Lang, I. Carvalho, E. P. Schenkel and L. S. C. Bernardes, Semi-synthesis of new glycosidic triazole derivatives of dihydrocucurbitacin B, *Tetrahedron Lett.*, 2015, **56**, 303–307.

32 Y. Manabe, *et al.*, Development of α1,6-fucosyltransferase inhibitors through the diversity-oriented syntheses of GDP–fucose mimics using the coupling between alkyne and sulfonyl azide, *Bioorg. Med. Chem.*, 2017, **25**, 2844–2850.

33 D. K. Sharma, *et al.*, Ammonium chloride mediated synthesis of alkyl glycosides and evaluation of their immunomodulatory activity, *RSC Adv.*, 2013, **3**, 11450–11455.

34 K. Izawa and T. Hasegawa, Tosylated and azidated inulins as key substrates for further chemical modifications to access inulin-based advanced materials: An inulin-based glycocluster, *Bioorg. Med. Chem. Lett.*, 2012, **22**, 1189–1193.

Faraday Discussions

PAPER

Biopolymer monolith for protein purification

Yoshiko Miura, *[a] Hirokazu Seto,[ab] Makoto Shibuya[a] and Yu Hoshino [a]

Received 11th February 2019, Accepted 24th April 2019

DOI: 10.1039/c9fd00018f

Porous glycopolymers, "glycomonoliths", were prepared by radical polymerization based on polymerization-induced phase separation with an acrylamide derivative of α-mannose, acrylamide and cross-linker in order to investigate protein adsorption and separation. The porous structure was induced by a porogenic alcohol. The pore diameter and surface area were controlled by the type of alcohol. The protein adsorption was measured in both batch and continuous flow systems. The glycomonoliths showed specific interaction with the sugar recognition protein of concanavalin A, and non-specific interaction to other proteins was negligible. The amount of protein adsorption to the materials was determined by the sugar density and the composition of the glycomonoliths. Fundamental knowledge regarding the glycomonoliths for protein separation was obtained.

Introduction

The development of biotechnology has become prominent in recent years. Since biological substances are usually obtained as a mixture, separation technology is necessary for any biotechnology applications.[1] The target biological substances are biomacromolecules with various sizes and properties such as cells, bacteria, proteins and nucleic acids.

Since biological substances are obtained as mixture in the aqueous solution, size-exclusion methods with membrane filtration are frequently used due to their practicality and rapidity of processing.[2] For example, bacterial separation is attained by filtration with a semi-micropore-sized membrane, a process which is known as filtration sterilization.[3] The sterilization can be accomplished only by filtration, and the processing speed is high. The size of proteins, bacteria, and cells are on the nanometer, sub-micrometer, and micrometer orders, respectively. Therefore, all separations can be attained by the size-exclusion method. However,

aDepartment of Chemical Engineering, Graduate School of Engineering, Kyushu University, 744 Motooka, Nishi-ku, Fukuoka, 819-0395, Japan. E-mail: miuray@chem-eng.kyushu-u.ac.jp

bDepartment of Chemical Engineering, Fukuoka University, 8-19-1 Nanakuma, Jonan-ku, Fukuoka 814-0180, Japan

the difficulty in membrane separation is not only due to the principle of size exclusion separation, but also due to the availability of practical filtration. The flux of filtration is described by fluid laws like the Hagen–Poiseuille law, where the flux changes in proportion to the fourth power of the pore radius.[4] Since the size of a protein is of the order of nanometers, a membrane pore for protein separation should be on the nanometer scale.[5] The flux of membrane filtration with nanometer pores is too slow for practical operation. Therefore, size exclusion separation is not suitable for nanometer-order substances like proteins and viruses. In addition, for membrane separation with nanometer-order pores it is difficult to perform a large amount of treatment due to fouling.[6] Actually, ultra-filtration (UF) with nanometer pores enables the separation of viruses and proteins.[1,5] However, it cannot separate large amounts of sample due to fouling, and it needs centrifugation to enlarge the flux.

Another important method is affinity separation with molecular recognition of ligands,[7] antibodies,[8] and proteins.[9] The best known affinity separation is antibody separation and purification using protein A and G, which is a key process of antibody medicine production.[10] Besides antibody purification, separation of biological substances using antibody–antigen,[9] sugar–protein,[11] DNA hybridization[12] and enzyme–substrate recognition[13] have also been reported. In affinity separation, membranes with small pores are not needed, and it is possible to separate proteins and viruses with sufficient flux and speed. Affinity separation is possible using materials with a molecular recognition unit.

Saccharide is one of the major ligands in living systems, and molecular recognition relates to various biological phenomena, including interaction with sugar recognition proteins (lectin), viruses, bacteria and cells.[14,15] The saccharide–protein interaction is usually weak, but can be amplified by multivalency.[16] Glycopolymers, polymers with pendant saccharides, are known to exhibit better molecular recognition based on their multivalency, and are useful biomaterials for molecular recognition.[17] The advantage of glycopolymers is the availability of polymer synthetic techniques. Various glycopolymers have been developed, such as glycopolymer-grafted substrates,[18] glycopolymer–gels[19] glycopolymer–gold nanoparticle counjugates,[20] and glycopolymer cryogels.[21,22] We have reported glycopolymers for the protein separation with the glycopolymer grafted membranes,[18,23] but their separation capacity was insufficient due to their limited surface area.

On the other hand, materials with large surface are appropriate for efficient affinity separation. Porous materials have a large surface area, and are utilized for molecular recognition unit immobilization. Protein separation with ligand-immobilized porous materials can be attained with continuous flow system, which is similar in operation to filtration techniques.[23,24] There are several representative porous polymers such as polymer monoliths and cryogels. The Fréchet group have reported a polymer monolith of polyacrylamide with porogenic alcohol, and they showed control of the pore structure by the porogenic solvent.[25,26] The Uyama group have reported polymer monoliths with phase separation.[27] The porous structure of monolith was determined by phase separation, which is induced by the porogenic solvent and monomer structure. In the polymer monoliths, phase separation is frequently utilized for the preparation of porous structures. Other well-known methods are cryogels[21,22] and porous polymers with precursor polymers.[28]

In this research, we investigated porous glycopolymers for protein affinity separation, where we call "glycomonoliths". The glycomonoliths were prepared by co-polymerization of acrylamide (AAm), α-mannose acrylamide derivative (p-acrylamido phenyl α-D-mannose) and a cross linker of N,N'-methylene bisacrylamide (BIS) using polymerization induced phase separation in the presence of porogenic solvents.[23-26] The porosity and surface area of the polymer monolith were measured by SEM and mercury porosimetry. Detailed protein adsorption was studied in both batch and flow systems in order to obtain fundamental knowledge of protein separation with the glycomonoliths. We have reported previously glycopolymer grafted membranes for protein separation, but the capacity of the protein separation was small.[24] Other groups have reported affinity separation of glycopolymer cryogels[21,22] where the surface area was also not large enough. Here we report the affinity separation of proteins with a monolith with a controlled structure.

Experimental

Reagents

AAm, 2,2'-azobisisobutyronitrile (AIBN), BIS, Bradford reagent, butanol (BuOH), ethanol (EtOH), and octanol (OctOH) were obtained from Fujifilm Wako Pure Chemical (Tokyo, Japan). AAm was recrystallized from chloroform. AIBN was recrystallized from acetone. A sugar monomer of p-nitrophenyl acrylamide α-mannose (**Man monomer**) was synthesized by following a previous report.[20] A porous glass membrane (diameter 20 mm, thickness 0.6 mm, pore 430 nm) (SPG Technology Co. Ltd, Miyazaki, Japan) was used for the reference material. Concanavalin A (ConA) (J-Oil Mills Inc, Tokyo Japan), peanut agglutinin (PNA) (Vector Laboratories Inc., Burlingame, CA) and bovine serum albumin (BSA, Sigma-Aldrich Co., St. Ouis, MO) were used without purification. The buffers used were 10 mM phosphate buffer saline with $CaCl_2$ and $MgCl_2$ (PBS(+), pH 7.4, 137 mM NaCl, 2.68 mM KCl, 1.47 mM KH_2PO_4, 8.1 mM Na_2HPO_4, 1.8 mM $CaCl_2$, and 0.49 mM $MgCl_2$), and 10 mM 4-(2-hydroxyethyl)-1-piperazineethanesulfonic acid (HEPES, pH 7.4).

Preparation of glycomonoliths

Glycomonoliths were prepared by solvent induced phase separation based on the previous report on polyAAm monoliths (Scheme 1). For example, a glycomonolith of 10 wt% was synthesized as follows: acrylamide (240 mg, 60 wt%), **Man monomer** (40 mg, 10 wt%) and BIS (120 mg, 30 wt%) were dissolved in the mixture of DMSO and alcohol (800 μL, 7 : 3). The solution was degassed with N_2 for 30 min and the polymerization was initiated by addition of AIBN solution (1 mol%). The polymerization was achieved at 70 °C for 30 min. The obtained monolith was immersed and washed with methanol and water for 1 day. Other glycomonoliths were prepared with different **Man monomer** and acrylamide ratios. Monoliths without **Man monomer** were also prepared by radical polymerization in the presence of porogenic alcohol.

Glycopolymer grafted membrane

The glycopolymer-grafted glass membrane was prepared by grafting-to method with thiol-terminated glycopolymer (Scheme 2).[24] The glycopolymer of poly(**Man**

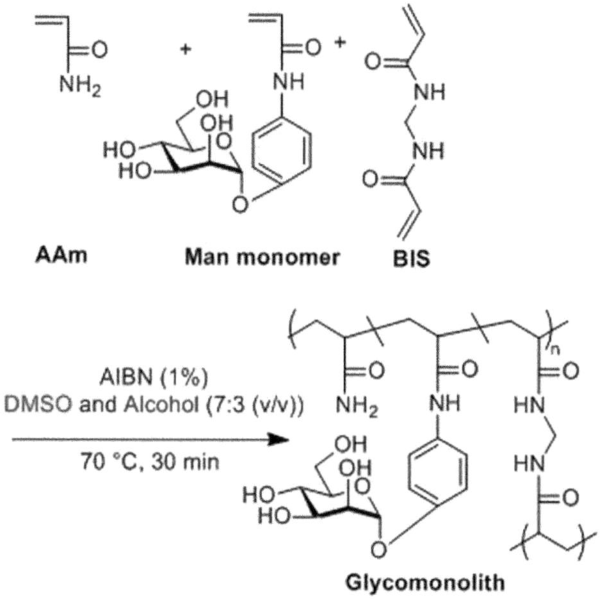

Scheme 1 Preparation of glycomonolith by radical polymerization in the presence of porogenic alcohol.

monomer) was synthesized by reversible addition fragmentation chain transfer (RAFT) polymerization. The obtained polymer was converted to a thiol-terminated molecule by hydrolysis. The glass membrane surface was modified with

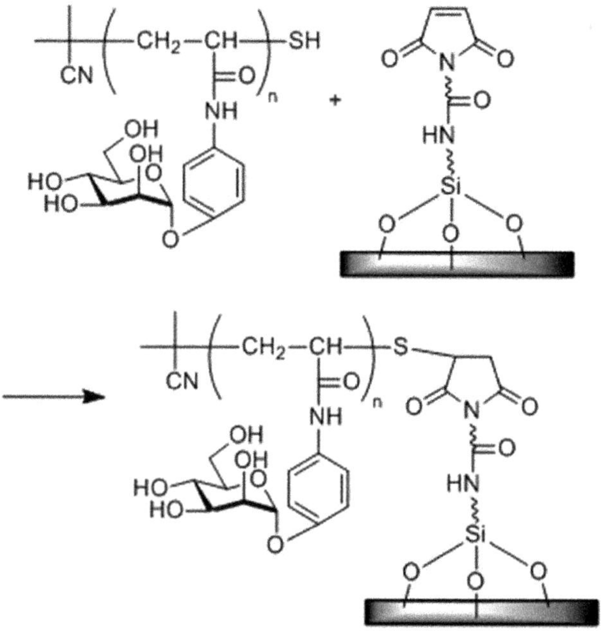

Scheme 2 Preparation of glycopolymer grafted glass membrane.

maleimide, and conjugated with the thiol-terminated glycopolymer. The modification of the glycopolymer was confirmed by X-ray photoelectron spectroscopy (data not shown).[24]

Characterization of the porous structure of glycomonolith

The morphology of the glycomonolith was observed by field emission scanning electron microscopy (FE-SEM, SU8000, Hitachi High-Technologies Co., Tokyo Japan). Before observation, the glycomonolith was coated with platinum (thickness: approx. 4 nm) with an auto fine coater (JFC-1600, JEOL Ltd., Tokyo, Japan). The pore size distribution in the glycomonolith was determined using mercury porosimetry (AutoPore IV9520, Micromeritics Instrument Co., Norcross, GA). The diameter and the pore volume of glycomonoliths were measured. The surface area was calculated from the pore volume.

Protein adsorption in the batch system

ConA was dissolved in 1 mM PBS(+) buffer at a concentration of 0.1–1.2 g L^{-1}. PNA and BSA were dissolved in the same concentration with 10 mM HEPES buffer. The monolith was cut into a piece of 3 mm square size, and mixed with protein solution in a solid–liquid ratio of 50 mg : 20 mL. The monolith was soaked in the protein solution and incubated at 25 °C for 20 h. The amount of protein absorption was measured by a UV spectrometer (Agilent 8453, Agilent Technologies, Santa Clara CA) at 280 nm. The protein adsorption was calculated as follows,

$$q = \frac{(C_0 - C) \times V \times d}{W}$$

where q, C_0, C, V, d, and W are the amount of adsorption (mg cm^{-3}), initial concentration (g L^{-1}), final concentration (g L^{-1}), fluid volume (L), density (mg cm^{-3}), and sample weight (g), respectively. The experiments were carried out with 7 different samples to check the reproducibility.

Protein adsorption by the flow system

Protein adsorption of the glycomonolith was evaluated by the continuous flow system in 10 mM PBS(+). Protein solution was passed through the glycomonolith at 10 mL h^{-1}. The effluent solution was collected and the concentration of protein in the effluent was determined with Bradford reagent.

Results and discussion

Glycomonolith preparation and morphology

Polymer monoliths were successfully prepared with high yield (>90%) (Scheme 1). In FE-SEM observation, the morphologies of the polymer were varied depending on the composition of the monomer and the porogen solvent (Fig. 1). When **Man monomer** was 0–10 wt%, pores were uniformly distributed in the materials as the polymer monolith. The morphologies of the polyAAm monolith and glycomonolith were similar. When the **Man monomer** ratio was increased, the pores were not uniformly distributed, and when **Man monomer** content was more than 70%, the porous structure could not be observed (Fig. 2). The morphology of the

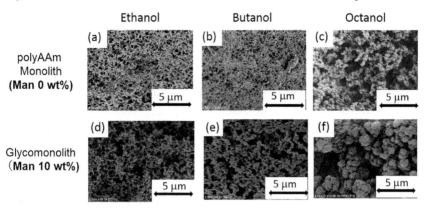

Fig. 1 SEM observation of polyAAm monolith and glycomonoliths with different monomer compositions and porogen alcohol; (a) polyAAm with EtOH, (b) polyAAm with butanol, (c) polyAAm with OcOH, (d) glycomonolith with EtOH, (e) glycomonolith with BuOH, and (f) glycomonolith with OcOH.

polymer was also different depending on the type of porogen. In the case of EtOH and BuOH, a fibrous structure was observed, and in the case of OcOH, agglomerate globules were observed.

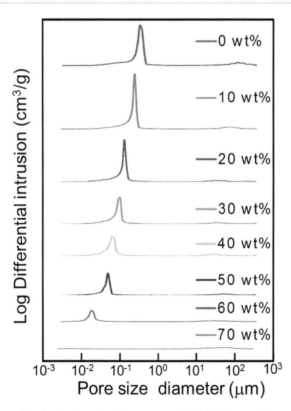

Fig. 2 The pore size distribution of polymer monolith with different **Man monomer** ratios evaluated by mercury porosimetry. The sample was prepared with BuOH.

Pore diameters and volumes of the dry monoliths were measured with mercury porosimetry, and the surface area was calculated (Fig. 2 and 3). The pore diameter at high **Man monomer** ratio was smaller than that at low **Man monomer** ratio. The incorporation of **Man monomer** also varied the pore diameter of the monolith (Fig. 2). The pore diameter of glycopolymer monoliths with **Man monomer** over 40 wt% was less than 100 nm. For the monolith containing 70% **Man monomer**, the pore structure could not be observed. Regarding polyAAm and the glyco-monolith with 10 wt% **Man monomer**, the pore diameter was 150–200 nm for EtOH, 250–300 nm for BuOH, and 1500–2000 nm for OcOH, respectively. In the surface area of the monolith, the pore diameter was greatly dependent on the kind of porogen alcohol, and the order of the surface area order was reversed, EtOH > BuOH > OcOH (Fig. 3).

In the polyAAm monolith, porous materials were obtained due to the solubility difference. The polymers have a lower solubility than the monomers, which

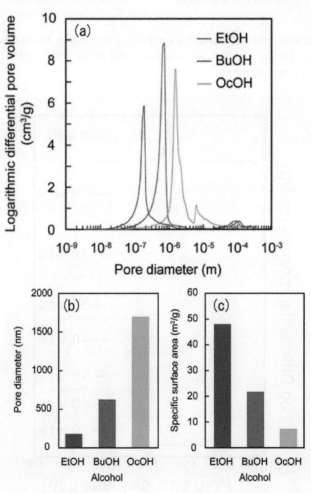

Fig. 3 Structure of glycopolymer monoliths having **Man monomer** 10 wt% with EtOH, BuOH and OcOH based on the mercury porosimetry test: (a) pore size distribution of glycomonoliths having **Man monomer** 10 wt% (b) mean pore size and (c) surface area.

caused polymerization-induced phase separation. Similar to the previous works in the literature.[25,26] Polymers are usually less soluble than monomers, and polyAAm and glycopolymers are water soluble, polar compounds. Besides, an alcohol with a short alkyl chain (EtOH) is more polar than one with a long alkyl chain (OcOH). The glycomonoliths in this study were less miscible to alcohol with long alkyl chains (OcOH), resulting in the formation of larger domains. The glycomonoliths showed rather better miscibility to alcohols with a short alkyl chain (EtOH), resulting in the formation of small domains. Therefore, glycomonoliths with the EtOH porogen had smaller pore diameter, and glycomonoliths with the OcOH porogen induced agglomerate globules with larger pore diameter.

The monomer property is also a decisive factor for the structure of glycomonoliths. The sugar unit (**Man monomer**) and glycopolymer are less soluble to both solvents (DMSO and alcohols). PolyAAm has a large solubility difference between alcohols and DMSO, and easily formed a porous structure with porogenic alcohols. Polymers with a high sugar content were not suitable for the preparation of porous polymers, because the solubility difference between the polymer and monomer was too small to induce phase separation. Polymers with high sugar monomer contents showed an aggregated structure with a small pore diameter. A glycomonolith of 10 wt% **Man monomer** formed an appropriate pore size as a glycopolymer monolith with a large surface area and sufficient flow flux. The pore size of glycomonoliths of much larger than protein size meaning that separation is conducted based on affinity. The pore size of glycomonoliths is much larger than protein size, meaning that separation is conducted based on affinity. The porous structure was maintained in the aqueous solution, which was confirmed by water permeation tests.

Protein adsorption in the batch system

Protein adsorption in the batch system was investigated with the glycomonoliths using BuOH and OcOH (Fig. 4). The protein adsorption was changed by the **Man monomer** ratio and the porogen alcohol. The protein adsorption with BuOH was larger than that with OcOH. In the case of BuOH, the amount of protein desorption was about 50 (mg cm^{-3}), which was the maximum for the monolith at 20–30 wt% **Man monomer** (Fig. 4(a)). The amount of protein adsorption over 30 wt% was decreased; the monolith of 40 and 50 wt% **Man monomer** was one fifth of this. In the glycomonolith with OcOH, the amount of protein adsorption did not change with **Man monomer** content; it was around 20 (mg cm^{-3}) (Fig. 4(b)). There was no clear relation between **Man monomer** ratio in the glycomonolith and the amount of protein adsorption.

The glycopolymer monolith of 10–30 wt% **Man monomer** was a suitable material for protein recognition. The linear glycopolymers have been reported to interact strongly with lectin by the cluster glycoside effect, and generally the glycopolymers with a high sugar ratio showed better molecular recognition.[17,20,29,30] Our results showed that the glycomonoliths with a high sugar ratio did not exhibit strong molecular recognition to lectin. Considering the mechanism of molecular recognition, molecular recognition is amplified by multivalent binding to the sugar recognition site and the increasing binding probability. The distance between each sugar binding site of ConA is 6.5 nm, and a 10% **Man monomer** ratio is sufficient for multivalent binding, considering the 3D structure

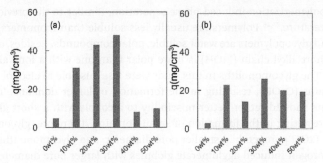

Fig. 4   Protein adsorption to glycomonoliths with different **Man monomer** ratios. Protein adsorption of glycomonoliths with (a) BuOH and (b) OcOH.

of the monolith. The binding probability is dependent on the protein diffusibility in the monolith. The protein diffusibility against pores that are too small (<100 nm) was limited due to insufficient flux, and little lectin was adsorbed.[30] The physical properties of the polymer monolith also could affect protein adsorption.[31-33] The detailed mechanism still needs more investigation.

The protein adsorptions of glycomonoliths (**Man monomer** 10 wt%) with different porogen alcohols were investigated (Fig. 5 and 6). Remarkably, protein adsorption in the glycomonoliths was observed only with ConA. Protein adsorption was not observed with PNA and BSA. PNA is galactose recognition protein, but it did not adsorb on the glycomonolith with Man. The polymer monolith without **Man monomer** (polyAAm monolith) did not adsorb any protein. The amount of protein was different with each porogen, in the order EtOH > BuOH > OcOH. The protein adsorption of the glycomonolith with EtOH was around 50 (g cm^{-3}) and almost double of that of the glycomonolith with OcOH. The protein adsorption of the glycomonolith with ConA was comparable to that of the glycopolymer-grafted glass membrane (Fig. 6). The amount of protein adsorption on the polymer-grafted membrane was the smallest, and that was approximately 30% of the glycomonolith with EtOH. The density of mannose on the materials

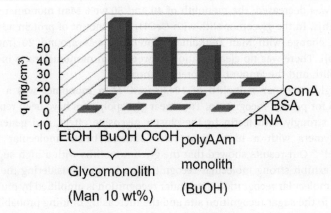

Fig. 5   Protein adsorption of polyAAm monolith and glycomonolith with different porogen alcohols with a protein concentration of 1.0 g L^{-1} for 20 h.

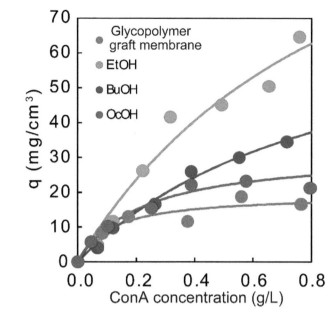

Fig. 6 Protein adsorption of glycomonoliths with different porogen alcohol and the glycopolymer-grafted membrane with varying protein concentration.

was calculated to be 113, 94, 100 and 0.27 $\mu g\ cm^{-3}$ in glycomonoliths with EtOH, BuOH, OcOH and glycopolymer-grafted membranes, respectively.

We confirmed a sufficient level of specific molecular recognition of the glycomonolith to ConA with no adsorption to PNA and BSA. Also, the absence of binding of polyAAm monolith indicated that the polyAAm based monolith did not induce non-specific protein adsorption, which makes it suitable as a biofunctional material, including applications in protein separation.[34] It has been reported that polyAAm and glycopolymer of the polyAAm backbone do not induce non-specific protein adsorption due to their hydrophilicity, and the polyAAm based monolith similarly showed specific protein adsorption. The amount of protein adsorption was strongly related to the surface area of the materials and the sugar density. Though the glycomonolith with 10 wt% **Man monomer** had the same composition, the amount of adsorbed protein was different. The polymer grafted membrane was also a porous materials with Man, but the amount of adsorbed protein was much smaller due to the low sugar density.[18] The sugar densities of the glycomonoliths were over 100 times larger than that of the polymer-grafted membrane. Since the porous materials are composed of glycopolymers, the glycomonoliths were more advantageous for the immobilization of biological ligands in materials and for the development of polymer separation materials. The glycomonoliths with 10–30 wt% from EtOH were the best materials for protein adsorption and separation in the batch system.

Protein adsorption by the flow system

Protein adsorption was measured with the continuous flow system, where the glycomonolith with 10 wt% **Man monomer** using BuOH protein was used. The

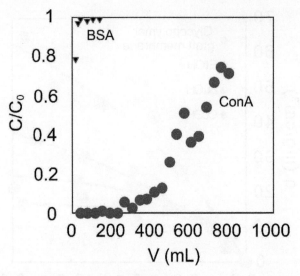

Fig. 7 Break through curve of protein adsorption on the glycomonolith with 10 wt% **Man monomer** using BuOH (C/C_0: the ratio of the eluted protein concentration to the initial protein concentration. V: the elution volume).

amount of protein adsorption with 100 mg L^{-1} at 60 mL h^{-1} and 10 mL h^{-1} feed was 0.32 mg g^{-1} and 68.8 mg g^{-1}, respectively. The protein adsorption was dependent on the flow rate, and the amount of protein adsorption at a slow flow rate was increased by over 100 times. The breakthrough curve of protein adsorption was measured with a flow rate of 10 mL min^{-1} using ConA and BSA (Fig. 7). ConA showed that the protein adsorption and breakthrough point was at around 380 mL effluent solution. BSA was quickly eluted from the glycomonolith, and showed almost no adsorption in the flow system.

Continuous flow separation, including filtration, was practical and speedy. Protein adsorption in the batch system was at near equilibrium, but that in the flow system was dynamic and in a non-equilibrium state. The amount of protein separation was dependent on the flow rate, which is the detention period. A longer detention period increased the amount of protein separation. On the other hand, BSA did not adsorb onto the glycomonolith because of the low non-specific adsorption on the polyAAm based monolith. The adsorbed ConA was separated from the glycomonolith with an acidic aqueous buffer (pH 3).[18] Since the glycomonolith induced specific binding to the mannose recognition protein, the purification of the protein mixture is possible.

Conclusion

Porous glycopolymers carrying α-mannose were prepared as glycomonoliths. The porous structure was induced by the solubility difference of the polymer using porogen alcohols. The diameter of the glycomonolith was controlled by the kind of porogen alcohol. The lower alcohol porogen gave a small diameter and large surface area, and the higher alcohol porogen provided a large diameter and small surface area. The protein adsorption of glycomonoliths showed high specificity to the sugar recognition protein of ConA. The amount of protein adsorption was determined by the surface area, but also by the monomer composition of the glycomonoliths. The protein adsorption was also measured by a continuous flow system, where the protein adsorption was greatly affected by the flow rate. Glycomonoliths showed a greater ability for protein adsorption and potential for protein separation. Since we have reported glycopolymers with affinity to viruses, toxic proteins and bacteria, glycomonoliths are applicable to the bioseparation of these. The large surface area and the high density of ligands are advantageous for various biomaterials and bioseparation applications.

Conflicts of interest

There are no conflict to declare.

Acknowledgements

We acknowledge financial support from a Grant-in-Aid for Scientific Research (B) (JP19H02766) and a Grant-in-Aid for Challenging Exploratory Research (JP16K140007). We appreciate assistance of Saga Ceramics Research Laboratory for the mercury intrusion test.

References

1 A. S. Grandison and M. J. Lewis, *Separation Processes in the Food and Biotechnology Industries: Principles and Applications*, Woodhead Publishing, 1996.
2 C. Charcosset, Purification of proteins by membrane chromatography, *J. Chem. Technol. Biotechnol.*, 1998, **71**, 95–110.
3 S. P. Denyer and N. A. Hodges, Sterilization: Filtration sterilization, in *Russell, Hugo & Ayliffe's Principles and Practice of Disinfection, Preservation & Sterilization*, 2004, pp. 436–472.
4 J. Pfitzner, Poiseuille and his law, *Anaesthesia*, 1976, **31**, 273–275.
5 R. van Reis and A. Zydney, Bioprocess membrane technology, *J. Membr. Sci.*, 2007, **297**, 16–50.
6 C. Güell and R. H. Davis, Membrane fouling during microfiltration of protein mixtures, *J. Membr. Sci.*, 1996, **119**, 269–284.
7 P. S. Stayton, T. Shimoboji, C. Long, A. Chilkoti, G. Ghen, J. M. Harris and A. S. Hoffman, Control of protein–ligand recognition using a stimuli-responsive polymer, *Nature*, 1995, **378**, 472–474.
8 M. Kim, K. Saito, S. Furusaki, T. Sugo and I. Ishigaki, Protein adsorption capacity of a porous phenylalanine-containing membrane based on a polyethylene matrix, *J. Chromatogr. A*, 1991, **586**, 27–33.
9 K. Ritter, Affinity purification of antibodies from sera using polyvinylidenedifluoride (PVDF) membranes as coupling matrices for antigens presented by autoantibodies to triosephosphate isomerase, *J. Immunol. Methods*, 1991, **137**, 209–215.
10 S. Hober, K. Nord and M. Linhult, Protein A chromatography for antibody purification, *J. Chromatogr. B: Anal. Technol. Biomed. Life Sci.*, 2007, **848**, 40–47.

11 A. Monzo, G. K. Bonn and A. Guttman, Lectin-immobilization strategies for affinity purification and separation of glycoconjugates, *TrAC, Trends Anal. Chem.*, 2007, **26**, 423–432.

12 O. Olsvik, T. Popovic, E. Skjerve, K. S. Cudjoe, E. Hornes, J. Ugelstad and M. Uhlen, Magnetic separation techniques in diagnostic microbiology, *Clin. Microbiol. Rev.*, 1994, **7**, 43–54.

13 P. Cuatrecasas, M. Wilchek and C. B. Anfinsen, Selective enzyme purification by affinity chromatography, *Proc. Natl. Acad. Sci. U. S. A.*, 1968, **61**, 636–643.

14 M. E. Taylor and K. Drickamaer, *Introduction to Glycobiology*, Oxford Press, London, 2002.

15 T. K. Dam and C. F. Brewer, Thermodynamic studies of lectin–carbohydrate interactions by isothermal titration calorimetry, *Chem. Rev.*, 2002, **102**, 387–430.

16 M. Mammen, S. K. Choi and G. M. Whitesides, Polyvalent interactions in biological systems: Implications for design and use of multivalent ligands and inhibitors, *Angew. Chem., Int. Ed.*, 1998, **37**, 2754–2794.

17 Y. Miura, Y. Hoshino and H. Seto, Glycopolymer nanobiotechnology, *Chem. Rev.*, 2016, **116**, 1673–1692.

18 H. Seto, Y. Ogata, T. Murakami, Y. Hoshino and Y. Miura, Selective protein separation using siliceous materials with a trimethoxysilane-containing glycopolymer, *ACS Appl. Mater. Interfaces*, 2012, **4**, 411–417.

19 Y. Kotsuchibashi, R. V. Agustin, J. Y. Lu, D. G. Hall and R. Narain, Temperature, pH, and glucose responsive gels *via* simple mixing of boroxole-and glyco-based polymers, *ACS Macro Lett.*, 2013, **2**, 260–264.

20 M. Takara, M. Toyoshima, H. Seto, Y. Hoshino and Y. Miura, Polymer-modified gold nanoparticles *via* RAFT polymerization: A detailed study for a biosensing application, *Polym. Chem.*, 2014, **5**, 931–939.

21 C. V. D. E. Ehe, T. Bus, C. Weber, S. Stumpf, P. Bellstedt, M. Hartlieb, U. S. Schubert and M. Gottschaldt, Glycopolymer-functionalized cryogels as catch and release devices for the pre-enrichment of pathogens, *ACS Macro Lett.*, 2016, **5**, 326–331.

22 S. Zhao, D. Wang, S. Zhu, X. Liu and H. Zhang, 3D cryogel composites as adsorbent for isolation of protein and small molecules, *Talanta*, 2019, **191**, 229–234.

23 H. Seto, M. Shibuya, H. Matsumoto, Y. Hoshino and Y. Miura, Glycopolymer monolith for affinity bioseparation of proteins in a continuous-flow system: Glycomonoliths, *J. Mater. Chem. B*, 2017, **5**, 1148–1154.

24 H. Seto, M. Takara, C. Yamashita, T. Murakami, T. Hasegawa, Y. Hoshino and Y. Miura, Surface modification of siliceous materials using maleimidation and various functional polymers synthesized by reversible addition–fragmentation chain transfer polymerization, *ACS Appl. Mater. Interfaces*, 2012, **4**, 5125–5133.

25 S. Xie, F. Svec and J. M. Fréchet, Preparation of porous hydrophilic monoliths: Effect of the polymerization conditions on the porous properties of poly(acrylamide-co-N,N′-methylenebisacrylamide) monolithic rods, *J. Polym. Sci., Part A: Polym. Chem.*, 1997, **35**, 1013–1021.

26 S. Xie, F. Svec and J. M. Fréchet, Rigid porous polyacrylamide-based monolithic columns containing butyl methacrylate as a separation medium for the rapid hydrophobic interaction chromatography of proteins, *J. Chromatogr. A*, 1997, **775**, 65–72.

27 K. Okada, M. Nandi, J. Maruyama, T. Oka, T. Tsujimoto, K. Konodoh and H. Uyama, Fabrication of mesoporous polymer monolith: A template-free approach, *Chem. Commun.*, 2011, **47**, 7422–7424.

28 G. Rohman, F. Lauprêtre, S. Boileau, P. Guérin and D. Grande, Poly(D,L-lactide)/poly(methyl methacrylate) interpenetrating polymer networks: Synthesis, characterization, and use as precursors to porous polymeric materials, *Polymer*, 2007, **48**, 7017–7028.

29 V. Ladmiral, G. Mantovani, G. J. Clarkson, S. Cauet, J. L. Irwin and D. M. Haddleton, Synthesis of neoglycopolymers by a combination of "click chemistry" and living radical polymerization, *J. Am. Chem. Soc.*, 2006, **128**, 4823–4830.

30 C. W. Cairo, J. E. Gestwicki, M. Kanai and L. L. Kiessling, Control of multivalent interactions by binding epitope density, *J. Am. Chem. Soc.*, 2002, **124**, 1615–1619.

31 K. P. S. Dancil, D. P. Greiner and M. J. Sailor, A porous silicon optical biosensor: Detection of reversible binding of IgG to a protein A-modified surface, *J. Am. Chem. Soc.*, 1999, **121**, 7925–7930.

32 Y. Hoshino, M. Nakamoto and Y. Miura, Control of protein-binding kinetics on synthetic polymer nanoparticles by tuning flexibility and including conformation changes of polymer chains, *J. Am. Chem. Soc.*, 2012, **134**, 15209–15212.

33 T. Hasegawa, S. Kondoh, K. Matsuura and K. Kobayashi, Rigid helical poly(glycosyl phenyl isocyanide)s: Synthesis, conformational analysis, and recognition by lectins, *Macromolecules*, 1999, **32**, 6595–6603.

34 M. Toyoshima, T. Oura, T. Fukuda, E. Matsumoto and Y. Miura, Biological specific recognition of glycopolymer-modified interfaces by RAFT living radical polymerization, *Polym. J.*, 2010, **42**, 172–178.

PAPER

Mimicking the endothelial glycocalyx through the supramolecular presentation of hyaluronan on patterned surfaces†

Xinqing Pang, [ID] ab Weiqi Li, [ID] ab Eliane Landwehr,[c] Yichen Yuan, [ID] ab Wen Wang[ab] and Helena S. Azevedo [ID] *ab

Received 11th February 2019, Accepted 29th March 2019

DOI: 10.1039/c9fd00015a

The glycocalyx is the immediate pericellular matrix that surrounds many cell types, including endothelial cells (ECs), and is typically composed of glycans (glycosaminoglycans, proteoglycans, and glycoproteins). The endothelial glycocalyx is rich in hyaluronic acid (HA), which plays an important role in the maintenance of vascular integrity, although fundamental questions about the precise molecular regulation mechanisms remain unanswered. Here, we investigate the contribution of HA to the regulation of endothelial function using model surfaces. The peptide sequence GAHWQFNALTVR, previously identified by phage display with strong binding affinity for HA and named Pep-1, was thiolated at the N-terminal to form self-assembled monolayers (SAMs) on gold (Au) substrates, and microcontact printing (μCP) was used to develop patterned surfaces for the controlled spatial presentation of HA. Acetylated Pep-1 and a scrambled sequence of Pep-1 were used as controls. The SAMs and HA-coated surfaces were characterized by X-ray photoelectron spectroscopy (XPS), contact angle measurements, and quartz crystal microbalance with dissipation (QCM-D) monitoring, which confirmed the binding and presence of thiolated peptides on the Au surfaces and the deposition of HA. Fluorescence microscopy showed the localization of fluorescently labelled HA only on areas printed with Pep-1 SAMs. Cell culture studies demonstrated that low molecular weight HA improved the adhesion of human umbilical vein endothelial cells (HUVECs) to the substrate and also stimulated their migration. This research provides insight into the use of SAMs for the controlled presentation of HA with defined size in cultures of HUVECs to study their functions.

[a]School of Engineering and Materials Science, Queen Mary University of London, London E1 4NS, UK. E-mail: h.azevedo@qmul.ac.uk

[b]Institute of Bioengineering, Queen Mary University of London, London E1 4NS, UK

[c]Department of Chemistry, University of Konstanz, Konstanz 78464, Germany

† Electronic supplementary information (ESI) available. See DOI: 10.1039/c9fd00015a

1. Introduction

Hyaluronic acid (HA), or hyaluronan, is a linear polysaccharide that consists of repeating disaccharide units of *N*-acetylglucosamine and glucuronic acid.[1] Despite its simple chemical structure, HA exhibits remarkable wide-ranging and often opposing biological functions, and these activities seem to be related to the HA molecular size.[2] High molecular weight HA is known to be space-filling, immunosuppressive, and anti-angiogenic. Molecules up to 20 kDa in size participate in the processes of ovulation, embryogenesis, and wound healing, while smaller HA oligosaccharides are known to be inflammatory, immune-stimulatory, and pro-angiogenic. HA is found in almost all living organisms, being degraded and resynthesized on a daily basis in the human body.[3] HA usually exists in the extracellular matrix (ECM), which provides cells with a physical and chemical microenvironment that determines their proliferation, migration, or differentiation.[4] The vascular endothelial glycocalyx, a brush-like layer located in the luminal surface of the vascular endothelium, is also rich in HA. Current studies suggest that the glycocalyx is a crucial component of many vascular activities, such as blood tissue exchange, inflammatory response, tissue homeostasis, fibrinolysis, coagulation, vascular regulation, vasodilation of various tissues, and angiogenesis.[5–11] HA is the only non-sulfated glycosaminoglycan (GAG) that binds to the cell surface receptor CD44 and a multitude of biological activities depend on its length. As a highly hydrophilic molecule, HA contributes to tissue hydrodynamics and the transport of water, and plays an important role in cell proliferation, migration, and maintaining vascular integrity.[12]

In order to dissect the key features of the ECM, researchers have developed synthetic platforms with defined chemistry that act as model surfaces for studying specific ECM–cell interactions.

Self-assembled monolayers (SAMs), or the spontaneous assembly of organosulfur compounds on metal surfaces, have been widely applied to prepare biocompatible substrates with defined chemical compositions for biomedical research, including wetting, protein adsorption, and cell adhesion studies.[13–15] In particular, gold (Au) has been the standard surface for creating SAMs because it is not toxic to cells and has high binding affinity to thiols, along with its inert characteristics.[16] Molecules used in SAMs typically consist of three parts: a head group (a thiol group), an alkyl chain, and a tail functional group ($-CH_3$, $-COOH$, $-PO_3^{2-}$, or $-OH$).[17]

The mechanism of SAM formation includes two steps: the rapid and strong chemisorption between the head groups and Au substrates, and the subsequent slow reassembly due to the interaction between the alkyl chains (van der Waals' forces).[18–20] The structure and quality of SAMs formed on Au substrates are affected by factors such as surface roughness, concentration and purity of the self-assembled molecules, immersion time, solvents, and temperature.[21–24] The formation, composition, and structure of SAMs have been characterized by complementary characterization techniques, such as X-ray photoelectron spectroscopy (XPS),[25] quartz crystal microbalance with dissipation (QCM-D) monitoring,[26,27] and contact angle measurements.[28–31]

Functional peptides attached to Au surfaces, forming well-arranged and reproducible SAMs, have been used in many biomedical studies.[32] For example, the work by Mrksich on using SAMs as ECM models has largely contributed to the elucidation of the role of peptide and protein ligands in cell–matrix interactions. In particular, SAMs presenting the Arg-Gly-Asp (RGD) peptide with different densities and spacing were used to investigate the adhesion and spreading of different cell types.[33] However, the application of peptides binding to specific components of the ECM has not yet been exploited.

Mummert *et al.* identified a HA-binding peptide (GAHWQFNALTVR) through phage display technology, named Pep-1, which presented specific binding to soluble, immobilized, and cell-associated forms of HA.[34] The ability of Pep-1 to bind to both HA-coated substrates and HA molecules expressed on the surfaces of endothelial cells was also demonstrated.[34]

In this study, we have modified Pep-1 with thiol functionality (Fig. 1A) to form Pep-1 SAMs on Au, which would result in surfaces displaying multiple peptide sequences with binding affinity for HA (Fig. 2). In addition, using microcontact printing (μCP), patterns of Pep-1 SAMs could be created on Au surfaces for the spatial localization of HA. μCP consists of transferring an ink solution from a patterned elastomeric mould, or stamp, to a substrate by contact with its surface.[35,36] The combination of μCP and SAMs is advantageous for obtaining good control over the surface chemistry and minimizing defects due to the molecular self-organization.[35]

We hypothesized that the supramolecular (non-covalent) immobilization of HA with defined sizes on surfaces could be used to probe how endothelial cells sense and respond to distinct HA sizes and would provide insight into the effect of HA on important cellular functions of the endothelium in terms of health and disease.

2. Materials and methods

Protection of 3-mercaptopropionic acid

In order to ensure the coupling of the acid group of 3-mercaptopropionic acid with the free amino in the peptide N-terminal, 3-(((4-methoxyphenyl)

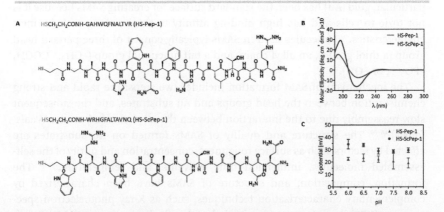

Fig. 1 Thiol-containing peptides used to create SAMs on Au surfaces and their characterization. (A) Chemical structure of the thiolated HA-binding peptide (HS-Pep-1) and thiolated scrambled Pep-1 (HS-ScPep1). (B) Circular dichroism (CD) spectra of the peptides at pH 7 and 0.1 mg mL^{-1}. (C) Zeta potential of the peptides at different pH values within the range 6 to 8.

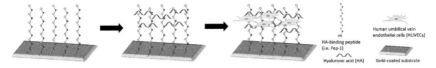

Fig. 2 Schematic representation of the fabrication process to obtain HA-coated surfaces for cell culturing using self-assembled monolayers of the HA-binding peptide.

diphenylmethyl)thio)propanoic acid was synthesized. *N*,*N*-Diisopropylethylamine (DIPEA, Sigma) and 3-mercaptopropionic acid were added dropwise into a stirred solution of 4-methoxytriphenylmethyl chloride (MMT, Sigma) in 1 : 1 dichloromethane (DCM, Sigma)/dimethylformamide (DMF, Sigma). The reaction mixture was concentrated by rotary evaporation, suspended in water, and then washed with diethyl ether. The organic layer was washed with brine, dried over magnesium sulfate (Thermo Scientific), and concentrated to oil by rotary evaporation. The oil was dried under high vacuum until a white powder remained. The chemical structure of the obtained product was confirmed using nuclear magnetic resonance spectroscopy (NMR, ESI, Fig. S1†).

Peptide synthesis and purification

Pep-1 (GAHWQFNALTVR) and a scrambled sequence (ScPep-1, WRHGFAL-TAVNQ)[37] were synthesized in an automated microwave peptide synthesizer (Liberty Blue, CEM, UK) on a 4-methylbenzhydrylamine (MBHA) rink amide resin (bead size: 100–200 mesh, Novabiochem), following the standard 9-fluorenylmethoxycarbonyl (Fmoc) solid phase peptide synthesis protocol. DCM was used to swell the resin and 20% (v/v) piperidine (Sigma) in DMF was used as the deprotection solution. The coupling was performed using 4 mol equivalents of Fmoc-amino acid (Novabiochem), 1-hydroxybenzotriazole hydrate (HOBt), and *N*,*N'*-diisopropylcarbodiimide (DIC). The 3-(((4-methoxyphenyl)diphenylmethyl) thio)propanoic acid tail was manually coupled to the N-terminal of the peptide under the same conditions as the Fmoc-amino acids. For the acetylated Pep-1 (Ac-Pep-1), the N-terminal was capped with an acetyl group by incubating the peptide-bound resin with 10% (v/v) acetic anhydride (Sigma) in DMF under shaking for 10 min. The coupling of the thiol tail or acetylation was confirmed using a Kaiser test kit (Sigma), where negative results (no free amine groups) indicated successful coupling and capping. The cleavage of the final peptide from the resin and the removal of the protecting groups was performed by shaking the resin with bound peptide and a mixture solution containing trifluoroacetic acid (TFA, Sigma)/thioanisole (Sigma)/1,2-ethanedithiol (EDT, Sigma)/anisole (Sigma) (90%/5%/2.5%/2.5%) for thiol-containing peptides and TFA/triisopropylsilane (TIS, Sigma)/water (95%/2.5%/2.5%) for the Ac-Pep-1, at room temperature for 3 hours. The peptides were concentrated using a rotary evaporator and subsequently precipitated in cold diethyl ether. The resulting suspension was centrifuged (Heraeus Multifuge X1, Thermo Scientific) at 4100 rpm for 20 minutes and the powder collected for freeze drying. The mass of the crude peptides was confirmed by electro-spray ionization mass spectrometry (ESI-MS, Agilent) and the purity was examined in an Alliance high-performance liquid chromatography (HPLC) system (Waters) coupled with an analytical reverse-phase C18 column (XBridge,

130 Å, 3.5 µm 4.6 × 150 mm, Waters). The peptide bond was used for detection through a UV/Vis detector (2489, Waters) set at 220 nm and using Empower software®. Peptide solutions (1 mg mL^{-1}, 100 µL) were injected into the column and eluted at 1 mL min^{-1} using a water/acetonitrile (ACN, Sigma) (0.1% TFA) gradient. An AutoPurification preparative scale HPLC system (Waters 2545 Binary Gradient HPLC system, Waters) containing a reverse-phase C18 column (X-bridge, 130 Å, 5 µm, 30 × 150 mm, Waters) was used to purify the peptides. Peptides were eluted at 20 mL min^{-1} using a gradient of water/ACN containing 0.1% TFA. Fractions were collected based on the mass detection performed by a SQ detector 2 (Waters) and the data were processed using MassLynx® software. After the purification process, the solvent was removed by rotary evaporation followed by freeze drying. Finally, the purity of the peptides was confirmed by ESI-MS and analytical HPLC, as described above.

Peptide characterization

Zeta potential. In order to investigate the overall charges of HS-Pep-1 and HS-ScPep-1 at different pH within the range 6 to 8, the ζ-potentials of the peptide aqueous solutions were measured using a Nano-ZS Zetasizer (Malvern Instruments). Briefly, the peptides were dissolved in ultrapure water (0.1 mM) and the pH was adjusted to 6, 6.5, 7, 7.5, and 8 by adding 0.1 M HCl or 0.1 M NaOH. Cuvettes containing gold electrodes (DTS1070, Malvern Panalytical) were used to load the peptide samples, and the ζ-potential was recorded at 25 °C.

Circular dichroism (CD) spectroscopy. The secondary structures of HS-Pep-1 and HS-ScPep-1 were characterized by CD. The peptides were dissolved in ultra-pure water (0.1 mg mL^{-1}) and the pH was adjusted to 7. A 1 mm path length quartz cuvette was used to load the peptide aqueous solutions, and the CD spectra were recorded at 25 °C from 190 to 300 nm using a PiStar-180 spectrometer (Applied Photophysics). Ultrapure water was measured to obtain a background spectrum, which was subtracted from the peptide sample spectra. Each represented spectrum is an average of 3 spectra. The molar ellipticity [θ] at wavelength λ was calculated using the following eqn (1):

$$[\theta] = \frac{100 \times \theta}{c \times d} \tag{1}$$

where θ is the observed ellipticity in mdeg, c is the molar concentration of the peptide solution, and d is the cuvette path length in cm.

Preparation of peptide SAMs and HA deposition

The gold-coated slides used in this study were either purchased from Dynasil (5 nm chrome, followed by 100 nm gold) for SAM characterization experiments, or coated with 5 nm chrome, followed by 20 nm gold, through evaporation in the School of Physics & Astronomy at Queen Mary University of London, to be used for microscopy in the cell culture assays. The HA used in all experiments was purchased from Lifecore Biomedical, Inc. (Chaska). Briefly, slides were submerged in an ethanolic solution (ethanol/water in a 9 : 1 ratio) containing 0.1 mM peptide (HS-Pep-1 or HS-ScPep-1) and incubated at room temperature overnight (Fig. 2). The slides were rinsed with ethanol, dried under N$_2$, and then incubated with 0.5 mg mL^{-1} aqueous solution of unmodified HA (molar mass of either 5 kDa, 60 kDa, or 700 kDa) for at least 24 hours at room temperature (Fig. 2). The HA-coated surfaces were rinsed with ultrapure water to remove weakly bound molecules, dried under N_2, and characterized or used in further studies.

Fluorescein-hyaluronic acid (HA)

HA was labelled with fluoresceinamine following the procedures previously described.[38,39] Briefly, 40 mL aqueous solution of 0.25% (w/v) unmodified 700 kDa HA was mixed with 40 mL DMF containing 10 mg of fluoresceinamine (Sigma). 200 mg of N-hydroxysuccinimide (NHS, Sigma) were added to the mixture and the pH adjusted to 4.75 using 0.1 M HCl. Then, 100 mg of N-(3-dimethylaminopropyl)-N′-ethylcarbodiimide hydrochloride (EDC, Sigma) were added and the pH maintained at 4.75. After 12 h, the solution was dialyzed against 100 mM NaCl using dialysis tubing (5000 Da MWCO, Sigma) for 2 days, followed by another 2 days of dialysis against ultrapure water, and then was freeze dried.

Preparation of PDMS stamps and micro-contact printing (μCP)

PDMS stamps were prepared following the procedure described by Qin et al.[36] Briefly, Sylgard 184 silicone elastomer base and the curing agent (Dow corning), mixed in a mass ratio of 10 : 1, were placed in a vacuum-connected desiccator. The degassed liquid mixture was poured onto the patterned template and then placed in an oven overnight to achieve a cured PDMS stamp. The micropatterns of the PDMS stamps were imaged by scanning electron microscopy (SEM, Inspect F) using a 5.0 kV beam after being coated with a gold layer.

HS-Pep-1 and HS-ScPep-1 were dissolved in ethanol (1.5 mM), swabbed onto the patterned side of the PDMS stamp using a cotton Q-tip, and then dried under a stream of nitrogen. The loaded stamp was brought into contact with the gold surface for 10 seconds. The patterned gold slides were then incubated with 0.5 mg mL^{-1} aqueous solution of fluorescein HA (700 kDa) overnight at room temperature. Bare Au incubated with 700 kDa fluorescein-HA solution (0.5 mg mL^{-1}) was used as a control. The samples were rinsed with ultrapure water and then dried under N_2. Images were then acquired using a Leica DMi8 Epifluorescence microscope (Leica) at 10× and 20× magnification.

Characterization of peptide SAMs and HA-coated surfaces

Contact angle. The contact angles of the bare Au surface, the Pep-1 and ScPep-1 SAMs, and the surfaces coated with HA were measured via the sessile drop technique using a Drop Shape Analyser (Model DSA100, Krüss). 2 μL of ultrapure water was dropped onto the surface and the contact angle was measured. Bare Au immersed in 60 kDa HA aqueous solution (0.5 mg mL^{-1}) was used as the control. The contact angle of each surface (>8 gold substrates) was measured in 3–5 different locations and the average was calculated.

Quartz crystal microbalance with dissipation (QCM-D) monitoring. SAM formation and HA deposition were monitored by a QCM-D (QS100, QSense). Before use, the gold-coated AT-cut quartz crystal (QSense) was cleaned with base piranha (30% ammonium hydroxide (Sigma)/30% H_2O_2 (Sigma)/water in a 1 : 1 : 3 ratio) at 60 °C, rinsed with ultrapure water, and then dried under N_2. The cleaned crystal was then UV-ozone treated (UVOCS T10X10 OES/E, Ultraviolet Ozone Cleaning Systems) for 20 minutes. For all experiments, baseline, deposition, and washing steps were performed at 37 °C in 150 mM sodium chloride (NaCl). Immediately after the baseline frequency of the crystal became stable, a solution of HS-Pep-1, Ac-Pep-1, or HS-ScPep-1 (0.1 mM in 150 mM NaCl) was injected into the crystal chamber for binding. The system was rinsed with NaCl to remove loosely bound molecules. A solution of 60 kDa HA (0.5 mg mL^{-1} in 150 mM NaCl) was then injected into the crystal chamber for binding. Again, once a stable frequency was acquired, the system was washed with NaCl solution to remove weakly associated HA molecules. The frequency (Δf) and dissipation (ΔD) changes were monitored in real time, and the results are shown for 34.7 MHz resonance. Mass changes (Δmass) were calculated using the Voigt model and the software QTools.

X-ray photoelectron spectroscopy (XPS). XPS analysis was performed on a Thermo Scientific™ XPS system. The analysis point area was 100 μm × 100 μm. The analyser pass energy for the survey spectra was 200.0 eV. The elemental spectra were acquired with an analyser pass energy of 50.0 eV. The spectra of Au4f, C1s, N1s, O1s, and S2p, and the survey spectra, were analysed using the software Advantage. The S peaks were fitted using two S2p doublets with 2 : 1 area ratios and splittings of 1.2 eV. The binding energies were calibrated by setting the Au4f$_{7/2}$ at 84.0 eV. Two replicates per group were measured and averaged.

Cell culture studies

Human umbilical vein endothelial cells (HUVECs, Lonza) were cultured in medium 199 (Invitrogen) supplemented with 10% foetal bovine serum (FBS), 1 ng mL^{-1} β-endothelial cell growth factor, 3 μg mL^{-1} bovine neural extract, 1.25 μg mL^{-1} thymidine, 10 U mL^{-1} heparin, 100 U mL^{-1} penicillin, and 100 μg mL^{-1} streptomycin. All supplements were purchased from Sigma. Cells were cultured in a 5% CO$_2$ incubator at 37 °C. The culture medium was exchanged every 2 days. Solutions of HS-Pep-1 (0.1 mM), HS-ScPep-1 (0.1 mM), and HA (0.1 mg mL^{-1}) used in the cell culture were sterilized under UV light for 30 min before SAM preparation.

Cell adhesion and migration assay

Gold-coated microscope slides (cut into ~0.5″ × 0.5″ pieces) were placed in a 12 well plate and SAMs were formed as described above. HUVECs were seeded at a density of 5 × 10^4 cells per well. In order to investigate the extent of cell adhesion, the spreading area of the cells was measured at 30, 60, and 90 min immediately after seeding. Images were obtained at 10× magnification using an optical microscope (DFC420 C, Leica). Only attached cells were considered for calculation of the cell area (Fig. 5A3, green arrow). Cells maintaining a round shape (not adhered to the substrate) were not considered (Fig. 5A3, yellow arrow). Cell areas were calculated using ImageJ. In order to track cell movement, time-lapse images were obtained every 10 min using a Lumascope 720 (Etaluma) at 37 °C and 5% CO$_2$. The cell trajectory and velocity were analysed using ImageJ.

Cell viability assay

8 well sticky-Slides (ibidi) were assembled on Dynasil gold coated slides (1″ × 3″) and SAMs were formed as described above. The metabolic activity of the HUVECs

over 24 and 48 h incubation periods was assessed using AlamarBlue™ cell viability reagent (ThermoFisher). AlamarBlue™ (10% volume of the well) was added to cell culture medium and then incubated at 37 °C and 5% CO_2 for 4 hours, protected from direct light. After incubation, the absorbance values were read at 570 nm and 600 nm on a BMG Labtech microplate reader. The reduction percentage was calculated following the manufacturer's instructions.

Data analysis and statistics

All data values are expressed as the mean ± standard deviation (SD). Statistical analysis was performed using GraphPad Prism 7.00 software. The statistical significance was evaluated using an unpaired t-test for the zeta potential data. The statistical differences of other experiments were determined using one-way analysis of variance (ANOVA) with a Tukey's honest significant difference (HSD) *post hoc* test. The statistical significant difference between groups was accepted at $P < 0.05$.

3. Results and discussion

Peptide SAMs on gold have been used to create defined surfaces for identifying substrates able to support cell adhesion[40] and proliferation,[41] or prevent protein adsorption (non-fouling surfaces).[32,42] Peptides can be attached to Au surfaces using the thiol functionality of cysteine, either at the N- or C-terminus.[32,43,44] However, using cysteine to anchor the peptide onto the Au surface does not allow for the obtaining of well-packed and dense monolayers due to the steric hindrance caused by the N-terminal. In order to circumvent this problem, researchers have conjugated peptides with alkanethiols at both termini,[40,42,45] or have incorporated a linker sequence of four proline residues linked to the terminal cysteine to confer rigidity and ensure closely packed monolayers.[32] Here, we synthesized peptides with a free thiol group at the N-terminus (Fig. 1A) using a bifunctional molecule (3-mercaptopropionic acid). The thiol functionality was protected first with an MMT group (ESI, Fig. S1†) to allow for the coupling of the carboxylic acid of the mercapto acid to the free amine of the peptide N-terminus. We expect that this peptide configuration will promote the formation of well-ordered SAMs.

Peptide characterization

In order to gain insight into the secondary structure adopted by the peptides used to form SAMs, CD spectroscopy was conducted on the peptides in solution. The CD spectrum of HS-Pep-1 showed a positive maximum at 194 nm and a negative maximum appeared at 218 nm (Fig. 1B). For HS-ScPep-1, the positive maximum was at 197 nm and the negative maximum at 217 nm. These are characteristic signatures of a β-sheet structure.[46] A β-sheet structure suggests the presence of peptide interchain interactions through hydrogen bonding. The zeta potentials of HS-Pep-1 and HS-ScPep-1 showed a positive charge for both peptides, as expected. There are two amino acids with ionizable side chains: the amine groups of arginine (R, $pK_a > 10$) and the imidazolium group of histidine (H, $pK_a = 6.1$). The amine group of R is protonated in the pH range studied, while H carries a positive charge at pH < 6. HS-ScPep-1 had a higher (significant difference in the t-test analysis) zeta potential compared to HS-Pep-1 in the pH range 6–8 (Fig. 1C), despite having exactly the same amino acid composition. Both peptides are highly hydrophobic, containing only 25% hydrophilic amino acids. However, their position in the peptide backbone is not the same, leading to a different distribution of hydrophobic and hydrophilic amino acids, resulting in different interactions among the peptide molecules and with the solvent. The CD data indicates a more pronounced β-sheet signal for HS-ScPep-1, which may lead to the formation of more stable aggregates with higher surface charges.

Patterning HA on micro-contact printed Pep-1 SAMs

A μCP technique was utilized to demonstrate the use of Pep-1 SAM to create HA patterns on Au surfaces (Fig. 3). PDMS moulds patterned with round holes and a 200 μm diameter were used to print HS-Pep-1 on the Au substrates. SEM images confirmed the hollow morphology and dimension of the patterns on the PDMS mould (Fig. 3D). Using HA labelled with fluorescein (green dye) and through fluorescence microscopy, HA was shown to be localized only on the Pep-1 printed areas (Fig. 3E). No fluorescent patterns were observed on either bare Au or ScPep-1 SAM after incubation with fluorescein-HA (ESI, Fig. S5†).

SAM characterization

In order to characterize changes in the hydrophilicity of the Au surfaces after modification, the water contact angle was measured and compared to that of a bare Au surface. The water contact angle on bare gold was 70.00 ± 5.59°, showing a highly hydrophobic surface, whilst that on Pep-1 SAMs was 60.92 ± 4.90°. After HA deposition, the substrates became more hydrophilic compared to

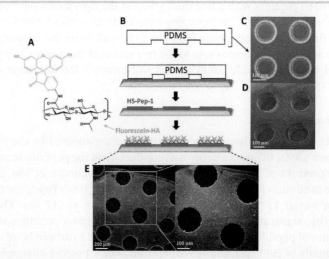

Fig. 3 Schematic illustration showing the fabrication of HA patterns by μCP. (A) Chemical structure of fluorescein-HA. (B) Flow diagram for μCP. HS-Pep-1 was printed on the gold surface using the PDMS stamp, and then the substrate was immersed in fluorescein-HA solution, allowing binding to the attached Pep-1. (C) Bright field microscopy images of the patterned PDMS mould. (D) SEM images of the patterned PDMS mould. (E) Fluorescence images showing the localization of fluorescein-HA (green) on the HS-Pep-1 printed areas.

the bare gold and those with immobilized Pep-1, suggesting the presence of HA on the surface. The surface coated with 60 kDa HA had a contact angle of 54.86 ± 3.65° (Fig. 4A). ScPep-1 SAMs (53.35 ± 4.06°) were shown to be more hydrophilic than those formed by Pep-1 and there were no significant differences between the ScPep-1 SAM and HA coated on ScPep-1-Au surfaces. The formation of the SAMs and the HA deposition were followed *in situ* by QCM-D monitoring. When an alternating potential is applied, the quartz crystal disk (QCM-D sensor) oscillates at its resonance frequency. A decrease in frequency was observed upon the addition of HS-Pep-1 and remained constant upon washing. A further decrease in frequency was observed after injection of HA (Fig. 4B). This decrease in frequency, combined with the increase in dissipation, indicates the binding of HS-Pep-1 and HA deposition on the surface of the Au crystal. When the thiol functionality was removed from the Pep-1 sequence (Ac-Pep-1), the binding of the peptide was diminished with the peptide being removed after washing, highlighting the need for the thiol group to form stable bonds with the Au. The binding of HS-ScPep-1 to

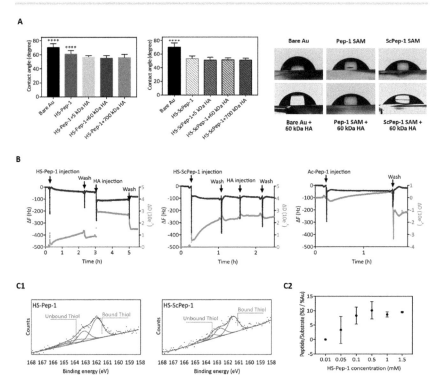

Fig. 4 Characterization of peptide SAMs formed on Au surfaces. (A) Contact angles of gold surfaces without and with peptide SAMs and after deposition of HA with different molecular weights (left: Pep-1 SAM; right: ScPep-1 SAM). (B) QCM-D monitoring of frequency changes (Δf, black) and dissipation changes (ΔD, orange) on the formation of Pep-1 and ScPep-1 SAMs, followed by addition of 60 kDa HA injection and adsorption of Ac-Pep-1. (C1) XPS S2p spectra of Pep-1 (left) and ScPep-1 (right) SAMs on gold surfaces. The S peaks were fitted using two S2p doublets with 2 : 1 area ratios and splittings of 1.2 eV. The position of the S2p$_{3/2}$ peaks assigned to bound thiolate and unbound thiol are shown in orange and green, respectively. (C2) HS-Pep-1 binding isotherm on Au, shown as a ratio of sulfur atomic percent to Au atomic percent (%S/%Au) for different concentrations of HS-Pep-1.

the Au crystal was confirmed, but when HA was injected on the ScPep-1 coated crystal, there was no significant frequency shift, indicating that HA did not bind to the ScPep-1 SAMs. These results confirm that the binding affinity of Pep-1 to HA is sequence dependent, and scrambling this sequence (ScPep-1) results in the loss of HA binding affinity. The resonance frequency changes upon mass deposition on the crystal surface and the viscoelastic properties can be analysed using the Voigt model.[47] Mass changes compared to the base line (after washing) further demonstrated the strong affinity of Pep-1 to bind to HA, the deposition mass of which increased sharply after HA injection (ESI, Fig. S6†). The bond formation between S and Au was confirmed by monitoring the $S2p_{3/2}$ binding energy, which was obtained by XPS.[48–50] The S2p spectra of the HS-Pep-1 modified gold surface showed two peaks at 161.9 eV and 163.1 eV, assigned to bound S atoms ($S2p_{3/2}$ and $S2p_{1/2}$), and a peak at 163.9 eV, corresponding to unbound thiols (Fig. 4C1). The position of the $S2p_{3/2}$ peak for the ScPep-1 modified gold surface was at 161.5 eV for bound thiolate and 163.4 eV for unbound thiol. The signal for unbound thiol on the ScPep-1 SAMs was smaller than the signal for Pep-1 SAMs. Different concentrations of HS-Pep-1 (0.01–1.5 mM) were tested in order to investigate the coverage of the gold surfaces and density of the SAMs (Fig. 4C2; ESI, Table S4†). When 0.01 mM of HS-Pep-1 was used, the sulfur composition was very low (%S/% Au $\sim$ 0%). However, the sulfur composition increased with increasing peptide concentration (8.27 $\pm$ 2.96% for 0.1 mM HS-Pep-1 and 9.24 $\pm$ 3.04% for 0.5 mM HS-Pep-1), with a decrease in the gold signal, indicating that the SAMs were more densely packed. For higher concentrations (1 mM and 1.5 mM) of HS-Pep-1, no significant changes in the sulfur composition were observed, suggesting saturation of the surface from 0.5 mM HS-Pep-1. XPS was also used to confirm that the binding of sulfur was responsible for the formation of SAMs on the gold surface. No sulfur was detected for bare Au in the XPS survey (ESI, Fig. S8†). Comparisons of the theoretical elemental composition of the peptides with the elemental percentage composition of the peptide SAMs formed on the surface obtained by XPS (ESI, Fig. S7†) show good correlation, further indicating the successful formation of the peptide SAMs on the Au surface.

Taken together, the results from the QCM-D analysis and XPS characterization showed the attachment of HS-Pep-1 on bare Au through a bond between sulfur and Au. Moreover, the HA binding affinity of Pep-1 was confirmed by frequency and dissipation shifts, and the deposition of HA led to more hydrophilic surfaces.

Cell culture studies

The effect of HA length on a culture of endothelial cells has been investigated in several studies,[51,52] but mainly using HA in solution (added to the culture medium). Covalent immobilization of HA on solid surfaces[53,54] has also been investigated, but the methods used require chemical modification of the HA. Using the Pep-1 SAM described in the previous sections, we have studied the effect of HA molecular weight on HUVECs, where HA is presented at surfaces in its native form without covalent immobilization. Through *in vitro* cell spreading experiments, low molecular weight HA (5 kDa and 60 kDa HA) was shown to stimulate cell spreading with higher cell surface areas (Fig. 5A). Cells cultured on substrates without HA (bare Au, Pep-1, and ScPep-1 SAMs) showed an advantage

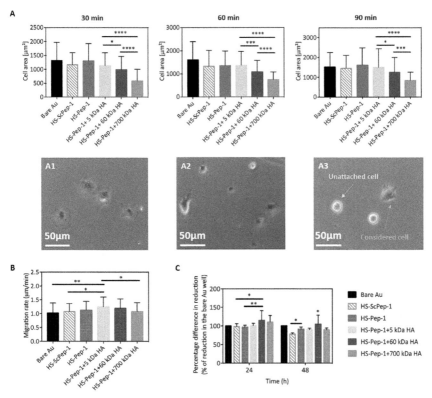

Fig. 5 (A) Spreading of HUVECs during different time periods when cultured on Pep-1-HA and HS-ScPep-1 SAMs coated with HA of different molecular weights. (A1–A3): cell spreading morphology when seeded on Pep-1 SAMs coated with 5, 60, and 700 kDa HA, respectively, at 60 min post-seeding. The green arrow points to cells considered for the calculation of cell area and the yellow arrow points to cells not considered for the calculation. (B) Migration and (C) viability of HUVECs cultured on the prepared surfaces.

regarding spreading in the first 30 min, but this advantage gradually disappeared after 60 min of culturing. 700 kDa HA slowed down the attachment of cells, with fewer attached cells observed at all time points, suggesting a suppression of cell adhesion and a significant reduction of cell surface areas. For example, after culturing for 60 min, the area of cells seeded on the 5 kDa HA-modified surfaces was $1375.72 \pm 597.13\ \mu m^2$, that of the 60 kDa HA-modified surfaces was $1091.05 \pm 492.69\ \mu m^2$, and that of the 700 kDa HA-modified surfaces significantly dropped to $751.24 \pm 336.00\ \mu m^2$ (Fig. 5A1–A3). Low molecular weight HA (5 kDa) also resulted in a noticeable enhancement in cell migration (Fig. 5B), with the highest migration rate at $1.25 \pm 0.35\ \mu m\ min^{-1}$. However, cells cultured on 700 kDa HA-modified surfaces had a slow migration rate of $1.07 \pm 0.32\ \mu m\ min^{-1}$, with only the cells on bare Au having a slower rate.

These results were consistent with the literature reporting that low molecular weight HA can stimulate cell motility, while high molecular weight HA inhibits it.[53] In the cell viability assay, cells seeded on bare Au were used as control cells, and the percentage difference between treated (seeded on peptide SAMs with and without HA) and control cells was calculated (Fig. 5C). Cells seeded on the Pep-1

SAM surfaces with 60 kDa HA showed the highest viability, with $115.12 \pm 26.00\%$ at 24 hours and $104.63 \pm 24.30\%$ at 48 hours, compared to the control cells.

4. Conclusion

In summary, we describe the development of self-assembled monolayers on gold using a HA-binding peptide (Pep-1) as a platform to mimic the function of the endothelial glycocalyx. For that, Pep-1 bearing an N-terminal thiol group was successfully synthesized. Water contact angle measurements indicated that surfaces modified using HS-Pep-1 and HA were more hydrophilic. QCM-D monitoring further demonstrated the strong affinity of Pep-1 to bind HA when immobilized on a solid surface. XPS showed that most of the sulfur atoms on the gold surface were bound thiolate species for both the Pep-1 and ScPep-1 SAMs. μCP enabled spatial control over HA localization. Cell culture experiments with HUVECs demonstrated that smaller sized HA (5 kDa and 60 kDa HA) stimulated improved cell spreading, migration, and viability when compared to high molecular weight HA. We expect that the knowledge obtained from these studies will take us a step closer to developing new HA-based biomaterials as potential therapeutic solutions for vascular diseases.

Conflicts of interest

There are no conflicts to declare.

Acknowledgements

Xinqing Pang acknowledges the School of Engineering and Materials Science at Queen Mary University of London for her PhD scholarship. Weiqi Li thanks the European Commission for his post-doctoral funding. Yichen Yuan thanks the China Scholarship Council for her PhD Scholarship (No. 201706630005). The authors thank Dominic Collis and Clare O'Malley for useful discussions relating to this project.

References

1 J. R. E. Fraser, T. C. Laurent and U. B. G. Laurent, *J. Intern. Med.*, 1997, **242**, 27–33.

2 B. P. Toole, *Nat. Rev. Cancer*, 2004, **4**, 528–539.

3 B. V. Nusgens, *Ann. Dermatol. Venereol.*, 2010, **137**(suppl. 1), S3–S8.

4 R. O. Hynes, *Science*, 2009, **326**, 1216–1219.

5 Y. Zeng, M. Waters, A. Andrews, P. Honarmandi, E. E. Ebong, V. Rizzo and J. M. Tarbell, *Am. J. Physiol. Heart Circ. Physiol.*, 2013, **305**, H811–H820.

6 R. Lindner and H. Y. Naim, *Exp. Cell Res.*, 2009, **315**, 2871–2878.

7 C. C. Michel and F. E. Curry, *Physiol. Rev.*, 1999, **79**, 703–761.

8 J. R. Levick and C. C. Michel, *Cardiovasc. Res.*, 2010, **87**, 198–210.

9 J. M. Tarbell, *Cardiovasc. Res.*, 2010, **87**, 320–330.

10 H. Vink and B. R. Duling, *Circ. Res.*, 1996, **79**, 581–589.

11 H. H. Lipowsky, *Ann. Biomed. Eng.*, 2012, **40**, 840–848.

12 T. Pavicic, G. G. Gauglitz, P. Lersch, K. Schwach-Abdellaoui, B. Malle, H. C. Korting and M. Farwick, *J. Drugs Dermatol.*, 2011, **10**, 990–1000.

13 E. Gatto and M. Venanzi, *Polym. J.*, 2013, **45**, 468.

14 G. M. Whitesides, J. K. Kriebel and J. C. Love, *Sci. Prog.*, 2005, **88**, 17–48.

15 M. Mrksich and G. M. Whitesides, *Annu. Rev. Biophys. Biomol. Struct.*, 1996, **25**, 55–78.

16 J. C. Love, L. A. Estroff, J. K. Kriebel, R. G. Nuzzo and G. M. Whitesides, *Chem. Rev.*, 2005, **105**, 1103–1169.

17 F. Schreiber, *Prog. Surf. Sci.*, 2000, **65**, 151–256.

18 D. K. Schwartz, *Annu. Rev. Phys. Chem.*, 2001, **52**, 107–137.

19 M. J. Pellerite, T. D. Dunbar, L. D. Boardman and E. J. Wood, *J. Phys. Chem. B*, 2003, **107**, 11726–11736.

20 R. G. Nuzzo and D. L. Allara, *J. Am. Chem. Soc.*, 1983, **105**, 4481–4483.

21 J. Christopher Love, D. B. Wolfe, R. Haasch, M. L. Chabinyc, K. E. Paul, G. M. Whitesides and R. G. Nuzzo, *J. Am. Chem. Soc.*, 2003, **125**, 2597–2609.

22 C. D. Bain, E. B. Troughton, Y. T. Tao, J. Evall, G. M. Whitesides and R. G. Nuzzo, *J. Am. Chem. Soc.*, 1989, **111**, 321–335.

23 R. H. Terrill, T. a. Tanzer and P. W. Bohn, *Langmuir*, 1998, **14**, 845–854.

24 N. Leventis and Y. C. Chung, *J. Electrochem. Soc.*, 1991, **138**, L21.

25 C. M. Whelan, M. R. Smyth, C. J. Barnes, N. M. D. Brown and C. a. Anderson, *Appl. Surf. Sci.*, 1998, **134**, 144–158.

26 F. Patolsky, M. Zayats, E. Katz and I. Willner, *Anal. Chem.*, 1999, **71**, 3171–3180.

27 F. Höök, Development of a Novel QCM Technique for Protein Adsorption Studies, PhD thesis, Department of Biochemistry and Biophysics and Department of Applied Physics, Chalmers University of Technology and Göteborg University, 1997.

28 S. H. Brewer, A. M. Allen, S. E. Lappig, T. L. Chasse, K. A. Briggman, C. B. German and S. Franzen, *Langmuir*, 2004, **20**, 5512–5520.

29 T. Wink, S. J. van Zuilen, A. Bult and W. P. van Bennekom, *Analyst*, 1997, **122**, 43R–50R.

30 D. Nedelkov and R. W. Nelson, *Trends Biotechnol.*, 2003, **21**, 301–305.

31 M. Balcells, D. Klee, M. Fabry and H. Höcker, *J. Colloid Interface Sci.*, 1999, **220**, 198–204.

32 A. K. Nowinski, F. Sun, A. D. White, A. J. Keefe and S. Jiang, *J. Am. Chem. Soc.*, 2012, **134**, 6000–6005.

33 M. Mrksich, *Acta Biomater.*, 2009, **5**, 832–841.

34 M. E. Mummert, M. Mohamadzadeh, D. I. Mummert, N. Mizumoto and A. Takashima, *J. Exp. Med.*, 2000, **192**, 769–779.

35 Y. N. Xia and G. M. Whitesides, *Angew. Chem., Int. Ed.*, 1998, **37**, 550–575.

36 D. Qin, Y. N. Xia and G. M. Whitesides, *Nat. Protoc.*, 2010, **5**, 491–502.

37 D. H. Jiang, J. R. Liang, J. Fan, S. Yu, S. P. Chen, Y. Luo, G. D. Prestwich, M. M. Mascarenhas, H. G. Garg, D. A. Quinn, R. J. Homer, D. R. Goldstein, R. Bucala, P. J. Lee, R. Medzhitov and P. W. Noble, *Nat. Med.*, 2005, **11**, 1173–1179.

38 J. Gajewiak, S. S. Cai, X. Z. Shu and G. D. Prestwich, *Biomacromolecules*, 2006, **7**, 1781–1789.

39 D. S. Ferreira, A. P. Marques, R. L. Reis and H. S. Azevedo, *Biomater. Sci.*, 2013, **1**, 952–964.

40 B. P. Orner, R. Derda, R. L. Lewis, J. A. Thomson and L. L. Kiessling, *J. Am. Chem. Soc.*, 2004, **126**, 10808–10809.

41 R. Derda, S. Musah, B. P. Orner, J. R. Klim, L. Y. Li and L. L. Kiessling, *J. Am. Chem. Soc.*, 2010, **132**, 1289–1295.

42 S. F. Chen, Z. Q. Cao and S. Y. Jiang, *Biomaterials*, 2009, **30**, 5892–5896.

43 R. McMillan, B. Meeks, F. Bensebaa, Y. Deslandes and H. Sheardown, *J. Biomed. Mater. Res.*, 2001, **54**, 272–283.

44 M. Boncheva and H. Vogel, *Biophys. J.*, 1997, **73**, 1056–1072.

45 L. Y. Li, J. R. Klim, R. Derda, A. H. Courtney and L. L. Kiessling, *Proc. Natl. Acad. Sci. U. S. A.*, 2011, **108**, 11745–11750.

46 Y. Shi, R. Lin, H. Cui and H. S. Azevedo, *Methods Mol. Biol.*, 2018, **1758**, 11–26.

47 M. V. Voinova, M. Rodahl, M. Jonson and B. Kasemo, *Phys. Scr.*, 1999, **59**, 391–396.

48 A. Francesko, D. S. da Costa, P. Lisboa, R. L. Reis, I. Pashkuleva and T. Tzanov, *J. Mater. Chem.*, 2012, **22**, 19438–19446.

49 D. G. Castner, K. Hinds and D. W. Grainger, *Langmuir*, 1996, **12**, 5083–5086.

50 C. Vericat, M. E. Vela, G. Benitez, P. Carro and R. C. Salvarezza, *Chem. Soc. Rev.*, 2010, **39**, 1805–1834.

51 W. Mo, C. Yang, Y. Liu, Y. He, Y. Wang and F. Gao, *Acta Biochim. Biophys. Sin.*, 2011, **43**, 930–939.

52 D. C. West and S. Kumar, *Exp. Cell Res.*, 1989, **183**, 179–196.

53 C. H. Antoni, Y. McDuffie, J. Bauer, J. P. Sleeman and H. Boehm, *Front. Bioeng. Biotechnol.*, 2018, **6**, 1–7.

54 S. Ibrahim, B. Joddar, M. Craps and A. Ramamurthi, *Biomaterials*, 2007, **28**, 825–835.

Faraday Discussions

DISCUSSIONS

Glycan interactions on glycocalyx mimetic surfaces: general discussion

Helena S. Azevedo, Adam B. Braunschweig, Ryan C. Chiechi, Yuri Diaz Fernandez, Jeffrey C. Gildersleeve, Kamil Godula, Laura Hartmann, Yoshiko Miura, Stephan Schmidt, W. Bruce Turnbull and Dejian Zhou

DOI: 10.1039/C9FD90063B

Adam Braunschweig opened discussion of the paper by Yoshiko Miura: Carbohydrate–carbohydrate interactions are poorly understood, even considered by some to be a myth because they are so weak. I was wondering if your materials could exploit or examine carbohydrate–carbohydrate interactions by using them for separations of carbohydrates, which still remain a major challenge.

Yoshiko Miura answered: The multivalent compounds are also applied to the carbohydrate–carbohydrate interaction. First of all, the carbohydrate–carbohydrate interactions are much weaker than sugar–protein interactions. It is still difficult to obtain quantitative binding constants of carbohydrate–carbohydrate interactions. The carbohydrate–carbohydrate interactions are amplified by multivalent glycopolymer compounds and carbohydrate–protein interactions.[1] In my opinion, from the quantitative measurement results, the carbohydrate–carbohydrate interactions are too weak to apply it as a functional material. (The binding constants of the glycopolymer and carbohydrate are in the order of 10^4 M^{-1}, which are too weak.) However, the continuous flow system is a useful method to estimate the carbohydrate–carbohydrate interaction.[2] Investigation of the glycopolymer monolith (porous materials) is still its infancy and should be expanded to new areas including carbohydrate–carbohydrate interactions.

1 K. Matsuura, H. Kitakouji, N. Sawada, H. Ishida, M. Kiso, K. Kitajima and K. Kobayashi, *J. Am. Chem. Soc.*, 2000, **122**, 7406–7407.
2 G. N. Misevic and M. M. Burger, *J. Biol. Chem.*, 1993, **268**, 4922–4929.

Laura Hartmann opened discussion of the paper by Kamil Godula: Did you look at different viruses attaching to the glycocalyx in your model?

Kamil Godula replied: We have not tested any viruses other than influenza A. However, we do see similar behaviour (*i.e.* enhanced clustering in the presence of a bulky glycocalyx) even for multimeric lectins, such as SNA. This suggests broader generality for the biophysical model we propose, which may extend to other viruses.

Laura Hartmann continued: Do you think, with the proposed two-step or gradient-like attachment of lectins or pathogens to the cell's glycocalyx, this can also have an impact on the specificity of these recognition events?

Kamil Godula responded: That is an interesting question and an intriguing possibility. One can imagine that two proteins with similar glycosylation may be targeted differently by a lectin or a pathogen, depending on their localization within the glycocalyx. Evidence suggests, for instance, that the influenza A virus may bind to the peripheral regions of the glycocalyx in an unproductive way and needs to translocate to glycan receptors in a different region of the glycocalyx in order to initiate internalization and infection.

Dejian Zhou opened discussion of the paper by Helena Azevedo: How does the molecular weight of hyaluronic acid affect its binding to the Pep-1 coated surface? Are there correlations between the molecular weight and amount bound?

Helena Azevedo replied: In another study,[1] we have investigated the deposition of hyaluronan with different molecular weights on the Pep-1 SAMs by QCMD, and we observed increasing mass deposition of HA with higher molecular weight.

1 X. Pang, C. O'Malley, J. Borges, M. M. Rahman, D. W. P. Collis, J. F. Mano, I. C. Mackenzie and H. S. Azevedo, Supramolecular Presentation of Hyaluronan onto Model Surfaces for Studying the Behavior of Cancer Stem Cells, *Adv. Biosyst.*, 2019, DOI: 10.1002/adbi.201900017.

Stephan Schmidt commented: Nature employs high molecular weight hyaluronic acid in the extracellular matrix to drive biological functions. How is this possible when cell assays typically show reduced interactions with high molecular weight hyaluronic acid in comparison to low molecular weight hyaluronic acid?

Helena Azevedo replied: That is a very good question. The conditions used in the *in vitro* studies do not fully replicate the *in vivo* scenario. In particular, the use of serum in cell cultures may play a role in how serum proteins interact with hyaluronic acid (HA) of different molecular weights. It may be worth investigating this in more detail.

Bruce Turnbull enquired: How do you interpret the QCMD data in terms of the arrangement of hyaluronan polysaccharide chains at the surface? Are they mostly lying parallel to the surface or extending away from the surface, and do you see different populations of weakly and strongly-bound chains?

Helena Azevedo responded: In another study[1] and using QCMD, we have determined the thickness of the HA layers using the Voigt model. We obtained values of 5–11 nm, indicating that the polymer is mostly lying parallel to the surface and not extending away from the surface. The thickness of a fully extended HA molecule of 1.5 MDa will give a thickness of $\approx$ 3.75 μm.

1 X. Pang, C. O'Malley, J. Borges, M. M. Rahman, D. W. P. Collis, J. F. Mano, I. C. Mackenzie and H. S. Azevedo, Supramolecular Presentation of Hyaluronan onto Model Surfaces for Studying the Behavior of Cancer Stem Cells, *Adv. Biosyst.*, 2019, DOI: 10.1002/adbi.201900017.

In the QCMD experiments, a washing step with saline solution (150 mM NaCl) is always done after the HA deposition. No significant changes in frequency (mass changes) are observed, indicating strongly bound chains.

Bruce Turnbull followed this by asking: What is the individual affinity of your synthetic peptides for hyaluronan?

Helena Azevedo replied: The apparent affinity of Pep-1 to hyaluronan (HA) was reported in the original paper[1] to be in the micromolar range (Kd = 1.65 μM, not very high affinity) using HA-coated beads. We have used surface plasmon resonance (SPR) with Pep-1 immobilized on a gold chip and flowing 20 kDa HA and found similar values (also in the micromolar range, unpublished data).

1 M. E. Mummert, M. Mohamadzadeh, D. I. Mummert, N. Mizumoto and A. Takashima, *J. Exp. Med.*, 2000, **192**, 769–780.

Helena Azevedo continued with the discussion of her paper: Regarding the use of 3D arrays *vs.* 2D arrays, despite the fact our work is based on 2D arrays (gold surfaces functionalized with self-assembled peptide monolayers), we have also attempted to move to 3D, using gold nanoparticles. However, Pep-1 (the peptide used in our work to bind hyaluronan) is quite hydrophobic and promoted aggregation of the gold nanoparticles, requiring further modification of the peptide sequence near the gold surface. Similarly, we are interested in translating these 2D model surfaces to develop 3D hydrogels and recreate tumor micro-environments. Many tumors are rich in hyaluronan (HA) and Pep-1 could be used as a binding motif to capture HA in a 3D hydrogel.

Yuri Diaz Fernandez reopened discussion of the paper by Yoshiko Miura: It is impressive that you can cover a considerable range of pore sizes by just changing the alcohol used during the polymerization process. Could you please comment on the mechanism behind the effect of the alcohol?

Yoshiko Miura responded: The porous glycopolymer was prepared based on the phase separation of the polymer. There are several methods of porous poly-mer preparation. Our current method is polymerization-induced phase separa-tion (PIPS), and Svec and Fréchet reported the preparation of porous polyacrylamide by PIPS in 1997.[1] The monomers have higher solubility before polymerization because the molecular weight of the monomer is small. The gly-copolymer has a lower solubility in the solvent. The solvent used was DMSO and alcohols were used as porogens. Since polyacrylamide is a water soluble polymer, the glycopolymer with polyacrylamide was also water soluble. The polymers were soluble in the polar solvent DMSO, but were not soluble in the less polar alcohol solvent. It was interesting that uniform porous polymers could be prepared with this simple preparation method. Considering the mechanism, there is a possi-bility that the interface affects the polymer phase separation, but it was not affected much due to the polymer structure.

1 S. Xie, F. Svec and J. M. J. Fréchet, *J. Polym. Sci. A*, 1997, **35**, 1013–1021.

Laura Hartmann asked: Can you also use PIPS to create micro- or nano-particles with a porous structure?

Yoshiko Miura responded: The PIPS mechanism is utilized for the preparation of various polymeric materials, including particles. The method is applied not only at the experimental level, but also to the commercial product. For example, see N. Tsujioka, N. Ishizuka, N. Tanaka, T. Kubo and K. Hosoya, *J. Polym. Sci. A.*, 2008, **46**, 3272–3281. Both the bulk and particle forms were obtained.

Jeffrey Gildersleeve continued the discussion of Kamil Godula's paper: Are the effects of crowding on rate and affinity different when a lectin or virus binds a glycolipid *versus* a glycoprotein? For example, glycolipids are much closer to the cell membrane, while glycans on a glycoprotein might be found closer to the external edge/boundary of a cell. In addition, glycolipid mobility can be quite different compared to a glycan on a glycoprotein.

Kamil Godula answered: We have not probed this effect experimentally but my expectation is that this may be the case. We are certainly observing differences in the rate and affinity of influenza A *vs.* lectin binding as a function of glycopolymer size and density in glycan array studies.

Adam Braunschweig addressed all the presenters: As we know, for glycopolymers, binding avidity and signal transduction following binding are responses that are sensitively dependent on backbone stiffness. This is a material property of which the role is notoriously difficult to understand. Could your synthetic systems be used to study how backbone stiffness affects avidity? Also, in what ways may your model polymers not accurately reflect how binding occurs biologically?

Yoshiko Miura responded: This is an important point in the design of glyco-polymers. The physical properties of the ligands are very important for molecular recognition, including in glycopolymer. The design of glycopolymers has been varied, and facile preparation is preferred. We have reported the effect of the physical properties of a glycopolymer nanogel.[1] As Professor Braunschweig suggested, the stiffness of the polymer affects the binding of the glycopolymer and its molecular recognition.

In the glycopolymer, molecular recognition is controlled by enthalpy and entropy, where the enthalpy gain is mainly from the sugar–protein interaction (*i.e.* the hydrogen bonding of sugar–protein). The polymer stiffness affects the molecular recognition due to an entropic effect. Although the comparison between the natural system and the artificial ligand (glycopolymer) is important, only using a limited method can differences be verified. For example, the measurement of the specificity and the binding affinity are representative examples. It is difficult to know the accuracy of the artificial glycopolymer system. However, if various glycopolymers with different multivalency and physical properties are synthesized and examined, the findings obtained from these materials should be considered as models to represent phenomena occurring in nature.

1 Y. Hoshino, M. Nakamoto and Y. Miura, *J. Am. Chem. Soc.*, 2012, **134**, 15209–15212.

Helena Azevedo added: We have synthesized hyaluronan (HA) glycopolymers whereby HA sugars were grafted on a synthetic rigid polymer backbone (manuscript in preparation). We have performed small angle X-ray scattering (SAXS) on solutions of the glycopolymers and their scattering patterns differ from those of native HA, which exhibits a pattern typical of salt-free poly-electrolyte solutions. We have also studied the binding of the HA glycopol-ymers to CD44, the major receptor for HA, using surface plasma resonance (SPR) and observed binding to CD44. I believe glycopolymers can be used as probes to dissect the role of backbone stiffness and monosaccharide composition/presentation in protein–carbohydrate interactions of natural systems.

Ryan Chiechi returned to the discussion of Kamil Godula's paper: How is the post-polymerization chemistry of the poly(epichlorohydrin)s characterized? The reaction schemes state that the yields are quantitative, however, there will be some fraction of missing pendant groups. Due to the nature of the polymers and that their effects are studied at the level of single polymer chains, might these vacancies be important?

Kamil Godula replied: The extent of glycosylation is determined by ^{1}H-NMR and by IR analysis for the conversion of pendant azide chains. We observed complete disappearance of the characteristic azide stretch in the IR spectra of glycopolymers, which we interpreted as complete conversion, seeing as we can detect the presence of even a single end-chain azide group in the precursor polymer backbone. The level of glycosylation will influence the physical proper-ties of the polymers (*i.e.* persistence length, stiffness, *etc.*)

Ryan Chiechi said: Polymer chains decorated with sugars were described as not rod like and yet not globular. Have their dynamics been modeled in detail?

Kamil Godula answered: Some studies on the effects of glycosylation on biopolymer stiffness and persistence length have been carried out in the past, particularly in the context of mucin glycoproteins and, recently, also in synthetic NCA glycopolypeptide polymers. The ability to model and predict these properties in synthetic polymers would be great to have in order to guide glycopolymer design.

Ryan Chiechi commented: When discussing the self-assembly of, presumably, very sterically crowded polymers at the surfaces of cells and lipid bilayers, the mechanistic explanations for their effects were very mechanical and invoked cartoons of macromolecules literally getting in each other's way (DOI: 10.1039/c9fd00024k). It would therefore seem that bulk descriptors, such as Young's modulus, do not capture the salient mechanical properties of the individual polymer chains. What is the best way to describe these properties?

Kamil Godula answered: I think a recent paper by Paszek, extending the physical principles applied to surface-bound polymer brushes, provides a nice quantitative framework for the mucinous glycocalyx.[1]

1 C. R. Shurer, J. C.-H. Kuo, L. M. Roberts, J. G. Gandhi, M. J. Colville, T. A. Enoki, H. Pan, J. Su, J. M. Noble, M. J. Hollander, J. P. O'Donnell, R. Yin, K. Pedram, L. Möckl, L. F. Kourkoutis, W. E. Moerner, C. R. Bertozzi, G. W. Feigenson, H. L. Reesink and M. J. Paszek, *Cell*, 2019, **177**, 1757–1770.

Jeffrey Gildersleeve remarked: You describe some interesting effects of crowding on the rates of binding and affinities. Have you evaluated the effects of proteins with significant differences in size, such as IgG *versus* IgM antibodies? Do you think crowding will influence the recognition of cells by different isotypes of antibodies?

Kamil Godula replied: We have not compared proteins of different sizes but we did compare lectins to whole influenza A viruses. The effects on virus binding are more pronounced in the presence of an extended glycocalyx (long polymers), while lectins are influenced more by the glycocalyx density (short polymers).

Conflicts of interest

There are no conflicts to declare.

Faraday Discussions

PAPER

Binary polymer brush patterns from facile initiator stickiness for cell culturing†

Lina Chen,[a] Peng Li,[a] Xi Lu,[a] Shutao Wang [b] and Zijian Zheng [*a]

Received 5th February 2019, Accepted 8th March 2019

DOI: 10.1039/c9fd00013e

We report a new initiator stickiness method to fabricate micropatterned binary polymer brush surfaces, which are ideal platforms for studying cell adhesion behavior. The atom transfer radical polymerization (ATRP) initiator, ω-mercaptoundecyl bromoisobutyrate (MUDBr), is found to adsorb on several hosting polymer brushes, including poly [oligo(ethylene glycol)methyl ether methacrylate] (POEGMA), poly(2-hydroxyethyl methacrylate) (PHEMA), and poly(glycidyl methacrylate) (PGMA) brushes. Based on the initiator stickiness, micropatterned initiator molecules are printed onto a layer of homogenous hosting polymer brushes via microcontact printing (µCP), and then, vertically, a patterned second layer of polymer brushes is grown from the initiator areas. With this simple, fast, and additive method, we demonstrate the fabrication of various binary polymer brushes, and show their applications for patterning cell microarrays and controlling cell orientation. This new approach to generating binary polymer brushes shows great potential for the manipulation of interfacial phenomena, facilitating a range of applications from semiconductors and lubrication to fundamental cell biology studies.

1. Introduction

The *in vivo* cellular microenvironment is the local surroundings of the cells, such as the extracellular matrix (ECM) and neighbouring cells.[1] Living cells sense and respond to the microenvironment through adhesion, protrusions, and intracellular forces, all of which are fundamental regulators of key cellular functions, such as migration, proliferation, and apoptosis.[2] In order to understand these complex processes, living cell assays have been developed that allow the delivery of experimental stimuli to cells and the measurement of the resulting cellular responses.[3–6] Among these assays, significant focus is given to cell microarrays due to their high-throughput adaptations.[7–9] A cell microarray is a laboratory tool

[a]Laboratory for Advanced Interfacial Materials and Devices, Institute of Textiles and Clothing, The Hong Kong Polytechnic University, Hong Kong SAR, P. R. China. E-mail: tczzheng@polyu.edu.hk

[b]CAS Key Laboratory of Bio-inspired Materials and Interfacial Science, CAS Center for Excellence in Nanoscience, Technical Institute of Physics and Chemistry, Chinese Academy of Sciences, Beijing 100190, P. R. China

† Electronic supplementary information (ESI) available. See DOI: 10.1039/c9fd00013e

that consists of the multiplex interrogation of small volumes of different biomolecules and cells on a solid support. With many revolutionary break-throughs in recent research in cell biology and regenerative medicine, it is well-recognized that cell microarrays can control the interfaces between biomaterials, and provide scientific insight into subcellular adhesion localization and the consequences of a series of cellular activities. *Via* cell microarrays, researchers can efficiently study cell sensitivity and the response to specific environmental cues,[2,10] which is extremely important in the development of biosensors, bioreactors, bioassays, tissue engineering, cell-based drug screening, and fundamental cell biology studies.[11–14]

The selective adhesion of cells on cell-adhesive micropatterns is a very effective method to generate cell microarrays.[6,15] The fabrication typically starts with the surface patterning of chemical or topological structures that can mimic ECM and lead to the selective adhesion of cells. Compared with self-assembled monolayers and thin films, surface-tethered polymer brushes[16–19] are particularly effective materials for surface modification due to their excellent environmental stability, as well as the versatility of regulating the surface hydrophilicity, surface charge, chemical reactivity, protein adhesion, mechanical properties, and surface topography.[20] In the past two decades, advances in various patterning technologies, including scanning probe lithography (SPL),[21] microcontact printing (μCP),[19] electron beam lithography (EBL),[22,23] and photolithography,[24] have enabled the precise construction of polymer brush patterns for cell microarrays. For example, Ober and co-workers[25] used cell-adhesive poly[2-(methacryloyloxy)ethyl] trime-thylammonium chloride (PMETAC) brush micropatterns separated by cell-inert short polyethylene glycol (PEG) chains by photolithography for spatially controlling the growth of rat hippocampal neurons. Burcu Colak *et al.*[26] synthesized cell-inert alkene-functionalized antifouling POEGMA brushes and coupled cell-adhesive peptides *via* thiol–ene bonds to study cell adhesion and spreading, and to form cell arrays. Huck's group[27] used μCP to prepare ECM protein micropatterns that were surrounded by POEGMA brushes for directing the spreading of single cells.

It is known that a pattern of binary polymer brushes with distinctive cell-adhesive and cell-inert contrast is more effective than a single-component poly-mer brush pattern.[28] However, only limited work on using binary polymer brushes for cell micropatterning has been reported, which is largely due to the difficulty in making the binary polymer brushes through a standard multi-step lithographic process. Typically, a 7-step process of photolithography and surface-initiated polymerization is needed to construct a binary polymer brush pattern.[29–34] On the other hand, Zhou and co-workers[35] reported a much easier method based on μCP to design multicomponent polymer brush micropatterns. They printed a thiol-functional initiator on gold and grew a first polymer brush. Then, the blank area was backfilled with the initiator to grow a second polymer brush.

Herein, we report a new initiator stickiness phenomenon that allows for the rapid fabrication of binary cell-adhesive and cell-inert polymer brushes through a simple two-step printing and surface-initiated atom transfer radical polymeri-zation (SI-ATRP) process. It is found that the ATRP initiator ω-mercaptoundecyl bromoisobutyrate (MUDBr) can stick efficiently on several polymer brushes, such as POEGMA, PHEMA, and PGMA brushes, through simple μCP. Sequent SI-ATRP could yield a second layer of patterned polymer brushes, which are vertically grown on and firstly adhered to the underlying layer. By rationally designing a suitable polymer combination, we readily fabricate a binary cell-adhesive/inert polymer surface for generating a cell microarray. As a proof-of-concept, human cervical carcinoma (HeLa) and NIH-3T3 cell micropatterning are demonstrated on the prepared binary polymer brush patterns.

2. Results and discussion

We prepared various polymer brush layers on Au substrates *via* SI-ATRP, including hydrophilic polymer brushes (PEGMA and PHEMA), comparatively hydrophobic polymer brushes [PNIPAm, PGMA, and poly(methyl methacrylate) (PMMA)], and polyelectrolytes [PMETAC, poly(methacrylic acid, sodium salt) (PMANa), and poly(3-sulfopropyl methacrylate) potassium salt (PSPMAK)]. The water contact angles of PMMA, POEGMA, PNIPAm, PGMA, and PHEMA at room temperature were 80°, 45°, 70°, 77°, and 46°, respectively. As shown in Fig. S1,† an attenuated total reflectance-Fourier transform infrared (ATR-FTIR) spectrometer was used to detect the surface chemistry of the various polymer brush coated Au substrates. A clear peak representing C=O stretching of the carboxyl group at 1717 cm^{-1} was detected, indicating the successful binding of the MUDBr initiator on the Au surface for the subsequent SI-ATRP. Furthermore, after SI-ATRP, the characteristic peaks of the POEGMA, PGMA, PNIPAm, PMMA, PMETAC, PHEMA, PSPMK, and PMANa brushes were observed at 1730 cm^{-1} (C=O stretching),[36] 907 cm^{-1} (C–O–C of epoxy group), 1565 cm^{-1} (C–N bond), 1736 cm^{-1} (C=O stretching),[37] 1473 cm^{-1} [(CH$_3$)$_4$N$^+$ bond bending vibration],[38] 3348 cm^{-1} (O–H bond),[39] 1248 and 1048 cm^{-1} (O–S bond), and 1710 cm^{-1} (C=O stretching), respectively, indicating successful polymerization. These polymer brush layers were deactivated by azide to prevent further polymerization.

Then, µCP was used to transfer the MUDBr initiator micropatterns on these polymer brushes surfaces (Fig. 1). µCP is a soft lithographic technique that uses the relief patterns on an elastomeric polydimethylsiloxane (PDMS) stamp to

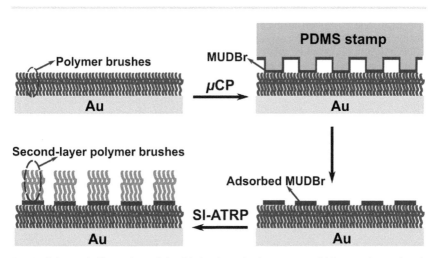

Fig. 1 Schematic illustration of the fabrication of micropatterned binary polymer brush systems by the initiator stickiness method.

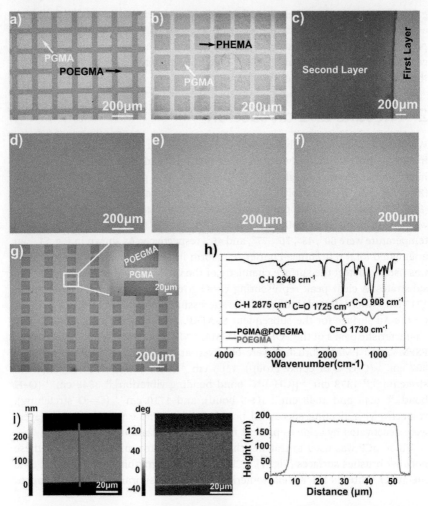

Fig. 2 (a–f) Optical microscopic images of (a) POEGMA, (b) PHEMA, (c) PGMA, (d) PMMA, (e) PMETAC, and (f) PNIPAm brush coated Au substrates after microcontact printing of the initiator and the SI-ATRP process. (g) Optical microscopic images of a PGMA grid on a POEGMA surface, called PGMA@POEGMA. (h) ATR-FTIR spectra of a POEGMA brush coated Au substrate and PGMA@POEGMA sample. (i) AFM topographic view, corresponding phase view, and cross-sectional profile of the PGMA grid on the POEGMA surface.

fabricate patterns on a substrate surface by conformal contact. Briefly, a MUDBr initiator-inked PDMS stamp was brought into contact with the various polymer brush surfaces for 60 s, where MUDBr micropatterns were formed on those polymer brushes that were sticky to the MUDBr (such samples were named MUDBr@polymer). In order to provide evidence of the initiator stickiness, we studied the X-ray photoelectron spectroscopy (XPS) spectra of the samples (Fig. 3). After printing the MUDBr patterns on the POEGMA brush surface, an evident S_{2p} peak located at 163.5 eV for the mercapto group[40] of the MUDBr molecules was detected (Fig. 3a), whereas there were no clear S_{2p} peaks in the S_{2p} spectra of the

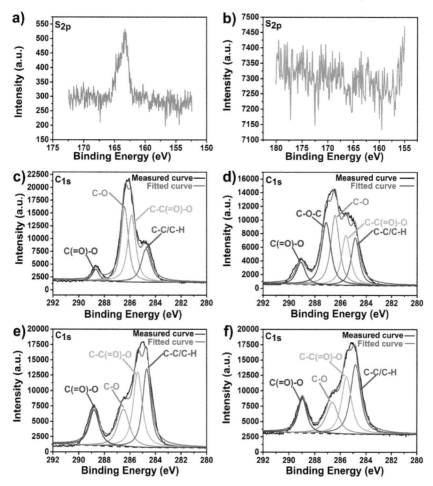

Fig. 3 The XPS S$_{2p}$ spectra for (a) MUDBr@POEGMA and (b) MUDBr@PMMA. Decomposition of the C$_{1s}$ peak of (c) POEGMA brushes, (d) PGMA@POEGMA binary brushes, (e) PMMA brushes, and (f) PGMA@PMMA.

MUDBr@PMMA sample. This indicates that, during the conformal contact printing process, the MUDBr was successfully transferred to the POEGMA brushes, but not to the PMMA. These initiator molecules, once transferred, were very stable even under ultrasonication.

Finally, SI-ATRP was performed again to grow the second layer of micropatterned polymer brushes, in order to form a binary polymer brush surface. The results were observed by optical microscopy (Fig. 2a–f). It is clearly seen that the PGMA brushes formed squares on the POEGMA brush surface (PGMA@POEGMA, Fig. 2a), the PGMA brushes formed a grid on the PHEMA brush surface (PGMA@PHEMA, Fig. 2b), and second-layer PGMA brushes formed on the PGMA brush surface (PGMA@PGMA, Fig. 2c). However, no brush was grown from the PMMA (Fig. 2d), PMETAC (Fig. 2e), and PNIPAm (Fig. 2f) brush surfaces. We studied the C$_{1s}$ spectra of the PGMA@POEGMA and PGMA@PMMA samples in more detail. Compared with pure POEGMA brushes, a clear peak representing the

C–O–C bond (287.1 eV)[41] of the epoxy group of the PGMA chains was detected from the PGMA@POEGMA sample (Fig. 3c and d), indicating the successful growth of PGMA brushes on the POEGMA brush surface. However, the C_{1s} spectra of the pure PMMA brushes[42] and the PGMA@PMMA sample had no significant differences (Fig. 3e and f), indicating that no PGMA brushes were grown on the PMMA sample. The ATR-FTIR spectrum also confirmed the successful growth of the polymer. Taking PGMA@POEGMA as an example (Fig. 2h), the C–O stretching of the epoxy group at 908 cm^{-1} is evidence of the grafting of the PGMA brushes. AFM characterization (Fig. 2i) showed that the PGMA microgrids on the POEGMA brush surface were spatially well-controlled with a 175 nm thick additional POEGMA layer.

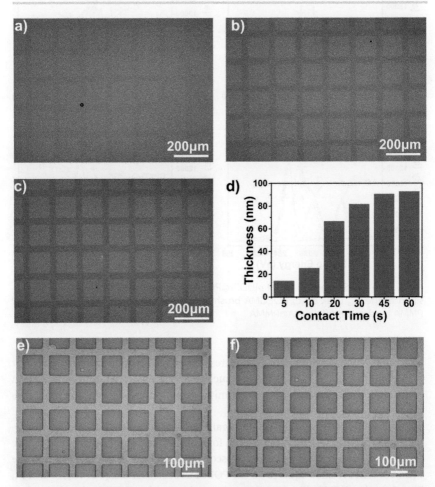

Fig. 4 (a–c) Optical microscope images of the patterned PGMA brush grid following different contact times of the PDMS stamp on the POEGMA brush coated Au substrate: (a) 5 s, (b) 20 s, and (c) 45 s. (d) The thickness of the PGMA brushes on the POEGMA surfaces with changing contact time. The thickness was measured by AFM in non-contact mode. (e and f) The stability of the binary PGMA@POEGMA substrate in the mix solution (DI water : DMF : ZnCl$_2$ = 3 : 5 : 2, w/w/w) with ultrasonication: (e) 10 min of ultrasonication and (f) 40 min of ultrasonication.

These results indicate that the binary polymer brush systems could be successfully prepared by our method on POEGMA, PGMA, and PHEMA brush surfaces, *i.e.* the POEGMA, PGMA, and PHEMA brushes had MUDBr initiator stickiness. On the other hand, the surfaces of other polymer brushes (PNIPAm, PMMA, PMETAC, PSPMK, and PMANa) did not have vertical micropatterns after μCP and the second SI-ATRP, meaning that these polymer brushes had very low or little initiator stickiness. At this moment, we are still not able to clearly conclude what the course of the selective initiator stickiness is. It is very likely to be associated with the hydrogen bond interaction between MUDBr and the polymer brushes.

It is also noted that the contact time during the μCP process affects the quantity of adsorbed MUDBr on the three kinds of polymer brush surfaces. As shown in Fig. 4a–d, the thickness of the PGMA grids on the POEGMA brush surface increased from 14 nm to above 90 nm with an increase in contact time (5 s to 60 s). Longer contact time resulted in more MUDBr adsorption. Importantly, the PGMA squares on the POEGMA brush surface were very robust, even when immersed into a mix solution (DI water : dimethyl formamide : $ZnCl_2$ = 3 : 5 : 2, w/w/w), which can destroy hydrogen bonds, and ultrasonicated for 40 min (Fig. 4e and f). Because the intertwinement of the two types of polymer chain was very strong, the PGMA squares still attached on the POEGMA surface were stable, even when the hydrogen bonds in the binary system were broken. This may be attributed to the intertwinement of the two types of polymer chain. Longer

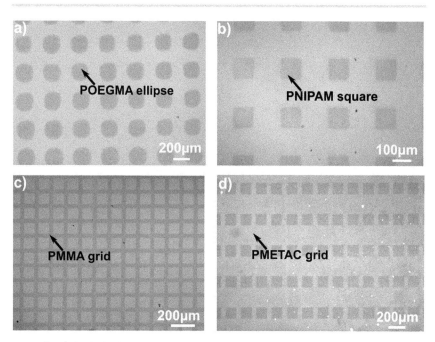

Fig. 5 (a–d) Optical microscope images of the various binary systems. (a) A POEGMA dot array on a POEGMA surface, called POEGMA@POEGMA, (b) a PNIPAm square array on a POEGMA surface, called PNIPAm@POEGMA, (c) a PMMA grid on a POEGMA surface, called PMMA@POEGMA, and (d) a PMETAC grid on a POEGMA surface, called PMETAC@POEGMA.

contact time leads to higher surface density of the initiator molecules, which leads to a more stretched polymer brush. Furthermore, we have used this simple method to fabricate binary POEGMA@POEGMA, PMMA@POEGMA, PNI-PAm@POEGMA, and PMETAC@POEGMA brush systems (Fig. 5a–d) successfully.

In order to demonstrate the application for patterning cell microarrays, we chose the binary PGMA@POEGMA surface. It is known that the POEGMA brushes are antifouling materials,[43] so they act as the cell-inert part in the binary system. On the other hand, although our previous experiments proved that the pure PGMA brushes were cell-inert,[21] the epoxide groups of the PGMA chains react very easily with the amino and carboxyl groups of proteins.[44,45] Hence, we modified the PGMA brushes with gelatin proteins, which facilitate cell adhesion, resulting in a cell-adhesive transformation of gelatin-PGMA. HeLa and NIH-3T3 cells were cultured on the gelatin-PGMA@POEGMA (micropatterned cell-adhesive gelatin-PGMA brushes separated by cell-inert POEGMA brushes) surface. After 24 h incubation, the HeLa cells selectively adhered on the gelatin-PGMA grid, forming a cell micropattern (Fig. 6b) that remained stable after 48 h (Fig. 6f). Inspired by cell orientation studies on microstripes,[31] we also designed 100, 40, and 5 μm wide gelatin-PGMA stripe patterns on the POEGMA brush surfaces (Fig. 7a–c). The protruded thicknesses of these stripes on the POEGMA layer were about 8 nm, 20 nm, and 45 nm, respectively. After being incubated for 24 h on these micro-striped gelatin-PGMA@POEGMA substrates, clear NIH-3T3 cell microstripes were formed on the 100 μm and 40 μm gelatin-PGMA stripe patterns, but not on the 5 μm microstripes. However, the cell orientation was well controlled on the 5 μm microstripes (Fig. 7d and g). This could be attributed to the fact that 5 μm was

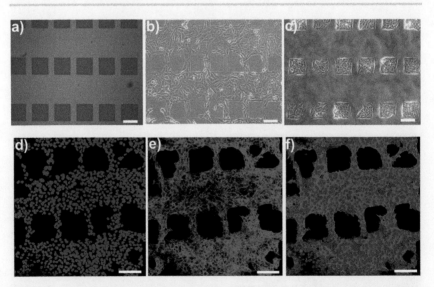

Fig. 6 (a–c) Optical microscope images of (a) the gelatin-modified PGMA@POEGMA substrate, called gelatin-PGMA@POEGMA, and (b and c) the cell micropatterning of HeLa cells on gelatin-PGMA grids on the POEGMA surface and gelatin-PGMA squares on the POEGMA surface for 24 h. (d–f) CLSM images of HeLa cells incubated on the gelatin-PGMA@POEGMA surface for 48 h. Red and blue colors represent the cell cytoskeleton and nucleus, respectively. The scale bars are 100 μm.

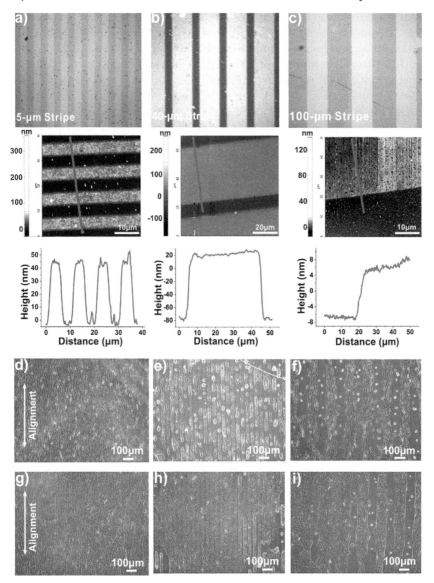

Fig. 7 (a–c) Optical images, AFM topography, and cross-section profiles of gelatin-PGMA stripes (5, 40, and 100 μm width) on POEGMA-glass. (d–i) Optical images of NIH-3T3 cells cultured on gelatin-PGMA stripes (5, 40, and 100 μm width) for 24 h (d–f) and 48 h (g–i).

slightly less than the cell's width, which applied a suitable amount of focal adhesion and was greatly beneficial to the regulation of cell behavior.

3. Conclusions

In conclusion, we have reported, for the first time, that ATRP initiator can be absorbed by several cell-inert polymer brushes, including POEGMA, PHEMA, and PGMA. By utilizing this initiator stickiness ability, MUDBr micropatterns could be readily made on a continuous layer of POEGMA, PHEMA, or PGMA brush by μCP. With subsequent SI-ATRP, a second layer of patterned polymer brushes, *e.g.* PGMA, was grown from the underlying polymer brush to form binary polymer brushes. Gelatin was further reacted with PGMA to produce a binary cell-adhesive (gelatin-PGMA) and cell-inert (POEGMA) surface. The binary surfaces with different pattern sizes were successfully applied to make cell microarrays and control the cell orientation. This initiator stickiness method is a very simple method to create micro- and nanoscale multicomponent surface patterns that are essential for a range of purposes, such as fundamental cell biology, making biosensors, and studying nanochemistry. More importantly, it has great potential to be applied to many different substrates due to the fact μCP is directly applied on the first polymer brush, which should overcome the most critical challenge in the previous research.

4. Experimental

Materials

The ATRP initiator MUDBr was kindly provided by Prof. Hongwei Ma, Suzhou Institute of Nano-Tech and Nano-Bionics. The following chemicals were all purchased from the listed distributors: phosphate buffered saline (PBS, pH 7.4, 1×; Gibco™), Dulbecco's Modified Eagle Medium (DMEM with high glucose; Gibco™), trypsin–EDTA (0.5%, Gibco™), fetal bovine serum (FBS, South American origin; Gibco™), penicillin–streptomycin (10 000 U mL^{-1}; Gibco™), 4-(1,1,3,3-tetramethylbutyl)phenyl-polyethylene glycol (Triton™ X-100, M_w 625; Sigma-Aldrich), albumin bovine V (BSA, from bovine serum, M_w 66 000; Sigma-Aldrich), rhodamine phalloidin (Rhod-phalloidin, Cytoskeleton Inc.), and 4',6-diamidino-2-phenylindole (DAPI, M_w 350.25; Sigma-Aldrich). NR9-8000 negative photoresist was brought from Futurrex Inc. All other chemicals were obtained from Sigma-Aldrich and used as received. Gold (Au) substrates were prepared by thermal evaporation of 25 nm Au/5 nm chromium (Cr) on $\langle 100 \rangle$ Si wafers with 500 nm SiO_2 on one side, and 15 nm Au/2 nm Cr on the glass wafer. The non-contact mode AFM tip was purchased from NanoSensors Inc.

Substrate modification

The cleaned Au substrates were simply immersed into ethanol solutions of MUDBr (2.5 mg mL^{-1}) at room temperature for 24 h to form initiator self-assembled monolayers (SAMs). The obtained SAM-modified Au substrates were rinsed with ethanol and dried *via* a nitrogen stream. Then, the MUDBr-modified Au substrates and activators (re)generated by electron transfer (AGET) SI-ATRP[46-48] were employed to graft functional polymer brushes (PGMA, PMMA, POEGMA, PNIPAm, PHEMA, PMANa, PSPMK, and PMETAC brushes) onto the substrate surfaces. Generally, the MUDBr-Au substrates were placed in centrifuge tubes with reaction solutions prepared according to the polymerization recipes as below, and the resultant substrates were then rinsed with methanol, deionized water, acetone, and toluene, dried *via* a nitrogen stream, and then terminated by sodium azide (NaN$_3$) (0.2 g NaN$_3$ dissolved in 30 mL dimethyl formamide) at 50 °C for 6 h. The reaction solution for the PGMA brushes was composed of GMA (7.8 g), methanol (4.8 g), deionized (DI) water (1.5 mL), ascorbic acid (150 mg), copper(ı) bromide (CuBr, 78 mg), and 2, 2′-bipyridyl (bpy, 210 mg). The reaction solution for the PMMA brushes was composed of MMA (5 g), methanol (3.2 g), DI water (1.5 mL), ascorbic acid (150 mg), CuBr (71 mg), and bpy (156 mg). The reaction solution for the POEGMA brushes was composed of OEGMA (4 g, M_w 300), methanol (4.8 g), DI water (1.5 mL), ascorbic acid (150 mg), CuBr (70 mg), and bpy (156 mg). The reaction solution for the PNIPAm brushes was composed of NIPAm (1.5 g), MeOH (12 g), deionized water (15 g), CuBr (0.048 g), PMDETA (0.165 µL), and ascorbic acid (0.2 g). The reaction solution for the PHEMA brushes was composed of HEMA (5 g), methanol (2 g), DI water (2.5 mL), ascorbic acid (150 mg), CuBr (147 mg), and bpy (400 mg). The reaction solution for the PMANa brushes was composed of NaMA (4.5 g), DI water (15 mL), ascorbic acid (150 mg), CuBr (144 mg), and bpy (310 mg). The reaction solution for the PSPMK brushes was composed of SPMAK (6.9 g), methanol (6.4 g), DI water (4 mL), ascorbic acid (150 mg), CuBr (66 mg), and bpy (260 mg). The reaction solution for the PMETAC brushes was composed of METAC (6.9 g, M_w 207.7, 80 wt% in H_2O), methanol (5.6 g), DI water (2 mL), ascorbic acid (200 mg), CuBr (135 mg), and bpy (260 mg). The polymerization time and reaction temperature were 2 h and 40 °C, respectively.

Preparation of vertically patterned binary polymer brushes based on facile initiator stickiness *via* µCP and SI-ATRP

According to the MUDBr initiator stickiness properties of some of the polymer brushes, the MUDBr micropatterns were generated on those polymer brush coated Au substrates *via* a µCP technique. The vertically patterned second-layer polymer brushes were then grown from the MUDBr micropatterns by SI-ATRP to obtain the binary polymer brush systems. A NR9-8000 negative photoresist was used in the PL process to prepare silicon (Si) molds.[49] The µCP process includes the preparation of the PDMS stamp and the ink transfer.

Fabrication of the PDMS stamp. A SYLGARD 184 silicone elastomer kit was used to prepare the PDMS stamp. The base and curing agent were mixed in a 10 (base) : 1 (curing agent) ratio by weight for manual mixing. Then, the mixture was degassed under vacuum, and placed onto the top of the fluorinated Si mold. After being cured at 80 °C overnight, the solidified PDMS stamp was carefully peeled off from the Si mold and washed with ethanol, followed by drying with compressed air.

Ink transfer. A 2.5 mM ethanol solution of MUDBr was coated onto the surface of the PDMS stamp using a cotton swab. After drying, the PDMS stamp was pressed on the polymer brush modified Au substrate for 60 s, and then peeled off and removed. The MUDBr pattern was transferred to the surface of the polymer brush modified Au substrate in the pressing process, and this sample was called MUDBr@polymer. Finally, the second layer of polymer brushes was then grown from the MUDBr@polymer surfaces *via* SI-ATRP and the patterned binary polymer brush structures were obtained. The conditions and reagents for the SI-ATRP reaction are the same as previously mentioned in the substrate modification subsection.

Gelatin immobilization

The PGMA micropatterns on the POEGMA brush coated Au substrate (PGMA@-POEGMA) were immersed in 10 mg mL^{-1} gelatin in PBS at 37 °C for 24 h to immobilize the gelatin. The resultant gelatin-modified PGMA@POEGMA (gelatin-PGMA@POEGMA) substrate was repeatedly washed with PBS solution to remove the unfixed gelatin, and dried *via* a nitrogen stream for cell culture experiments.

Cell microarrays

HeLa and NIH-3T3 cells were incubated in TCPS dishes in high glucose DMEM media with 5% FBS and 1% penicillin–streptomycin, and placed in an incubator with 75% humidity and 5% CO_2 at 37 °C. For the cell behavior studies, the micropatterned gelatin-PGMA@POEGMA was put into 90 mm TCPS dishes and seeded with cells at a density of 3×10^5 cells per mL.

Fluorescence dying of cell micropatterning

Cells on the micropatterned gelatin-PGMA@POEGMA surface were washed with PBS and fixed with 4% paraformaldehyde in PBS for 10 min at room temperature, and then permeabilized with 0.1% Triton™ X-100 in PBS for 3 min at room temperature and washed with PBS. These steps were repeated three times. Next, the permeabilized cells were blocked with 1% BSA in PBS for 20 min. After being washed with PBS, the cells were incubated with 5 μg mL^{-1} of Rhod-phalloidin in a dark place at 37 °C for 1 h. Finally, the cells were rinsed with PBS and treated with 2.5 μg mL^{-1} DAPI for 15 min at 37 °C for nuclei staining.

Characterization

An ATR-FTIR (PerkinElmer Spectrum 100) spectrometer was used to detect the surface chemistry of various polymer brush coated Au substrates and the vertically patterned binary polymer brushes. A drop shape analyzer (EasyDrop, Krüss) was applied to detect the water contact angle of the various polymer brush coated Au substrates. An optical microscope (Nikon Eclipse 80i) with a bright field was employed to observe the binary polymer brush micropatterns and cell micropatterns. AFM topography in air was measured by XE-100 AFM in non-contact mode under ambient conditions. The XPS analysis of these experiments was carried out with a Thermo ESCALAB 250 spectrometer. Confocal laser scanning microscopy (CLSM, Leica TCS SPE) was used to observe the stained cells.

Conflicts of interest

The authors declare no competing financial interests.

Acknowledgements

We acknowledge the financial support of the Hong Kong Polytechnic University (G-YBV9).

References

1 X. Jiang and G. M. Whitesides, *Eng. Life Sci.*, 2003, **3**, 475–480.
2 M. Théry, *J. Cell Sci.*, 2010, **123**, 4201–4213.
3 Y. Fang, A. M. Ferrie, N. H. Fontaine, J. Mauro and J. Balakrishnan, *Biophys. J.*, 2006, **91**, 1925–1940.

4 D. R. Albrecht, V. L. Tsang, R. L. Sah and S. N. Bhatia, *Lab Chip*, 2005, **5**, 111–118.

5 K. R. King, S. Wang, D. Irimia, A. Jayaraman, M. Toner and M. L. Yarmush, *Lab Chip*, 2007, **7**, 77–85.

6 R. Jonczyk, T. Kurth, A. Lavrentieva, J.-G. Walter, T. Scheper and F. Stahl, *Microarrays*, 2016, **5**, 11.

7 J. F. Pascoal, T. G. Fernandes, G. J. Nierode, M. M. Diogo, J. S. Dordick and J. M. Cabral, in *Cell-Based Microarrays*, Springer, 2018, pp. 69–81.

8 X. Yan, L. Zhou, Z. Wu, X. Wang, X. Chen, F. Yang, Y. Guo, M. Wu, Y. Chen and W. Li, *Biomaterials*, 2019, **198**, 167–179.

9 M. B. Oliveira and J. F. Mano, in *Cell-Based Microarrays*, Springer, 2018, pp. 11–26.

10 E. D'Arcangelo and A. P. McGuigan, *BioTechniques*, 2015, **58**, 13–23.

11 J. Nakanishi, T. Takarada, K. Yamaguchi and M. Maeda, *Anal. Sci.*, 2008, **24**, 67–72.

12 J. S. Choi and T. S. Seo, *Biomaterials*, 2016, **84**, 315–322.

13 M. Mrksich and G. M. Whitesides, *Annu. Rev. Biophys. Biomol. Struct.*, 1996, **25**, 55–78.

14 D. Falconnet, G. Csucs, H. M. Grandin and M. Textor, *Biomaterials*, 2006, **27**, 3044–3063.

15 R. G. Willaert and K. Goossens, *Fermentation*, 2015, **1**, 38–78.

16 T. Fujie, H. Haniuda and S. Takeoka, *J. Mater. Chem.*, 2011, **21**, 9112–9120.

17 A. F. Hirschbiel, S. Geyer, B. Yameen, A. Welle, P. Nikolov, S. Giselbrecht, S. Scholpp, G. Delaittre and C. Barner-Kowollik, *Adv. Mater.*, 2015, **27**, 2621–2626.

18 Z. Zhou, P. Yu, H. M. Geller and C. K. Ober, *Biomacromolecules*, 2013, **14**, 529–537.

19 K. Y. Tan, H. Lin, M. Ramstedt, F. M. Watt, W. T. Huck and J. E. Gautrot, *Integr. Biol.*, 2013, **5**, 899–910.

20 S. Edmondson, V. L. Osborne and W. T. Huck, *Chem. Soc. Rev.*, 2004, **33**, 14–22.

21 L. Chen, Z. Xie, T. Gan, Y. Wang, G. Zhang, C. A. Mirkin and Z. Zheng, *Small*, 2016, **12**, 3400–3406.

22 I.-T. Hwang, C.-H. Jung, C.-H. Jung, J.-H. Choi, K. Shin and Y.-D. Yoo, *J. Biomed. Nanotechnol.*, 2016, **12**, 387–393.

23 Q. Yu, L. M. Johnson and G. P. López, *Adv. Funct. Mater.*, 2014, **24**, 3751–3759.

24 Y. Kumashiro, J. Ishihara, T. Umemoto, K. Itoga, J. Kobayashi, T. Shimizu, M. Yamato and T. Okano, *Small*, 2015, **11**, 681–687.

25 R. Dong, R. P. Molloy, M. Lindau and C. K. Ober, *Biomacromolecules*, 2010, **11**, 2027–2032.

26 B. Colak, S. Di Cio and J. E. Gautrot, *Biomacromolecules*, 2018, **19**, 1445–1455.

27 J. E. Gautrot, B. Trappmann, F. Oceguera-Yanez, J. Connelly, X. M. He, F. M. Watt and W. T. S. Huck, *Biomaterials*, 2010, **31**, 5030–5041.

28 I. Bos, H. Merlitz, A. Rosenthal, P. Uhlmann and J.-U. Sommer, *Soft Matter*, 2018, **14**, 7237–7245.

29 J. Hou, R. Chen, J. Liu, H. Wang, Q. Shi, Z. Xin, S.-C. Wong and J. Yin, *J. Mater. Chem. B*, 2018, **6**, 4792–4798.

30 A. Johnson, J. Madsen, P. Chapman, A. Alswieleh, O. Al-Jaf, P. Bao, C. R. Hurley, M. L. Cartron, S. D. Evans and J. K. Hobbs, *Chem. Sci.*, 2017, **8**, 4517–4526.

31 H. Takahashi, M. Nakayama, K. Itoga, M. Yamato and T. Okano, *Biomacromolecules*, 2011, **12**, 1414–1418.

32 F. Zhou, L. Jiang, W. Liu and Q. Xue, *Macromol. Rapid Commun.*, 2004, **25**, 1979–1983.

33 L. Li, T. Nakaji-Hirabayashi, H. Kitano, K. Ohno, Y. Saruwatari and K. Matsuoka, *Colloids Surf., B*, 2018, **161**, 42–50.

34 P. Chapman, R. E. Ducker, C. R. Hurley, J. K. Hobbs and G. J. Leggett, *Langmuir*, 2015, **31**, 5935–5944.

35 F. Zhou, Z. Zheng, B. Yu, W. Liu and W. T. S. Huck, *J. Am. Chem. Soc.*, 2006, **128**, 16253–16258.

36 X. Li, M. Wang, L. Wang, X. Shi, Y. Xu, B. Song and H. Chen, *Langmuir*, 2013, **29**, 1122–1128.

37 T. K. Mandal, M. S. Fleming and D. R. Walt, *Nano Lett.*, 2002, **2**, 3–7.

38 C. Combellas, F. Kanoufi, S. Sanjuan, C. Slim and Y. Tran, *Langmuir*, 2009, **25**, 5360–5370.

39 T. Perova, J. Vij and H. Xu, *Colloid Polym. Sci.*, 1997, **275**, 323–332.

40 D. G. Castner, K. Hinds and D. W. Grainger, *Langmuir*, 1996, **12**, 5083–5086.

41 K. K. Lau and K. K. Gleason, *Thin Solid Films*, 2008, **516**, 674–677.

42 S. Ben Amor, G. Baud, M. Jacquet, G. Nansé, P. Fioux and M. Nardin, *Appl. Surf. Sci.*, 2000, **153**, 172–183.

43 G. Gunkel, M. Weinhart, T. Becherer, R. Haag and W. T. Huck, *Biomacromolecules*, 2011, **12**, 4169–4172.

44 F. Xu, Q. Cai, Y. Li, E. Kang and K. Neoh, *Biomacromolecules*, 2005, **6**, 1012–1020.

45 Z. Xie, C. Chen, X. Zhou, T. Gao, D. Liu, Q. Miao and Z. Zheng, *ACS Appl. Mater. Interfaces*, 2014, **6**, 11955–11964.

46 A. Simakova, S. E. Averick, D. Konkolewicz and K. Matyjaszewski, *Macromolecules*, 2012, **45**, 6371–6379.

47 B. T. Cheesman, J. D. Willott, G. B. Webber, S. Edmondson and E. J. Wanless, *ACS Macro Lett.*, 2012, **1**, 1161–1165.

48 K. Matyjaszewski, *Macromolecules*, 2012, **45**, 4015–4039.

49 A. Han, O. Wang, M. Graff, S. K. Mohanty, T. L. Edwards, K.-H. Han and A. B. Frazier, *Lab Chip*, 2003, **3**, 150–157.

Faraday Discussions

PAPER

Probing the nanoscale organisation and multivalency of cell surface receptors: DNA origami nanoarrays for cellular studies with single-molecule control†

William Hawkes,‡[abc] Da Huang,‡[c] Paul Reynolds,[d] Linda Hammond,[e] Matthew Ward,[b] Nikolaj Gadegaard, [d] John F. Marshall,*[e] Thomas Iskratsch *[b] and Matteo Palma *[c]

Received 19th February 2019, Accepted 18th March 2019

DOI: 10.1039/c9fd00023b

Nanoscale organisation of receptor ligands has become an important approach to study the clustering behaviour of cell-surface receptors. Biomimetic substrates fabricated via different nanopatterning strategies have so far been applied to investigate specific integrins and cell types, but without multivalent control. Here we use DNA origami to surpass the limits of current approaches and fabricate nanoarrays to study different cell adhesion processes, with nanoscale spatial resolution and single-molecule control. Notably, DNA nanostructures enable the display of receptor ligands in a highly customisable manner, with modifiable parameters including ligand number, ligand spacing and most importantly, multivalency. To test the adaptability and robustness of the system we combined it with focused ion beam and electron-beam lithography nanopatterning to additionally control the distance between the origami structures (i.e. receptor clusters). Moreover, we demonstrate how the platform can be used to interrogate two different biological questions: (1) the cooperative effect of integrin and growth factor receptor in cancer cell spreading, and (2) the role of integrin clustering in cardiomyocyte adhesion and maturation. Thereby we find previously unknown clustering behaviour of different integrins, further outlining the importance for such customisable platforms for future investigations of specific receptor organisation at the nanoscale.

[a]Randall Centre of Cell and Molecular Biophysics, King's College London, UK

[b]Division of Bioengineering, School of Engineering and Materials Science, Queen Mary University of London, UK. E-mail: t.iskratsch@qmul.ac.uk

[c]School of Biological and Chemical Sciences, Queen Mary University of London, UK. E-mail: m.palma@qmul.ac.uk

[d]School of Engineering, University of Glasgow, UK

[e]Centre for Tumour Biology, Barts Cancer Institute, Queen Mary University of London, UK. E-mail: j.f.marshall@qmul.ac.uk

† Electronic supplementary information (ESI) available. See DOI: 10.1039/c9fd00023b

‡ Authors contributed to this work equally.

Introduction

Mammalian cells are embedded in an extracellular matrix (ECM), composed of proteoglycans, glycosaminoglycans and glycoproteins. The ECM components serve as anchoring points for the cells, but the binding of integrins and other cell surface receptors also induces cellular signalling cascades which regulate fundamental processes from cell growth, to differentiation, motility or cell death.[1] The contact interface between the ECM and cell-binding domains, depends on the receptors' nanoscale spatial organisation[2,3] and clustering.[4–7] Integrin–ECM interactions are also regulated by cellular (cytoskeletal) and extracellular forces (such as the passive stiffness of the ECM) that affect the stability of the bonds.[1,8–10] Importantly integrin clustering distributes the forces between the receptor–ligand complexes, increasing the maximum force per bond and enabling further integrin adhesion assembly.[4–6,9,11] Additionally, it is known that integrins work synergistically with other membrane receptors to modulate cell behaviour.[5,12–14]

Different strategies have been developed to fabricate biomimetic substrates for the investigation of ECM geometries in cell spreading and focal adhesion formation.[6,15–20] Achieving nanoscale control of adhesion receptors, with single molecule resolution, is essential for such investigations; micellar diblock copolymer self-assembly[21] and nanopatterning approaches[4,17,18,22] have been the most notable strategies to meet this requirement. This has been achieved *via* the arrangement of metal nanodots in arrays and their use as tethering points for cell-binding domains. These studies demonstrated that in order to establish stable focal adhesions and initiate integrin clusters, the examined cells ($\alpha v \beta 3$ integrin rich fibroblast, osteoblast and cancer cells) require a preferential spacing of $\sim$60 nm between single RGD peptides.[3,4,6,15] However, a recent study with peptidomimetic ligands indicated that major differences exist in the clustering behaviour between $\alpha v \beta 3$ and $\alpha 5 \beta 1$ integrins[23] and information about other integrin subtypes or from different cellular systems is lacking. Most importantly, these strategies have shown only partial application for multivalent investigations with single-molecule control.[16,19]

In this regard, we recently presented a strategy for the fabrication of biomimetic nanoarrays, based on the use of DNA origami that permit the multivalent investigation of ligand–receptor molecule interactions (simultaneous binding of multiple different ligands) in cancer cell spreading with nanoscale spatial resolution and single-molecule control.[20] Here we demonstrate the extension of this approach to different cell adhesion processes. Nanoscale spatial resolution and single-molecule control have been achieved by selectively functionalising DNA nanostructures with specific cell surface receptor ligands for human cutaneous melanoma cells and neonatal rat cardiomyocytes. We combined this with both an electron-beam and focused ion beam nanopatterning approach to assemble DNA origami in nanoarray configurations. The platforms so developed were then employed for cancer cell spreading investigations and to investigate cardiomyocyte attachment and maturity in response to different peptide configurations. This demonstrates the general applicability and validity of DNA origami nanoarrays as biomimetic substrates for the study of (multivalent) ligand–receptor interactions in the regulation of cellular adhesion and function, with nanoscale spatial resolution and single-molecule control.

Results and discussion

Design of the arrays and nanostructures

Cellular adhesion to – and probing of – the extracellular environment involves recruitment, activation and clustering of integrins.[1,2,11,24] Sufficient ligand density is key for cell spreading and viability.[6,25,26] For certain cell types and integrins (especially $\alpha v \beta 3$), the preferential ligand spacing required for spreading is 60 nm,[4,15,22] but a recent study suggested that differences might exist between integrin subtypes.[23] However, the clustering behaviour of integrin subtypes in different cellular systems is still elusive. Furthermore, the role of the nanoscale organisation for the crosstalk between different cell surface receptors is not well understood. To address these challenges, we designed DNA origami nano-structures for two specific applications, namely the investigation of the cooper-ativity of growth factor and integrin signalling in cancer cells, as well as the role of integrin ligand spacing and densities in the regulation of cardiomyocyte function.

For the cancer cell investigations, DNA origami were engineered and func-tionalised with A20FMDV2 peptides, selective for $\alpha v \beta 6$ integrin subtypes, and epidermal growth factor (EGF). The $\alpha v \beta 6$ integrin subtype is overexpressed in almost one third of carcinomas,[27] correlates with poor cancer survival[28] and has been identified as a key marker for metastasis[29] *via* its modulation of proliferative signalling pathways.[30] EGF receptors (EGFR) are known to work cooperatively with integrins and play a key role in regulating the signalling mechanisms at cell adhesion sites.[31] EGFR is also upregulated in many cancers and, together with integrins, regulates cancer cell proliferation, migration and metastasis.[32] For this reason, the A20FMDV2 peptides and EGF were chosen for multivalent investiga-tions of cooperative substrate bindings effects of $\alpha v \beta 6$ integrins and EGF (respectively) in human melanoma cells (A375P, human malignant cutaneous melanoma cell line). DNA origami were designed to present 6 peptides/EGF along their outer edges, at 60 nm intervals.

To demonstrate the versatility of DNA origami substrate nanopatterning, we sought to extend this approach to the investigation of integrin clustering dynamics for primary neonatal rat cardiomyocytes (NRCs), which due to their contractile nature differ from other cell types in adhesion structure and cyto-skeletal organisation.[10,33] For this purpose, DNA origami were engineered to present the cyclic RGD (cRGDfC) peptide. Neonatal cardiomyocytes contain relatively high levels of the RGD binding $\alpha 5 \beta 1$ integrin subtype which is down-regulated shortly after birth (in exchange for laminin binding integrins), but is re-expressed in cardiomyocytes after myocardial infarction and in other cardiac diseases, where it is strongly associated with the activation of pathological sig-nalling pathways.[34] Because of the absence of $\alpha v \beta 3$ integrin from NRCs, cRGDfC peptides were utilised to investigate the effect of ligand distance and density on $\alpha 5 \beta 1$ integrin clustering and downstream cardiomyocyte behaviour. DNA origami configurations of 6, 12 and 18 peptides were fabricated, constituting inter-peptide spacings of approximately 60 nm, 30 nm and 20 nm (respectively) along the outer edge of each DNA origami structure.

Synthesis and assembly of functionalised DNA origami nanostructures

To functionalise DNA origami with A20FMDV2 and cRGDfC, two different attachment chemistries were employed (Fig. 1). A20FMDV2 was conjugated to ssDNA *via* a maleimide–thiol reaction, bridging the cysteine thiol on the peptide and a deprotected maleimide on ssDNA (5′ end OH group of the phosphate). The cRFDfC peptides were conjugated to ssDNA with a 1 hour, UV mediated thiol-ene reaction.[35] For all ssDNA–peptide products, Reverse-Phase High Performance Liquid Chromatography (RP-HPLC) was used to verify successful conjugation. Using this approach, the conjugation efficiency was estimated *via* the relative change in UV adsorption (260 nm) of the unconjugated ssDNA compared to the emerging conjugated product (ESI Fig. 1†). Additional validation of the yield was obtained using a Microvolume spectrophotometer (see Methods). Purified peptide–ssDNA conjugates were then hybridised to DNA origami in solution *via* complementary "sticky-end" strands incorporated into the design, enabling selective positioning of peptides on the origami structure with ~6 nm precision.

Individual EGF moieties were selectively positioned and attached to DNA origami *via* a streptavidin–biotin conjugation technique (Fig. 1 and 2). ssDNA was modified to present the biotin which were firstly hybridised to the origami structure. Streptavidin modified EGF was then incubated in solution overnight to achieve successful conjugation to DNA origami structures.

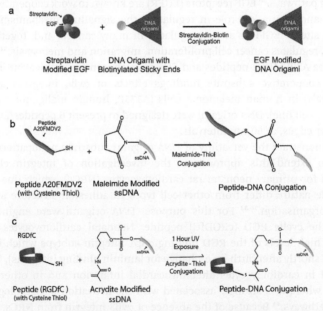

Fig. 1 The three linker chemistries used to conjugate ligands to DNA origami. (a) Streptavidin–biotin reaction. DNA origami were labelled with biotin and then streptavidin labelled EGF was incubated with labelled origami in solution. (b) The first thiol–ene strategy utilised a free thiol on the A20FMDV2 peptide and maleimide modified ssDNA. (c) The second thiol–ene strategy utilised a free thiol on the cRGDfC peptide and an acrydite modified ssDNA.

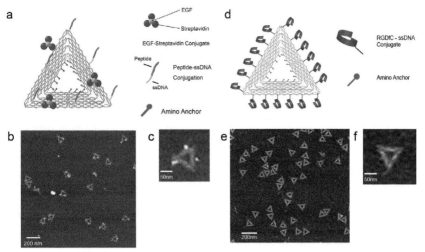

Fig. 2 Schematic of functionalised DNA origami: (a) DNA origami with peptide and EGF modification. Peptide, EGF and amino anchor modifications were positioned along the inner and outer edges of the DNA origami, and in-plane to the DNA nanostructure; this ensure the availability of the ligands regardless to the orientation of the DNA origami once immobilized on surfaces. (b) AFM characterisation of EGF functionalised DNA origami. (c) Zoom-in AFM image of EGF modified DNA origami. (d) Schematic of DNA origami functionalised with 18 cRGDfC peptides. (e) AFM characterisation of peptide modified DNA origami. (f) Zoom-in AFM characterisation of peptide functionalised DNA origami.

The successful design of both types of DNA nanostructures was further verified by casting diluted solution on muscovite mica substrates. Fig. 2 shows atomic force microscopy (AFM) images of the triangular DNA origami employed in this study.

Nanoscale surface lithography for selective positioning of functionalised DNA origami

Combining nanoscale lithographic substrate fabrication and functionalised DNA origami presents a powerful tool for multivalent investigations into ligand–receptor clustering dynamics. Furthermore, the use of three peptide concentrations (6, 12 and 18 peptides) and two DNA origami spacing configurations (200 nm and 300 nm spacing) permits unparalleled control over both local and global peptide concentration and stoichiometry at physiologically relevant sales (30–300 nm).

Two approaches were implemented to selectively position individual DNA origami over arrays of up to 3.68 mm^2 using focussed ion beam (FIB) and electron-beam (EBL) lithography. The first approach implemented was a one-step FIB lithographic process (Fig. 3) to fabricate holes in metal-coated SiO$_2$ substrates.[20,36] Chromium and gold were initially evaporated onto SiO$_2$ substrates to form 2 nm and 8 nm layers, respectively. Using FIB, 200 × 200 nm squares were milled into the metal, at 300 nm intervals. A 300 nm spacing per nanoaperture and therefore per DNA origami, was chosen for these nanoarrays to achieve a density of at least 87 ligands per μm^2.[20] The exposed SiO$_2$ was then silanised with a carboxylic acid terminated silane through which origami were covalently cross-linked to the

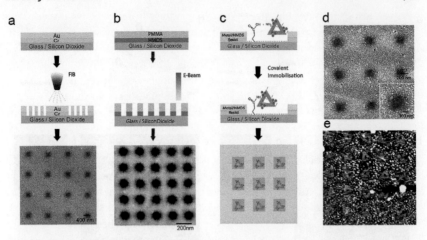

Fig. 3 Schematics of FIB: (a) and E-beam (b) nanopattern fabrication. (c) Covalent immobilisation of DNA origami *via* carbodiimide reaction chemistry between amino modifications on the central void within the DNA origami and carboxylic acid silane. Characterisation of successful cross-linking of origami to the FIB (d) and E-beam (e) nanopatterns was carried out with AFM.

surface using an EDC/sulfo-NHS carbodiimide crosslinking reaction, and the amino modifications in the central void of the origami (Fig. 3). The stability of the metallic surface offers a reusable, one-step fabrication of patterned arrays with the ability to selectively position single, functionalised DNA origami, with a yield of 82% ± 0.17.[20]

The second approach utilised EBL to define the DNA origami bindings sites.[37] Here, 150 nm in diameter holes were patterned in a hexamethyldisilazane layer (1–2 nm thickness) and poly(methyl methacrylate) (∼100 nm thickness) resist at intervals of 200 or 300 nm, exposing the underlying SiO_2 (Fig. 3). Glass coverslips were patterned with a total of six, 1.5 × 1.5 mm nano-patterned areas (3 × 200 nm and 3 × 300 nm arrays). An isotropic oxygen plasma etch was then used to generate silanol groups in the holes which were correspondingly silanised with a carboxylic silane. DNA origami were then cross-linked to the surface *via* the same carbodiimide chemistry as above. While lacking the reusability of the FIB nanopattern, this approach offers the advantages of a higher yield of single DNA origami binding events (100% of sites filled, with *ca.* 90% single origami attachment) on transparent substrates.

Multivalent activation of EGF and αvβ6 integrins in human cutaneous melanoma cells

To investigate the synergistic effect of integrin/EGF binding in cancer cell interaction with the ECM, human melanoma cells (A375P) were seeded on FIB nano-patterned substrates functionalised with DNA origami for 1.5 hours before fixation. The cooperative effects of αvβ6 integrin and EGF binding was evaluated by varying the ratio of A20FMDV2 peptides and EGF ligands (number of peptides : EGF = 1 : 1, 3 : 1 and 3 : 3). The degree of cell attachment and spreading was evaluated with phase contrast and immunofluorescence microscopy. Control experiments of blank FIB nanopatterns did not mediate the attachment of any cells. Melanoma cells adhered to all substrates but the degree of spreading, as indicated by phase images and phalloidin staining, was greater with increasing ligand density (see Fig. 4). The EGF receptor (EGFR) sub-family of receptor tyrosine kinases has well validated roles in cancer progression and metastasis.[38] Immunofluorescence of phosphorylated tyrosines (pTyr) was therefore used as a marker of general tyrosine kinase activity, including EGFR activity. Fig. 4 shows that the highly motile melanoma cell line A375Pb6 does not form strong focal adhesions on the origami. Moreover, the ratio of peptide : EGF dramatically changes cell signalling and cell behaviour. At a peptide : EGF ratio of 1 : 1 the cells are all well spread, actin is distributed diffusely and pTyr signalling is weak. When the ratio of peptide : EGF is 3 : 1 the cells remain well spread, pTyr signalling increases and some cells show sub-membranous actin on one side of the cell, consistent with potential development of a migratory leading edge. Interestingly, additional EGF ligands (peptide : EGF = 3 : 3) reverse the cells to a more rounded phenotype and reduce the pTyr signalling. These data show clearly that nanoscale modulation of the ligand density of interactive membrane receptors can induce dramatic changes in cells that warrant more detailed investigation.

Peptide functionalised DNA origami mediates cell adhesion and spreading in neonatal rat cardiomyocytes

To investigate integrin receptor clustering in primary neonatal rat cardiomyocyte (NRC) cultures we first tested the ability of the DNA origami to mediate cell attachment and spreading in a dose dependent manner on random, non-patterned arrays. DNA origami functionalised with 6, 12 and 18 cRGDfC

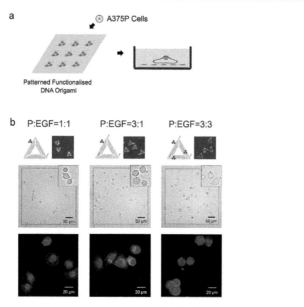

Fig. 4   Cancer cell investigation on patterned DNA origami nanoarrays: (a) Schematic of cell investigation; (b) A375P cell-spreading study on patterned DNA origami substrates with 300 nm spacing include DIC images of cell spreading on different substrates (peptide : EGF = 1 : 1, peptide : EGF = 3 : 1 and peptide : EGF = 3 : 3) with 6× zoom-in inset images, and fluorescent staining images with DAPI (blue), actin (red) and pTyr (green).

peptides per origami (corresponding to approximate peptide distances of 60, 30, and 20 nm, respectively) were cast randomly on 16 mm round glass coverslips and covalently cross-linked *via* the EDC/s-NHS carbodiimide chemistry described above. DNA origami without any peptides attached were used as a control. An optimum concentration of 2 nM DNA origami was used to achieve a uniform surface coverage of 10 ± 3 DNA origami per μm^2 (Fig. 5). Surfaces were passivated with an methoxy PEG silane and blocked with 5% bovine serum albumin to minimise non-specific adhesion. NRCs were seeded on the substrates and cultured for 24 hours before the degree of NRC attachment, spreading and phenotype was evaluated. Compared to the 0-peptide control, the number of cells attached to the substrate and the degree of cell spreading increased when 12 or more peptides (30 nm distance between peptides, $\sim120 \pm 36$ ligands per μm^2) were present on each DNA origami, however cells mainly displayed a continuous filamentous instead of a striated F-actin staining pattern, indicating an immature phenotype (Fig. 5). When 18 peptides were present (180 ± 54 ligands per μm^2), a further increase in cell attachment and spreading was observed and sarcomeric striations were clearly visible (see ESI Fig. 2†), thus confirming the ability of functionalised DNA origami to mediate cell attachment and verifying the successful synthesis of the different origami designs.

Nanoscale control of global and local ligand clustering dynamics in cardiomyocytes

Random positioning of peptide functionalised DNA origami demonstrated that a minimum of 12 peptides per DNA origami ($\sim120 \pm 36$ ligands per μm^2) were required to initiate cardiomyocyte adhesion and spreading. However, random positioning of DNA origami provides no control over ligand geometry above the level of individual origami (~85 nm^2) and offers limited information on global and local clustering dynamics, which are key for cell adhesion and spreading.[5,21] To investigate this, DNA origami functionalised with 0, 6 or 12 cRGDfC peptides were bound to EBL nanopatterned substrates at 200 nm and 300 nm intervals (Fig. 5), and this allowed us to further investigate the role of cluster spacing, due to the differences in receptor clustering behaviour between cardiomyocytes and fibroblasts. EBL nanopatterned substrates were blocked against non-specific adhesion with a methoxy PEG silane and 5% BSA before seeding NRCs for 24 hours. All cells adhered to the nanopattern were imaged and the degree of cell spreading and cytoskeletal maturity was evaluated by calculating the cell area and F-actin intensity. On 0 peptide controls, a small degree of nonspecific cell adhesion was observed (Fig. 6), although almost all NRCs were rounded up, presumably in a state of detachment. Negative controls of nanopatterns without DNA origami did not facilitate any cardiomyocyte adhesion. Functionalised DNA origami with 6 cRGDfC peptides (60 nm peptide spacing) resulted in the initiation of cell spreading when origami were spaced at 200 nm intervals (150 peptides per μm^2), but not at 300 nm intervals (54 peptides per μm^2); the cell area at 200 nm intervals was measured to be (mean $\pm$ SD) 599 ± 178 μm^2 compared to 442.5 ± 125 μm^2 at 300 nm intervals, equivalent to the area of rounded unspread cardiomyocytes.

Increasing the number of peptides to 12 per DNA origami (30 nm peptide spacing) resulted in a marked increase in cell attachment and clear initiation of

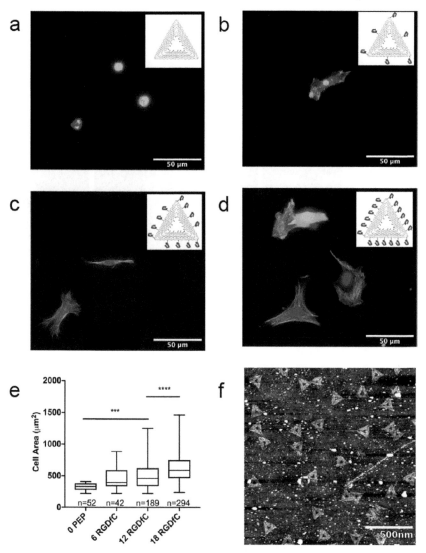

Fig. 5 Verification of neonatal rat cardiomyocyte (NRC) cell adhesion and spreading mediated by randomly positioned DNA origami functionalised with 0 (a), 6 (b), 12 (c) and 18 (d) cRGDfC peptides; this corresponds to peptide distances of 60, 30, and 20 nm, respectively. Insets display the DNA origami design used at a given condition. N numbers represent the number of cells. (e) DAPI and F-actin staining was used to measure cell area. (f) AFM characterisation of 2 nM DNA origami cast randomly on SiO_2 substrates. Fluorescent channels in (a–d) represent F-actin (green) and DAPI (blue). p values from ANOVA with Turkey correction for multiple comparisons: *** $= p < 0.001$, **** $= p < 0.0001$.

spreading at both 300 nm (108 peptides per μm^2) and 200 nm (300 peptides per μm^2) origami spacings, although differences in area, and F-actin content and organisation persisted (Fig. 6). At 300 nm, the cell area increased to 645 ± 265 μm^2, although a sub-population of rounded cells was still observed. At 200 nm origami spacing, average cell area increased to 755 ± 192 μm^2 and almost all cells were spread, indicating stable adhesion formation.

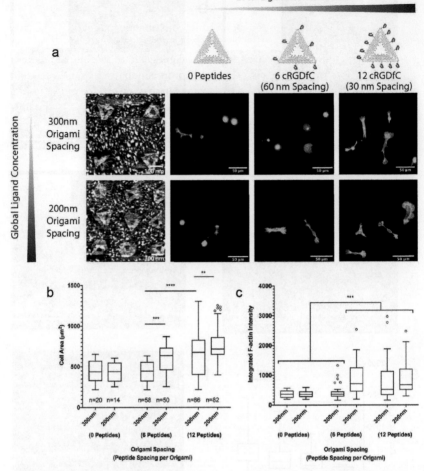

Fig. 6 Increasing global and local ligand concentration mediates NRC attachment and spreading. (a) AFM images of 200 nm and 300 nm DNA origami spacing configurations (far left). DAPI (blue) and F-actin (green) staining was used to evaluate cell area (b) and integrated F-actin intensity (c). p values from ANOVA with Turkey correction for multiple comparisons: ** = p < 0.01, *** = p < 0.001, **** = p < 0.0001. Box and whisker diagrams: box limits represent the 25th to 75th percentiles and whiskers represent the 1.5× interquartile range. N numbers represent the number of cells.

This difference in spreading and F-actin content (Fig. 6) depending on origami distance, while maintaining the same ligand spacing on the origami, surprisingly indicated that, in contrast to fibroblasts, the density between the ligands influences cardiomyocyte adhesion. 60 nm has been reported to be the optimal spacing to form stable focal adhesions and initiate efficient cell spreading in fibroblasts, osteoblasts and cancer cell lines which, in contrast to NRCs, all express high levels of αvβ3 integrins.[4,15,21,22,26] However, the observed variation in cardiomyocyte spreading at both 200 and 300 nm spacing is indicative of differences in adhesion formation between integrin subtypes in agreement with a previous study using peptidomimetic ligands.[23]

Conclusions

Our results demonstrate the adaptability of biomimetic nanoarray platforms based on the use of DNA origami to study the spreading and adhesion of different cell types. We show that DNA nanostructures conjugated to receptor ligands can be placed with high efficiency and precision onto nanopatterns fabricated with different methods (FIB, EBL). These substrates can then be used to display a wide range of different receptor ligands in a highly controlled manner – alone or in combination with other cell surface receptors. We demonstrate the usefulness of the platform to study different cell types, namely cancer cells and cardiomyocytes. Our findings further show that different receptor types vary strongly in their clustering behaviour; this demonstrates that previous findings from studies employing biomimetic arrays to study RGD/αvβ3 integrin interactions at the nanoscale are not universally applicable. Therefore, further work is needed to investigate the nanoscale organisation for different receptors and receptor combinations in detail. By providing nanoscale spatial resolution, single-molecule control, and multivalent capability, we believe the DNA origami biomimetic nanoarrays presented here are the ideal platform for such studies.

Experimental methods and materials

Synthesis of DNA origami

DNA origami was assembled by combining M13mp18 (5 nM) and staple strands (50 nM) in 50 μL of TAE buffer with 12.5 mM Mg^{2+} (DNA sequences can be obtained at ref. 20). M13mp18 is a bacteria phage vector strand with 7249 bases long. An appropriate quantity of ions, such as magnesium here, or sodium, are required for efficient DNA hybridization. This acts to equilibrate electrostatic repulsion between highly negatively charged DNA molecules. An amount of 12.5 mM Mg^{2+} is sufficient to achieve a high yield of DNA origami and limit any aggregation effects. DNA origami are synthesised by annealing from an initial temperature of 94 °C to completely melt all dsDNA. Temperature step-controlled annealing was carried out in a PCR machine. Samples were cooled from 94 to 65 °C at a rate of ∼0.3 °C per minute. A cooling rate of 0.1 °C per minute is employed from 65 °C to room temperature. The self-assembled DNA origami were then purified using Millipore Amicon Ultra 100 kDa spin columns in a centrifuge at 2000 rcf for 6 min, three times, to remove excess staple strands. DNA origami were adjusted to a concentration ∼20 nM and stored in Lo-Bind Eppendorf tubes at 4 °C. A NanoDrop spectrophotometer is used to detect the approximate concentration of DNA origami products based on the constant of a molecular weight of 330 g mol^{-1} per base and an extinction coefficient of 33 mg mL^{-1} for A260; the actual result is close to the estimated numbers. To ensure efficient assembly and labelling of DNA origami with peptide conjugates, unmodified staple strands and amino anchors were added at a 5× excess to the M13mp18 backbone and peptide conjugates were added at a 10× excess.

AFM characterization of DNA origami

AFM (Bruker, Dimension Icon) was used to image the DNA origami structures. DNA origami was cast on either silicon dioxide, glass, or mica surfaces for imaging. The DNA origami solution is diluted by TAE buffer with 30 mM $MgCl_2$ to around 1 nM in order to get a good separation of the DNA nanostructures once immobilized on surface. Magnesium is required in the procedure as an ion charged bridge, immobilizing DNA origami to the substrate surfaces. Mica samples were cleaved twice by solid scotch tape immediately prior to casting. 5 μL of diluted DNA origami solution was directly deposited on freshly cleaned mica and left to adsorb on the surface for 2 min. Subsequently, the substrate was washed by distilled water to remove non-adsorbed origami and then blown dry by compressed air. Silicon nitride ScanAsyst-Air tips with a 0.4 N m^{-1} spring constant were used to scan the sample by AFM under ScanAsyst Mode. A resolution of 512 pixels per line with 1 Hz scan rate was chosen for appropriate imaging of the DNA nanostructure.

The highly hydrophobic hexamethyldisilazane (HMDS) layer on EBL patterned surfaces made it very difficult to accurately track the surface when scanning in air. Therefore, to characterise DNA origami on the EBL nanopattern, AFM characterisation was carried out in fluid tapping mode. All samples were imaged in 5 mM Tris, pH 8.3, 40 mM $MgCl_2$. Silicon nitride ScanAsyst-Fluid+ with a 0.7 N m^{-1} spring constant were used at a resolution of 512 pixels per line and a 1 Hz scan rate. Amplitude setpoint and gain were optimised for each sample and tip.

Modifications of DNA origami

Peptide A20FMDV2 was assembled on DNA origami using the maleimide–thiol linkage on ssDNA complementary to the sticky ends. The numbers of peptides on a single origami depends on how many complementary sticky ends that origami has. Up to 18 positions were chosen in this study (see modified sequences in ESI†). Commercially available protected maleimide modified ssDNA need to be deprotected before conjugation. 50 nmol of the protected maleimide modified ssDNA was freeze-dried at first. Deprotection and conjugation will be inefficient in the presence of water. Freeze-dried samples were washed adding 2 mL of anhydrous acetonitrile, and the solvent was evaporated by a rotary evaporator. Two millilitres of anhydrous toluene was added to the vial, mixed, and then evaporated. This rinsing procedure was repeated three times. After the second evaporation, toluene was added and heated to 90 °C for 4 h to deprotect the maleimide modifier. After incubation, the toluene was evaporated. The vial containing the deprotected maleimide-modified ssDNA was immediately mixed with the reduced peptide. The thiol group on the cysteine of the peptide needed to be reduced from the oxide form right before conjugation. The conjugation reaction occurs after the mixing. The mixture was put on a shaker for 1 h to complete the conjugation. Peptide–ssDNA conjugation was validated and purified by reversed-phase HPLC (RP-HPLC) with TEAA buffer (triethyl amine acetic acid). The purified products were freeze-dried and resuspended in TAE buffer. Peptide–ssDNA conjugation was mixed with staple strands and M13mp18, and standard DNA-origami synthesis was carried out. A20FMDV2 is a linear peptide with 21 amino acid. Heating up to 94 °C will not affect the structure and function of this peptide. Epidermal growth factor (EGF) was modified on DNA origami *via* streptavidin–biotin conjugation. The EGF was biotinylated and assembled with streptavidin, which is commercially available. Streptavidin-modified EGF was attached on origami structure *via* incubation with a DNA origami solution which had biotinylated sticky ends. AFM characterization was carried out to confirm the EGF modification.

For cRGDfC peptides, conjugation to ssDNA was performed through a UV mediated thiol–ene reaction, on acrydite modified ssDNA (Integrated DNA Technologies). Peptides were diluted to 100 µM in H_2O containing $100\times$ TCEP pH 7.0, to reduce any disulfide bonds and present the cysteine groups for conjugation. Peptides were then mixed with the acrydite modified ssDNA at a final concentration of 20 µM and 200 µM, respectively. Reactions were carried out in 120 mM Tris buffer with 11 µM photoinitiator (2-hydroxy-4'-(2-hydroxyethoxy)-2-methylpropiophenone). Samples were exposed to 260 nm UV light for 1 hour and the conjugates were purified by RP-HPLC and freeze drying, as described above.

Fabrication of FIB and EBL substrates

Fabrication of the FIB substrates was carried out as previously described.[36] Briefly, glass coverslips were cleaned in Piranha solution before chromium and gold were evaporated on the surface in 2 nm and 8 nm thick layers (respectively). Substrates were then baked on a hotplate at 300 °C for 15 minutes. 200 × 200 nm squares were then milled into the substrate using the FIB over an array of a desired size. Prior to cross-linking DNA origami, substrates were cleaned using oxygen plasma for 5 minutes (Harrick, 18 W, room air).

Further substrates were produced by direct-write EBL on coverslips. Coverslips were solvent cleaned in an ultrasonic bath (acetone followed by methanol, IPA, and water) and dehydrated in a 180 °C oven overnight. An oxygen plasma treatment for 60 s at 100 W power prepared the surface for silane deposition, which was conducted from HMDS vapour in a closed container at 150 °C. A 950k molecular weight poly(methyl methacrylate) (PMMA) film was spin-coated on the coverslips at 5k rpm, followed by a short solvent bake on a 180 °C hotplate for 30 s. A 10 nm aluminium charge conduction layer was evaporated on the coverslips, and they were mounted on a 4-inch silicon wafer for EBL processing using crystalbond 555 adhesive. EBL exposure was carried out as described previously[39] to define arrays of 200 nm and 300 nm pitch holes, diameter 150 nm in the PMMA layer. After exposure, the aluminium charge conduction layer was removed in a 2.6% TMAH solution (MF-CD26) for 30 s followed by rinsing with water. The surface roughness for both kinds of patterned surfaces was calculated from AFM topographical heights (Z) by measuring the root mean square $R_q = \sqrt{\dfrac{\sum Z_i^2}{N}}$. The FIB and EBL substrates exhibited R_q values of 3.2 ± 0.1 and 0.7 ± 0.1 nm, respectively.

DNA origami patterning and cross-linking

For patterning of DNA origami FIB nanofabricated substrates, 60 µL of 1 nM of DNA origami were incubated on the surface for 1.5 hours in Tris buffer (5 mM; pH 8.3) with 30 mM $MgCl_2$. Substrates were placed in a six-well plate with a moist Kimwipe to prevent dewetting and agitated on a shaker. Samples were then washed with Tris buffer (5 mM; pH 8.3) containing 30 mM $MgCl_2$ and a 0.01% solution of carboxyethylsilane (CTES) and incubated for 2 min on a shaker. The buffer was then exchanged for MOPS buffer (10 mM; pH 8.0) with 30 mM $MgCl_2$.

The initial wash with MOPS buffer removed all primary amines from the Tris washes ahead of cross-linking. Next, the MOPS buffer was exchanged with an equal volume of MOPS buffer, pH 8.0, containing 100 mM EDC (1-ethyl-3-(3-(dimethylamino)propyl)carbodiimide), and 50 mM sulfo-NHS (N-hydroxysulfosuccinimide). Samples were incubated for 10 minutes at room temperature with gentle agitation. Finally, samples were then washed with the MOPS buffer without $MgCl_2$ and rinsed with DPBS containing 125 mM NaCl. PBS precipitates with Mg and should remove all non-covalent, physisorbed origami structures. Origami positioning was verified under AFM.

For patterning of DNA origami on e-beam patterned substrates, patterned substrates were first developed in a 2 : 1 solution of propanol-2 : methyl isobutyl ketone for 60 seconds at 23 °C, followed by a 30-second wash in 100% propanol-2. DNA origami binding sites were then etched using oxygen plasma treatment (100 W, 70 seconds, room air) followed by silanisation with 0.1% CTES in Tris buffer (5 mM, pH 8.0). The PMMA resist was then removed by immersion in NMP at 50 °C and sonicated or 10 minutes. DNA origami were incubated for 1 hour at room temperature at a concentration of 1 nM in Tris buffer (5 mM, pH 8.3, 50 mM $MgCl_2$, see the ESI† Methods section for more information on placement conditions). Due to the hydrophobic nature of the HMDS layer and the size of the patterned area, 100 μL of origami solution was required to cover the surface sufficiently. Samples were then washed with MOPS buffer (10 mM, pH 8.1, 50 mM $MgCl_2$) to remove primary amines, followed by a wash with MOPS buffer (10 mM, pH 8.1, 50 mM $MgCl_2$) containing 50 mM EDC and 25 mM sulfo-NHS. Finally, samples were then washed with the MOPS buffer without $MgCl_2$ and rinsed with DPBS containing 125 mM NaCl. Before AFM characterisation, samples were placed into deionised H_2O.

Cancer cell-spreading study

Cells were grown at 37 °C and 8% (v/v) carbon dioxide/air condition in a humidified incubator. Cells were maintained as adherent monolayers on tissue culture plastic. The growth medium contained DMEM accompanied by 10% FBS. Cells were sub-cultured approximately every 3 days. After incubation with 0.25% (w/v) trypsin/EDTA solution, adhered cells were removed from culture plastic and neutralized by adding three folds of growth medium. Cells were then suspended to fresh tissue culture flasks with fresh growth medium. Cells were removed from tissue culture plastic by trypsin, neutralized, pelleted, and resuspended in fresh growth medium. The substrates were blocked with 1% BSA before cells plating. Growth medium (2 mL) was added to each well and then incubated for 1.5 h. The substrates were carefully washed by PBS twice to remove non-adhered cells. The patterned area with cells was imaged using DIC to observe the cell behaviours. ImageJ was used for counting the cell numbers and average cell area analysis. For this outline of cells were drawn manually after thresholding. The combined cell area were then measured and divided by the cell numbers.

Cardiomyocyte isolation and culture

Neonatal rat cardiomyocytes were prepared as previously described.[33] Briefly, newborn rat hearts were dissected into ice-cold ADS buffer (116 mM NaCl, 20 mM Hepes, 0.8 mM NaH_2PO_4, 5.6 mM glucose, 5.4 mM KCL, 0.8 mM $MgSO_4$) and washed once with ADS buffer. The ADS buffer was then removed and hearts were incubated with of 5 mL enzyme solution in ADS (ES, 246U collagenase and 0.6 mg pancreatin per mL), for 5 min, at 37 °C under vigorous shaking. Tissues were then triturated with a pipette and the supernatant was discarded. This step was followed by 5–6 digests, until the hearts were completely digested. Each time 5 mL fresh ES was added to the hearts and incubated 15 min at 37 °C, under shaking. Hearts were pipetted up and down 30 times using a Pasteur pipette. The supernatant was then transferred into plating medium (65% DMEM, 17% M199, 10% horse serum, 5% FCS, 2% Glutamax, 1% penicillin/streptomycin (P/S)). Two digests each were combined in one tube with 20 mL plating medium, then cleared through a 100 mm cell strainer and spun down at 1200 rpm for 5 min at RT, before resuspended in 10 mL plating medium. Cells were pooled together and preplated for 90 min to enrich the cardiomyocytes. Cardiomyocytes were then plated onto the respective substrates in serum free medium (70% DMEM, 22% M199, 5% FCS, 2% Glutamax, 1% penicillin/streptomycin (P/S)). Medium was changed the next day to serum free maintenance medium (78% DMEM, 19% M199, 2% horse serum, 2% Glutamax, 1% P/S).

Immunofluorescence

Immunofluorescent investigations were carried out as described previously.[33] Briefly cells were fixed with 4% paraformaldehyde for 10 minutes followed by a wash with PBS. Cells were then permeabilised with 0.2% Triton-X for 5 minutes followed by additional PBS washes. Nonspecific binding was blocked with a 1 hour incubation in 5% BSA before incubation with Alexa 488 labelled phalloidin and DAPI for 1 hour at room temperature to visualise F-actin and cell nuclei, respectively. Substrates were washed with PBS (3 × 5 minutes) and mounted on a coverslip in Mowiol mounting medium.

Statistical analysis

All statistical analysis was carried out using Graphpad Prism 8. For random origami experiments, a one-way ANOVA with Turkey's correction for multiple comparisons was used to test for significant differences. For nanopatterned origami experiments, statistical analyses were carried out using a two-way ANOVA with a Turkey's correction for multiple comparisons.

Conflicts of interest

There are no conflicts to declare.

Acknowledgements

We would like to acknowledge the James Watt Nanofabrication Centre and its staff at the University of Glasgow for the nanofabrication work. W. H. is recipient of a BBSRC LIDO Studentship. D. H. was supported by the China Scholarship Council. T. I. was supported by a British Heart Foundation Intermediate Basic Science Research Fellowship (FS/14/30/30917) and a BBSRC new investigator award (BB/S001123/1). N. G. also acknowledges ERC funding through the FAKIR 648892 Consolidator Award.

References

1　T. Iskratsch, H. Wolfenson and M. P. Sheetz, *Nat. Rev. Mol. Cell Biol.*, 2014, **15**, 825–833.

2　K. Zhang, H. Gao, R. Deng and J. Li, *Angew. Chem., Int. Ed.*, 2018, **58**, 2–12.

3　E. A. Cavalcanti-Adam, T. Volberg, A. Micoulet, H. Kessler, B. Geiger and J. P. Spatz, *Biophys. J.*, 2007, **92**, 2964–2974.

4　M. Schvartzman, M. Palma, J. Sable, J. Abramson, X. Hu, M. P. Sheetz and S. J. Wind, *Nano Lett.*, 2011, **11**, 1306–1312.

5　F. Karimi, A. J. O'Connor, G. G. Qiao and D. E. Heath, *Adv. Healthcare Mater.*, 2018, **7**, e1701324.

6　R. Changede, H. Cai, S. Wind and M. Sheetz, *bioRxiv*, 2018, DOI: 10.1101/435826.

7　H. Wolfenson, T. Iskratsch and M. P. Sheetz, *Biophys. J.*, 2014, **107**, 2508–2514.

8　J. D. Hood and D. A. Cheresh, *Nat. Rev. Cancer*, 2002, **2**, 91–100.

9　R. Oria, T. Wiegand, J. Escribano, A. Elosegui-Artola, J. J. Uriarte, C. Moreno-Pulido, I. Platzman, P. Delcanale, L. Albertazzi, D. Navajas, X. Trepat, J. M. Garcia-Aznar, E. A. Cavalcanti-Adam and P. Roca-Cusachs, *Nature*, 2017, **552**, 219–224.

10　M. Ward and T. Iskratsch, *Biochim. Biophys. Acta, Mol. Cell Res.*, 2019, DOI: 10.1016/j.bbamcr.2019.01.017.

11　P. Roca-Cusachs, T. Iskratsch and M. P. Sheetz, *J. Cell Sci.*, 2012, **125**, 3025–3038.

12　M. Mammen, S. K. Choi and G. M. Whitesides, *Angew. Chem., Int. Ed. Engl.*, 1998, **37**, 2754–2794.

13　J. D. Humphries, A. Byron and M. J. Humphries, *J. Cell Sci.*, 2006, **119**, 3901–3903.

14　M. A. Wozniak, K. Modzelewska, L. Kwong and P. J. Keely, *Biochim. Biophys. Acta*, 2004, **1692**, 103–119.

15　M. Arnold, E. A. Cavalcanti-Adam, R. Glass, J. Blummel, W. Eck, M. Kantlehner, H. Kessler and J. P. Spatz, *ChemPhysChem*, 2004, **5**, 383–388.

16　H. Cai, D. Depoil, M. Palma, M. P. Sheetz, M. L. Dustin and S. J. Wind, *J. Vac. Sci. Technol., B: Nanotechnol. Microelectron.: Mater., Process., Meas., Phenom.*, 2013, **31**, 6F902.

17　H. Cai, J. Muller, D. Depoil, V. Mayya, M. P. Sheetz, M. L. Dustin and S. J. Wind, *Nat. Nanotechnol.*, 2018, **13**, 610–617.

18　Y. Keydar, G. Le Saux, A. Pandey, E. Avishay, N. Bar-Hanin, T. Esti, V. Bhingardive, U. Hadad, A. Porgador and M. Schvartzman, *Nanoscale*, 2018, **10**, 14651–14659.

19　G. Le Saux, A. Edri, Y. Keydar, U. Hadad, A. Porgador and M. Schvartzman, *ACS Appl. Mater. Interfaces*, 2018, **10**, 11486–11494.

20　D. Huang, K. Patel, S. Perez-Garrido, J. Marshall and M. Palma, *ACS Nano*, 2019, **13**, 728–736.

21　J. Huang, S. V. Grater, F. Corbellini, S. Rinck, E. Bock, R. Kemkemer, H. Kessler, J. Ding and J. P. Spatz, *Nano Lett.*, 2009, **9**, 1111–1116.

22　M. Schvartzman, K. Nguyen, M. Palma, J. Abramson, J. Sable, J. Hone, M. P. Sheetz and S. J. Wind, *J. Vac. Sci. Technol., B: Microelectron. Nanometer Struct.-Process., Meas., Phenom.*, 2009, **27**, 61–65.

23 V. Schaufler, H. Czichos-Medda, V. Hirschfeld-Warnecken, S. Neubauer, F. Rechenmacher, R. Medda, H. Kessler, B. Geiger, J. P. Spatz and E. A. Cavalcanti-Adam, *Cell Adhes. Migr.*, 2016, **10**, 505–515.

24 H. Wolfenson, B. Yang and M. P. Sheetz, *Annu. Rev. Physiol.*, 2019, **81**, 585–605.

25 R. Changede, X. Xu, F. Margadant and M. P. Sheetz, *Dev. Cell*, 2015, **35**, 614–621.

26 R. Zarka, M. B. Horev, T. Volberg, S. Neubauer, H. Kessler, J. P. Spatz and B. Geiger, *Nano Lett.*, 2019, **19**(3), 1418–1427.

27 A. Saha, D. Ellison, G. J. Thomas, S. Vallath, S. J. Mather, I. R. Hart and J. F. Marshall, *J. Pathol.*, 2010, **222**, 52–63.

28 A. L. Wong and S. C. Lee, *Int. J. Breast Cancer*, 2012, **2012**, 415170.

29 Z. Li, S. Biswas, B. Liang, X. Zou, L. Shan, Y. Li, R. Fang and J. Niu, *Sci. Rep.*, 2016, **6**, 30081.

30 Z. Li, P. Lin, C. Gao, C. Peng, S. Liu, H. Gao, B. Wang, J. Wang, J. Niu and W. Niu, *Tumour Biol.*, 2016, **37**, 5117–5131.

31 N. Balanis, M. Yoshigi, M. K. Wendt, W. P. Schiemann and C. R. Carlin, *Mol. Biol. Cell*, 2011, **22**, 4288–4301.

32 J. S. Desgrosellier and D. A. Cheresh, *Nat. Rev. Cancer*, 2010, **10**, 9–22.

33 P. Pandey, W. Hawkes, J. Q. Hu, W. V. Megone, J. Gautrot, N. Anilkumar, M. Zhang, L. Hirvonen, S. Cox, E. Ehler, J. Hone, M. Sheetz and T. Iskratsch, *Dev. Cell*, 2018, **44**, 326–336.

34 S. Israeli-Rosenberg, A. M. Manso, H. Okada and R. S. Ross, *Circ. Res.*, 2014, **114**, 572–586.

35 J. Torres-Kolbus, C. Chou, J. Liu and A. Deiters, *PLoS One*, 2014, **9**, e105467.

36 D. Huang, M. Freeley and M. Palma, *Sci. Rep.*, 2017, **7**, 45591.

37 A. Gopinath and P. W. Rothemund, *ACS Nano*, 2014, **8**, 12030–12040.

38 E. Raymond, S. Faivre and J. P. Armand, *Drugs*, 2000, **60**(1), 15–23; discussion 41–12.

39 N. Gadegaard, S. Thoms, D. Macintyre, K. Mcghee, J. Gallagher, B. Casey and C. Wilkinson, *Microelectron. Eng.*, 2003, **67–68**, 162–168.

Faraday Discussions

PAPER

Magneto- and photo-responsive hydrogels from the co-assembly of peptides, cyclodextrins, and superparamagnetic nanoparticles†

Benedikt P. Nowak and Bart Jan Ravoo ⓘ *

Received 29th January 2019, Accepted 25th February 2019

DOI: 10.1039/c9fd00012g

Dual response to external non-invasive stimuli, such as light and magnetic field, is a highly desirable property in soft nanomaterials with potential applications in soft robotics, tissue engineering, and life-like materials. Within this class of materials, hydrogels obtained from the self-assembly of low molecular weight gelators (LMWGs) are of special interest due to their ease of preparation and modification. Herein, we report a modular co-assembly strategy for a magneto- and photo-responsive supramolecular hydrogel based on the arylazopyrazole (AAP) modified pentapeptide gelator Nap-GFFYS, and β-cyclodextrin vesicles (CDVs) with superparamagnetic cobalt ferrite nanoparticles embedded in their membranes. Upon application of a magnetic field, a reversible increase in the storage modulus is observed during rheological measurements. Additionally, a gel rod could be manipulated with a weak permanent magnet, resulting in macroscopic bending of the rod. Furthermore, through irradiation with UV and visible light, respectively, the host–guest interaction between the AAP moiety and the hydrophobic cavity of the β-CD can be deactivated on demand, thus lowering the stiffness of the hydrogel reversibly.

Introduction

Hydrogels are an emerging subclass of soft nanomaterials driven by their unique mechanical properties as well as their inherent biocompatibility,[1,2] and therefore find widespread applications in tissue engineering,[3,4] drug delivery systems,[5] and further types of smart materials.[6–8] In general, hydrogels are either polymeric gels or supramolecular gels based on the self-assembly of low molecular weight gelators (LMWGs).[4] Among the latter, short peptides with aromatic protecting groups are especially attractive due to their ease of modification and preparation *via* solid phase peptide synthesis (SPPS).[9,10] Through variation of the amino acid

Center for Soft Nanoscience and Organic Chemistry Institute, Westfälische Wilhelms Universität Münster, Busso Peus Straße 10, 48149 Münster, Germany. E-mail: b.j.ravoo@uni-muenster.de

† Electronic supplementary information (ESI) available: Experimental procedures and supporting data. See DOI: 10.1039/c9fd00012g

side chains, as well as their sequence in the peptide, the LMWG can be readily optimized for various applications.[11] Over the last few years, stimuli-responsive hydrogels with on-demand tunable mechanical properties have gained increasing attention. Amongst a wide variety of potential stimuli,[12] such as pH,[13,14] temperature,[15] redox,[16] or enzymes,[17] light[18] and magnetic field[19] are of particular interest due to their non-invasive character, as well as their high temporal and spatial resolution. In order to confer magnetic properties on hydrogels, researchers have focused mainly on embedding superparamagnetic particles into the hydrogel matrix.[20] Commonly, the superparamagnetic particles are blended into a polymer gel or are formed by co-precipitation inside the polymer gel, and various impressive systems have been described using this type of magneto-rheological polymer gels.[21–23] Nevertheless, magnetic hydrogels based on LMWGs are rarely reported.[24]

On the other hand, photo-responsive LMWGs, based on molecular photo-switches such as azobenzenes[25,26] or arylazopyrazoles (AAP),[27] are well established. Recently, our group reported a hydrogel exploiting the photo-responsive host–guest interaction between an AAP-modified LMWG and β-cyclodextrin (β-CD) to tune the cross-linking density and, thus, the mechanical properties of the supramolecular gel.[27] However, to the best of our knowledge, a self-assembled LMWG-based hydrogel exhibiting both magneto- and photo-induced changes in the rheological properties is unprecedented.

Herein, we present a modular strategy to prepare magneto- and photo-responsive hierarchical hydrogels based on the co-assembly of an AAP-modified Nap-GFFYS gelator and cyclodextrin vesicles (CDVs) with superparamagnetic nanoparticles embedded in their membranes. Recently, we have described magneto-responsive supramolecular colloids based on CDVs with embedded oleic acid-coated Fe_3O_4 nanoparticles.[28] In the present study, $CoFe_2O_4$ nanoparticles were used, since they have a higher magnetic susceptibility and are reported to be biocompatible.[29,30] In combination with a photo-switchable guest moiety in the peptide gelator, both magneto- and photo-rheological properties can be readily achieved in an unprecedented dual responsive supramolecular hydrogel (see Fig. 1a). In the context of the Faraday Discussion on nano-lithography of biointerfaces, we note that glycans can potentially be embedded in the gel or displayed at the surface of the gel using host–guest chemistry,[31–34] so that both the material stiffness and its biorecognition properties can be tuned using a combination of co-assembly components and orthogonal physical stimuli.

Results and discussion

The amino acid sequence of the gelator was chosen for its high gelation efficiency and was taken from the literature with slight modifications.[35] Nap-GFFYS (Fig. 1b) was synthesized *via* SPPS using a standard Fmoc-protection group strategy and napthyloxyacetic acid protected phenylalanine as the N-terminal amino acid. The photo-responsive AAP moiety was introduced at the C-terminal amino acid serine using an established approach, in which AAP-TEG-azide was coupled to the propargyl-functionalized Fmoc-L-serine *via* a copper-catalyzed azide–alkyne click reaction (CuAAC).[27] Using this AAP-modified amino acid at the C-terminus during SPPS, the photo-responsive gelator Nap-GFFYS-AAP (Fig. 1b) was obtained after

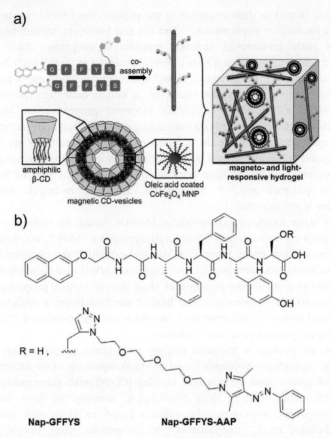

Fig. 1 (a) Schematic representation of the co-assembly of the LMWGs Nap-GFFYS and Nap-GFFYS-AAP in the presence of superparamagnetic cyclodextrin vesicles to yield a magneto- and photo-responsive hydrogel. (b) Structures of the gelator Nap-GFFYS and the photo-responsive analogue Nap-GFFYS-AAP.

stepwise elongation. Details on the gelator synthesis and analysis are provided in the ESI.†

Superparamagnetic $CoFe_2O_4$ nanoparticles with a hydrodynamic diameter of 9 nm in cyclohexane were synthesized using a thermolytic method starting from the respective iron(III) and cobalt(II) acetylacetonate salts in the presence of oleic acid, oleylamine, and 1-octadecanole as ligands. Utilizing a technique previously reported by our group, the particles were immobilized in the membranes of CDVs.[28] In short, oleic acid-stabilized $CoFe_2O_4$ nanoparticles and amphiphilic β-CD were dissolved in chloroform and the solvent slowly evaporated in an argon stream. After hydration with water overnight and repeated extrusion through a 100 nm pore size polycarbonate membrane, magnetic CDVs (M-CDVs) with a narrow size distribution and a hydrodynamic diameter *ca.* 100 nm in water were obtained. Details on the synthesis and characterization of the nanoparticles and M-CDVs are provided in the ESI.†

The gelation properties of co-assemblies consisting of Nap-GFFYS, of which 20% was substituted by Nap-GFFYS-AAP and 250 μM M-CDV (corresponding to

1 mg ml^{-1} CoFe$_2$O$_4$-nanoparticles), were analyzed by standard inverted vial tests using an acid triggered method based on the hydrolysis of glucono-δ-lactone (GdL).[13] Photographs of the gels are reported in the ESI (see Fig. S1†). A very low critical gelation concentration of less than 0.06 wt% total gelator amount was obtained, even though it is noteworthy that 2.2 wt% of non-gelating organic compounds (gluconic acid and M-CDV) are present in the gel. Furthermore, a shrinkage of the hydrogels with gelator concentrations lower than 0.07 wt% was observed after storage for more than 12 h. We assume that this is due to non-covalent cross-linking of the self-assembled AAP-modified nanofibers with the M-CDVs in these low concentration samples. For a hydrogel with 0.1 wt% total gelator content, a threshold GdL concentration of less than 0.25 mg ml^{-1} could be determined (see Fig. S2†). This observation indicates the highly efficient hydro-gelation of the supramolecular co-assembly of peptides and M-CDVs.

The photophysical properties of the AAP-modified gelator Nap-GFFYS-AAP were investigated by UV/vis spectroscopy (Fig. 2a). The characteristic absorption bands at 325 nm for the π → π* transition and at 426 nm for the n → π* transition are consistent with our previous reports for various AAP derivatives.[27,36] Under UV (λ = 365 nm) irradiation, a decrease in the π → π* band is observed, while a significant increase in the absorption band at 426 nm for the n → π* transition can be found. This observation reveals that the E- to Z-isomerization of the AAP moiety is completed after 10 min of irradiation. Through irradiation with visible light (λ = 520 nm), the original spectrum is restored, indicating back-isomerization to the thermodynamically stable E-isomer. The slightly increased band for the π → π* transition compared to the original spectrum is probably caused by minimal amounts of the Z-isomer present in the thermodynamic equilibrium of the non-irradiated (as-prepared) sample. This photo-switching between the E- and Z-isomers is fully reversible over at least three cycles (Fig. 2b).

Furthermore, the structure of the self-assembled nanofibers was analyzed using circular dichroism (CD) spectroscopy (see Fig. S3†).[37] The spectrum of the unmodified gelator Nap-GFFYS shows a strong negative band around 217 nm, which indicates β-sheet secondary structures. This β-sheet type of assembly structure remains unchanged when 20% of the unmodified gelator is substituted by Nap-GFFYS-AAP, as just a slight decrease in the band intensity but no change in the characteristic signal around 217 nm is observed. Thus, it can be concluded that the AAP modification does not interrupt or alter the assembly structure.

In order to investigate the influence of an external magnetic field on the mechanical properties of a hydrogel with 0.15 wt% total gelator content (20% Nap-GFFYS-AAP) and 250 μM M-CDVs (corresponding to 0.5 mg ml^{-1} magnetic nanoparticles), magneto-rheological experiments with 0.1% deformation and 1 Hz frequency were executed while applying a magnetic field (0.78 T) *in situ* for 1 min intervals (Fig. 3a). A higher storage modulus is observed when the field is switched on, whereas a lower storage modulus is observed when the field is switched off and reversible switching can be observed over at least four cycles. This observation can be explained by the alignment of the M-CDVs in a magnetic field,[28] in this case applied perpendicular to the shearing direction, thus increasing the interior resistance towards the flowing of the hydrogel and therefore, its stiffness. We note that reversibility is not given over the first half cycle, which is presumably caused by the rearrangement of the hydrogel structure directly around the M-CDVs during the first alignment process. Negative control

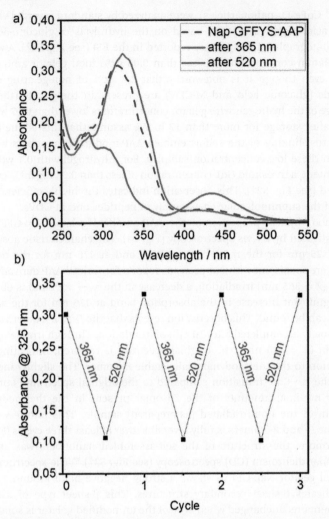

Fig. 2 (a) UV/vis spectra of the *E*- and *Z*-isomers of Nap-GFFYS-AAP (0.1 mM) in 20 mM HEPES buffer at pH = 8.01. (b) Photo-isomerization of Nap-GFFYS-AAP over three fully reversible cycles.

experiments without M-CDVs did not show a significant change in the mechanical properties during magnetization. This supports our hypothesis that M-CDV alignment during magnetization is the origin of the magneto-response of the hydrogel.

Moreover, the magneto-response can be visualized by the magnetic bending of a hydrogel rod (Fig. 4). Hydrogel rods were prepared starting from a solution of the gelator components and the M-CDVs in alkaline milieu (pH ≈ 12). GdL was added to the solution to slowly acidify the system and thus trigger the hydrogel formation. This solution was immediately transferred into the tip of a Pasteur pipette. After gelation overnight, the gel rod inside the pipette was carefully immersed into a water-filled vial. When placing a weak permanent magnet against the vial wall, a controlled and fast movement of the hydrogel rod towards

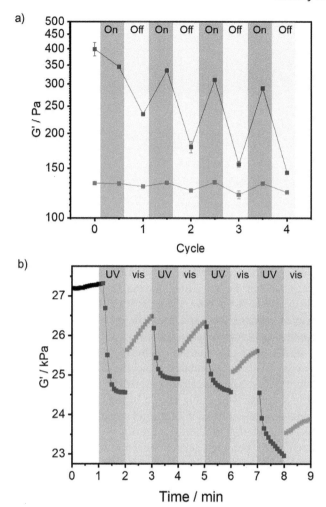

Fig. 3 (a) Impact of a magnetic field (0.78 T) on the storage modulus of the hydrogels over four cycles; red: 0.15 wt% Nap-GFFYS; blue: 0.15% wt% Nap-GFFYS with 20% Nap-GFFYS-AAP and 250 μM M-CDVs (corresponding to 1 mg ml^{-1} MNPs); On: 0.78 T, Off: 0.01 T. (b) Effect of UV (365 nm) or vis (520 nm) irradiation on the storage modulus of a 1 wt% Nap-GFFYS hydrogel with 20% of the LMWG substituted with Nap-GFFYS-AAP in the presence of 250 μM M-CDVs (corresponding to 0.5 mg ml^{-1} MNPs).

the magnet was observed (Fig. 4). The removal of the magnet resulted in a quick recovery of the initial unbent shape. This process could be repeated several times without any obvious damage or dissolution of the hydrogel. By placing the magnet in different positions against the vial wall, various bending directions could be obtained (see Fig. S4†). This finding underlines the high magneto-response, as well as the mechanical stability, of the supramolecular hydrogel.

The mechanical response of the hydrogel during irradiation was tested using a 1 wt% total gelator content (20% Nap-GFFYS-AAP) and 250 μM M-CDVs (corresponding to 0.5 mg ml^{-1} MNPs) hydrogel (Fig. 3b). Oscillatory rheology was performed with 0.1% deformation and 1 Hz frequency while irradiating *in situ*

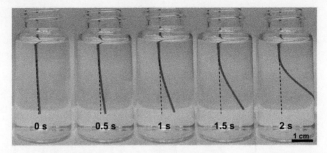

Fig. 4 Time-dependent bending of a 0.5 wt% Nap-GFFYS hydrogel rod with 20% of the LMWG substituted by Nap-GFFYS-AAP and 500 μM M-CDVs (corresponding to 2 mg ml^{-1} MNPs) in aqueous solution. A weak permanent magnet was placed on the right hand side of the sample.

with UV ($\lambda = 365$ nm) and vis ($\lambda = 520$ nm) light, respectively. Under UV irradiation, the hydrogel showed a fast decrease in the storage modulus of approximately 10%. This can be attributed to the dissociation of the host–guest interaction between the Z-AAP moiety and the M-CDVs, which decreases the cross-linking density of the nanofibers and, thus, the gel stiffness. Upon reconstruction of these cross-linking points under vis-light irradiation, causing the isomerization of the AAP unit back to its thermodynamically stable E-isomer, the storage modulus increases. This photo-induced switching of the gel stiffness is reversible over at least four cycles, even though an overall decrease in the storage modulus is observed. This decrease is probably caused by the conditions of the rheological measurements, since the formation of the gel structure is gradually inhibited by the continuously oscillating measurement setup. We note that an optimal photo-response is achieved when high gelator concentrations (>1 wt%) are used, while the highest magneto-response is observed when low gelator concentrations (≈ 0.1 wt%) are utilized. Furthermore, we note that while the magneto-response of the storage modulus is significantly higher than the photo-response for the hydrogels shown in Fig. 3, we contend that each response can be optimized by varying the composition of the hydrogel or the molecular structure of the peptide gelator.

Conclusion and outlook

In summary, we reported the first example of a magneto- and photo-responsive supramolecular hydrogel based on the self-assembly of peptides combined with the host–guest chemistry of cyclodextrins. The magneto-response was achieved *via* the incorporation of superparamagnetic CoFe$_2$O$_4$ nanoparticles in the membranes of β-cyclodextrin vesicles, which act as multivalent cross-links with the Nap-GFFYS-AAP gelator peptide. Furthermore, the photo-isomerization of AAP confers a photo-response to the gel. Indeed, the supramolecular hydrogel showed reversible magneto- and photo-rheological properties. Furthermore, the magneto-response could be visualized by bending a rod of the hydrogel using a weak permanent magnet. These results will aid the development of adaptive nanomaterials with applications in soft robotics, tissue engineering, and life-like materials. We note that functional guest molecules can potentially be embedded in the gel or displayed at the surface of the gel using host–guest chemistry[31–34] so that both the material stiffness and its biorecognition properties can be tuned using co-assembly and physical stimuli. We envisage that, in this way, bio-interfaces with tunable and responsive stiffness and ligand display can be assembled.

Methods and materials

The synthesis and analysis of the peptides and superparamagnetic nanoparticles are described in the ESI.† The M-CDVs were prepared by slow hydration and extrusion.[28] Oleic acid and oleylamine-capped cobalt ferrite NPs and amphiphilic β-CD were dissolved in chloroform. The solvent was slowly evaporated from the solution in an argon stream while continually rotating the flask. The residual solvent was removed from the film under strong vacuum. The film was hydrated with the addition of MilliQ water and vigorous stirring overnight. This solution was sonicated for 10 s and then passed repetitively through a 100 nm pore size polycarbonate membrane in a manual extruder. The hydrogels were prepared by gradual acidification using GdL.[11] Stock solutions of Nap-GFFYS and Nap-GFFYS-AAP were prepared in MilliQ water at pH ≈ 12 *via* stepwise addition of NaOH. The stock solutions were diluted to the respective concentrations and a solution of M-CDVs was added. This precursor solution was transferred into a vial with the respective amount of GdL present and mixed by gently shaking the vial. After leaving the vial unmoved overnight, hydrogel formation was tested by the standard inverted vial test.

Conflicts of interest

There are no conflicts to declare.

Notes and references

1 D. O. Wichterle and D. Lim, *Nature*, 1960, **185**, 117.
2 J. Kopeček and J. Yang, *Angew. Chem., Int. Ed.*, 2012, **51**, 7396–7417.
3 M. P. Lutolf and J. A. Hubbell, *Nat. Biotechnol.*, 2005, **23**, 47.
4 D. M. Ryan and B. L. Nilsson, *Polym. Chem.*, 2012, **3**, 18–33.
5 T. R. Hoare and D. S. Kohane, *Polymer*, 2008, **49**, 1993–2007.
6 X. Hu, D. Zhang and S. S. Sheiko, *Adv. Mater.*, 2018, **30**, 1707461.
7 A. Harada, R. Kobayashi, Y. Takashima, A. Hashidzume and H. Yamaguchi, *Nat. Chem.*, 2011, **3**, 34–37.
8 Z. Li, G. Davidson-Rozenfeld, M. Vázquez-González, M. Fadeev, J. Zhang, H. Tian and I. Willner, *J. Am. Chem. Soc.*, 2018, **140**, 17691.
9 E. K. Johnson, D. J. Adams and P. J. Cameron, *J. Mater. Chem.*, 2011, **21**, 2024–2027.
10 M. Zelzer and R. V. Ulijn, *Chem. Soc. Rev.*, 2010, **39**, 3351–3357.
11 D. J. Adams, K. Morris, L. Chen, L. C. Serpell, J. Bacsa and G. M. Day, *Soft Matter*, 2010, **6**, 4144.
12 X. Yang, G. Zhang and D. Zhang, *J. Mater. Chem.*, 2012, **22**, 38–50.

13 D. J. Adams, M. F. Butler, W. J. Frith, M. Kirkland, L. Mullen and P. Sanderson, *Soft Matter*, 2009, **5**, 1856.

14 L. Xu, L. Qiu, Y. Sheng, Y. Sun, L. Deng, X. Li, M. Bradley and R. Zhang, *J. Mater. Chem. B*, 2018, **6**, 510–517.

15 H. Xie, M. Asad Ayoubi, W. Lu, J. Wang, J. Huang and W. Wang, *Sci. Rep.*, 2017, **7**, 8459.

16 Y. Zhang, B. Zhang, Y. Kuang, Y. Gao, J. Shi, X. X. Zhang and B. Xu, *J. Am. Chem. Soc.*, 2013, **135**, 5008–5011.

17 K. J. C. van Bommel, M. C. A. Stuart, B. L. Feringa and J. van Esch, *Org. Biomol. Chem.*, 2005, **3**, 2917–2920.

18 E. R. Draper and D. J. Adams, *Chem. Commun.*, 2016, **52**, 8196–8206.

19 Y. Li, G. Huan, X. Zhang, B. Li, Y. Chen, T. Lu, T. J. Lu and F. Xu, *Adv. Funct. Mater.*, 2013, **23**, 660–672.

20 N. J. François, S. Allo, S. E. Jacobo and M. E. Daraio, *J. Appl. Polym. Sci.*, 2007, **105**, 647–655.

21 M. A.-P. O. Shefi, *Nano Lett.*, 2016, **16**, 2567–2573.

22 S.-H. Hu, T.-Y. Liu, D.-M. Liu and S.-Y. Chen, *Macromolecules*, 2007, **40**, 6786–6788.

23 X. Lin, B. N. Quoc and M. Ulbricht, *ACS Appl. Mater. Interfaces*, 2016, **8**, 29001–29014.

24 J. Hoque, N. Sangaj and S. Varghese, *Macromol. Biosci.*, 2018, 1800259.

25 T. M. Doran, D. M. Ryan and B. L. Nilsson, *Polym. Chem.*, 2014, **5**, 241–248.

26 Y. Huang, Z. Qiu, Y. Xu, J. Shi, H. Lin and Y. Zhang, *Org. Biomol. Chem.*, 2011, **9**, 2149–2155.

27 C.-W. Chu and B. J. Ravoo, *Chem. Commun.*, 2017, **53**, 12450–12453.

28 J. H. Schenkel, A. Samanta and B. J. Ravoo, *Adv. Mater.*, 2014, **26**, 1076–1080.

29 A. Sunny, K. S. Aneesh Kumar, V. Karunakaran, M. Aathira, G. R. Mutta, K. K. Maiti, V. R. Reddy and M. Vasundhara, *Nano-Struct. Nano-Objects*, 2018, **16**, 69–76.

30 S. Munjal, N. Khare, C. Nehate and V. Koul, *J. Magn. Magn. Mater.*, 2016, **404**, 166–169.

31 A. Samanta, M. C. A. Stuart and B. J. Ravoo, *J. Am. Chem. Soc.*, 2012, **134**, 19909–19914.

32 J. Voskuhl, M. C. A. Stuart and B. J. Ravoo, *Chem.–Eur. J.*, 2010, **16**, 2790–2796.

33 R. V. Vico, J. Voskuhl and B. J. Ravoo, *Langmuir*, 2011, **27**, 1391–1397.

34 U. Kauscher and B. J. Ravoo, *Beilstein J. Org. Chem.*, 2012, **8**, 1543–1551.

35 L. Lv, H. Liu, X. Chen and Z. Yang, *Colloids Surf., B*, 2013, **108**, 352–357.

36 L. Stricker, E.-C. Fritz, M. Peterlechner, N. L. Doltsinis and B. J. Ravoo, *J. Am. Chem. Soc.*, 2016, **138**, 4547–4554.

37 G. Gottarelli, S. Lena, S. Masiero, S. Pieraccini and G. P. Spada, *Chirality*, 2008, **20**, 471–485.

Faraday Discussions

PAPER

Hierarchically patterned striped phases of polymerized lipids: toward controlled carbohydrate presentation at interfaces†

Tyson C. Davis,‡[a] Jeremiah O. Bechtold,‡[a] Tyler R. Hayes,[a]
Terry A. Villarreal[a] and Shelley A. Claridge [ID] *[ab]

Received 19th February 2019, Accepted 7th March 2019

DOI: 10.1039/c9fd00022d

Complex biomolecules, including carbohydrates, frequently have molecular surface footprints larger than those in broadly utilized standing phase alkanethiol self-assembled monolayers, yet would benefit from structured orientation and clustering interactions promoted by ordered monolayer lattices. Striped phase monolayers, in which alkyl chains extend across the substrate, have larger, more complex lattices: nm-wide stripes of headgroups with 0.5 or 1 nm lateral periodicity along the row, separated by wider (~5 nm) stripes of exposed alkyl chains. These anisotropic interfacial patterns provide a potential route to controlled clustering of complex functional groups such as carbohydrates. Although the monolayers are not covalently bound to the substrate, assembly of functional alkanes containing an internal diyne allows such monolayers to be photopolymerized, increasing robustness. Here, we demonstrate that, with appropriate modifications, microcontact printing can be used to generate well-defined microscopic areas of striped phases of both single-chain and dual-chain amphiphiles (phospholipids), including one (phosphoinositol) with a carbohydrate in the headgroup. This approach generates hierarchical molecular-scale and microscale interfacial clustering of functional ligands, prototyping a strategy of potential relevance for glycobiology.

Introduction

Interfaces with precisely constructed chemical environments at the micrometer and nanometer scales are required for applications ranging from the design of electronic devices to the controlled display of complex biomolecules.[1] Increasingly, the goals of controlling interfacial structure may include not only

[a]Department of Chemistry, Purdue University, West Lafayette, IN 47907, USA. E-mail: claridge@purdue.edu
[b]Weldon School of Biomedical Engineering, Purdue University, West Lafayette, IN 47907, USA

† Electronic supplementary information (ESI) available: Detailed experimental methods, large-scale SEM images. See DOI: 10.1039/c9fd00022d

‡ These two authors made equal contributions.

positioning functional groups on the surface, but also controlling their orientation, clustering, or placement relative to other functional groups, mimicking complex structures such as those in cell membranes.

Monolayers of molecules such as alkanethiols have been broadly utilized to structure interfacial chemistry, particularly on coinage metals.[2] In alkanethiol monolayers, ordered lattices of alkyl chains position terminal functional groups with nearest-neighbor distances ∼0.5 nm, tilted at angles influenced by the bond between the thiol and the substrate.[2] Lattices displaying simple functional groups (*e.g.* carboxylic acids) influence further assembly at the interface (*e.g.* selecting for specific crystal facets of calcite); microcontact printing enables geometrically patterned assembly over microscopic (or large nanoscopic) areas.[3–5]

Controlling presentation of more complex, biologically relevant functionalities raises new challenges. In biological environments, polysaccharides, peptides, and other entities are presented in controlled orientations, with both nanoscale and microscale spatial ordering. To mimic elements of these environments for applications such as high-throughput screening of biomolecular interactions,[6,7] it would be useful to present microstructured areas of surface containing nanostructured clusters of specific ligand chemistries, enabling multivalent binding similar to molecular recognition events in the glycocalyx.[8–17]

However, even monosaccharides occupy interfacial footprints substantially greater than that of an alkyl chain in an alkanethiol monolayer (∼0.25 nm^2). Thus, creating simple lattices of these larger moieties becomes less straightforward. Designing complex clusters of functional groups at biologically relevant scales— with linear dimensions large relative to alkyl chain nearest neighbor distances in standing phases (>0.5 nm) but small relative to those typically achieved through microcontact printing (significantly <100 nm)—becomes especially challenging.

One complementary strategy for clustering structures with larger footprints arises from a transformation to the monolayer structure.[1] Since at least the 1960s, it has been known that long-chain alkanes can adopt lying down orientations on graphite and other layered materials such as MoS$_2$ and WS$_2$.[18,19] More recently, the surface chemistry of 2D materials (particularly graphite and graphene) has been regulated using striped phases of functional alkanes,[18–23] in which the alkyl chains extend horizontally across the substrate. Scanning probe microscopy studies[18–23] have shown that this arrangement produces nm-wide stripes of headgroups with 0.5 or 1 nm lateral periodicity along the row (for single-chain and dual-chain amphiphiles, respectively), separated by wider (∼5 nm, dependent on chain length) stripes of exposed alkyl chains. Assembly of functional alkanes containing an internal diyne allows the monolayer to be photopolymerized, creating a conjugated ene–yne polymer backbone that has been studied extensively in the context of molecular electronics.[21,23,24] Polymerization also stabilizes the non-covalently adsorbed monolayer, increasing potential utility of patterns of functional groups displayed at the interface.[25–27]

Just as clustering of functional groups at biological active sites creates unique chemical environments to promote specific interactions, precise positioning of functional groups in striped phases also creates unique chemical environments (Fig. 1). We have observed that striped phases of diyne phospholipids[26] exhibit distinct characteristics in comparison with striped phases composed of other amphiphiles.[28] Phospholipids can adopt a 'sitting' orientation in which the terminal amine in the headgroup protrudes a few Ångströms from the interface.[26]

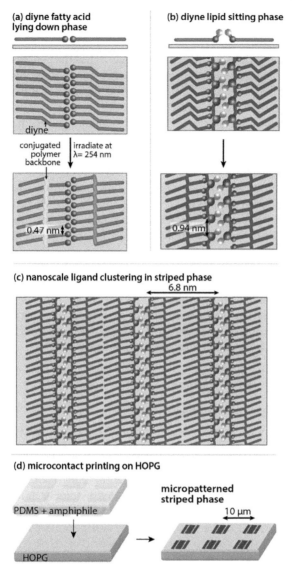

Fig. 1 Illustrations of: (a) striped phase of diynoic acids on HOPG, showing a 0.47 nm distance between functional groups along the stripe direction; (b) striped phase of diyne phospholipids, showing a 0.94 nm distance between functional groups along the row; (c) multiple rows of the striped phase, showing lamellar periodicity, a route to nanoscale ordering of complex functional groups; and (d) illustration of poly(dimethylsiloxane) (PDMS) transfer of amphiphiles to HOPG to form striped phases.

The phosphate and ester functional groups create a tailored chemical environment around the amine. Both head and chain structures influence nano- and micro-scale assembly of striped phases,[27,29,30] and chain elements including the position of the polymer backbone can be used to modulate solvent availability of the polar headgroups.[28] Flexible 1D zwitterionic arrays formed by the striped phase also impact further assembly of inorganic and organic nanostructures at the interface.[31,32] More broadly, the striped phospholipid polymer architecture represents a potential means for flexible yet controlled presentation of ligands at the interface.

Microcontact printing of striped phases (Fig. 1d) has the potential to combine microscopic geometric control over surface chemistry with molecular-scale control over ligand presentation, a capability of potential use in glycobiology. However, the strong focus on molecular-scale structure in noncovalent striped-phase monolayers on highly oriented pyrolytic graphite (HOPG) has meant that such monolayers are typically ordered and characterized at length scales < 100 nm.[33–35] Recently, we have shown that some amphiphiles order into striped phases with edge lengths > 10 μm,[27] scales relevant to controlling interactions with biological entities, and that monolayer ordering can be characterized by scanning electron microscopy (SEM), making it possible to characterize surface functionalization up to mm scales.[29,30,36] Some noncovalent monolayers can also be robust enough to survive vigorous solution processing and other environmental interactions.[27,31]

Here, we demonstrate microcontact printing of striped phases of amphiphiles on HOPG, utilizing both diyne amphiphiles (e.g. diynoic acids, diyne phospholipids) and a saturated phosphoinositol. This approach generates hierarchical molecular-scale and microscale interfacial clustering of functional ligands, including carbohydrates, prototyping a strategy of potential relevance for controlled presentation of carbohydrates at interfaces.

Results and discussion

Preparation of striped monolayers on HOPG

Striped monolayers of both single-chain amphiphiles (e.g. 10,12-pentacosadiynoic acid (PCDA), Fig. 2a and b) and dual-chain amphiphiles (e.g. 1,2-bis(10,12-tricosadiynoyl)-sn-glycero-3-phosphocholine (diyne PC), Fig. 2a and c) are typically prepared via drop-casting or Langmuir–Schaefer (LS) conversion,[23,26,27,31,37,38] then polymerized via UV irradiation and characterized by atomic force microscopy (AFM) (Fig. 2d and e). In AFM images, striped lamellar patterns are oriented at 120° angles, in epitaxy with the HOPG lattice; each stripe represents a row of lying-down molecules. SEM images of striped phases (Fig. 2f–i) typically exhibit brighter areas representing the molecular domains, against a darker background of HOPG. Long linear features along the image diagonals in Fig. 2f and g represent step edges in the HOPG substrates. Higher-resolution SEM images (Fig. 2h and i) reveal linear defects within the ordered molecular domains, highlighting the directionality of the molecular rows.[29] Use of this combination of techniques enables us to characterize both microscopic and nanoscopic ordering in striped phases, including those with carbohydrate headgroups (vide infra).

Preparation of patterned striped monolayers on HOPG by microcontact printing

Microscopic patterns of striped phase monolayers were prepared on HOPG by microcontact printing,[14] as shown in Fig. 3. Stamps used for microcontact printing of alkanethiols on gold are commonly prepared with a 10 : 1 ratio of elastomer base to crosslinker, resulting in a nominal elastic modulus of ∼2.6 ± 0.02 MPa at commonly used curing conditions (65 °C, 1 h).[39] For transfer to

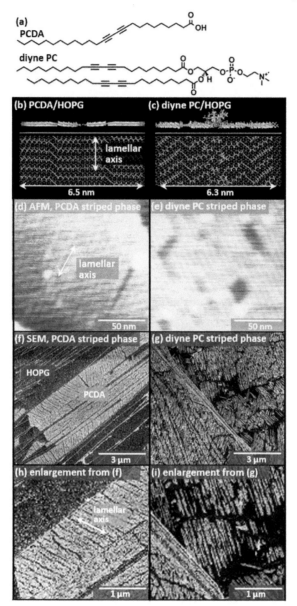

Fig. 2 (a) Structures of PCDA and diyne PC. (b, c) Molecular models of striped phases of (b) PCDA and (c) diyne PC on HOPG. (d, e) AFM images of striped phases of (d) PCDA and (e) diyne PC, illustrating the lamellar pattern. (f–i) SEM images of striped phases of (f, h) PCDA and (g, i) diyne PC, illustrating long-range ordering.

HOPG, which has relatively low local surface roughness, we often found that stamps prepared with a 10:2 ratio of base to crosslinker (nominal elastic modulus 3.6 ± 0.1 MPa)[39] improved transfer fidelity, while still enabling conformal contact.

A number of studies have previously examined factors relating to ink delivery to the substrate, with the goals of limiting diffusion of the ink outside the stamp

Fig. 3 (a) SEM image of microscopic areas of PCDA striped phases assembled on HOPG by microcontact printing. (b) Higher-resolution SEM image illustrating coverage in the square interior and the small fractional coverage of molecular domains assembled outside the stamp contact area. An AFM image (inset in (b)) shows the striped phase structure.

contact area,[40–42] and limiting delivery of impurities from the PDMS stamp.[41,43] Delivering a controlled amount of diyne amphiphile to the substrate is especially important in assembling noncovalent monolayers; screening several possible methods for controlling diyne amphiphile delivery, we found that immersing the stamp in a solution of amphiphile in carrier solvent (1.1 mM for PCDA and single-chain amphiphiles, 0.55 mM for diyne PC and dual chain amphiphiles, maintaining the concentration of alkyl chains) generally maximized coverage of striped phase inside the contact area while minimizing coverage outside the contact area.

Ink concentrations used here are similar to those typically utilized for assembly of standing phases of alkanethiols on Au (1–10 mM),[2] although fewer molecules are required to fill a given area of the surface: the molecular footprint of an alkyl chain in a lying-down phase (1.5 nm^2 for PCDA) is much larger than for a standing phase ($\sim$0.25 nm^2). Fig. 3a and b show SEM images of a pattern of squares transferred to HOPG using the stamp preparation and inking conditions described above. Fig. 3b shows a higher-resolution image of the square pattern.

High coverage is observed within the squares; AFM is used to verify that molecular coverage is comprised of striped domains (Fig. 3b, inset, and ESI†). Areas between the square stamp contact areas (channel regions) contain low number-densities of long, narrow molecular domains characteristic of submonolayer island nucleation and growth under conditions of low surface monomer concentrations.[44] Areas between squares also contain material that appears in dark contrast in SEM images. Similar features appear on substrates brought into contact with stamps wetted with the carrier solvents in the absence of amphiphile (see ESI†). Deposition of impurities is also common in microcontact printing of alkanethiols on gold. Previous studies suggest that the deposited material is the oligomeric PDMS crosslinker, in which hydrosilyl groups undergo oxidation to form more polar species exhibiting increased solubility in the ink or carrier solvent.[43,45]

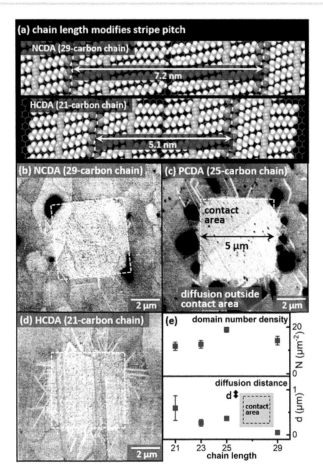

Fig. 4 (a) Molecular models of diynoic acid striped phases with the longest (29 carbon) and shortest (21 carbon) chains utilized in these experiments. (b–d) SEM images of 10,12-diynoic acids: (b) nonacosadiynoic acid (NCDA, 29-carbon chain), (c) pentacosadiynoic acid (PCDA, 25-carbon chain), (d) henicosadiynoic acid (HCDA, 21-carbon chain). (e) Average domain number density per μm², N, and average distance molecular layer extends outside stamped area, d, for chain lengths from 21–29 carbons.

Transfer characteristics of single-chain amphiphiles based on chain length

In using a striped phase to pattern functionality at an interface, shorter chain lengths correspond to smaller stripe pitch values, and thus shorter distances between linear clusters of functional groups on the surface (Fig. 4a). However, chain length also impacts dynamics in the self-assembly process. In previous demonstrations of microcontact printing to form standing phases (*e.g.* alkanethiols on Au), others have observed that molecular diffusion around the stamp contact area increases for molecular inks with shorter chains.[46–48] Here, we tested the transfer and assembly of 10,12-diynoic acids with chain lengths from 21 to 29 carbons to form noncovalently adsorbed striped phases to better understand the range of pitches that can reasonably be established, and the fidelity of patterning (Fig. 4b–d). In the Figure, areas exhibiting linear defects typical of striped phases (similar to those in Fig. 2h) have been colored yellow as a guide to the eye. Image segmentation was used to estimate the average distance over which each amphiphile spread outside the stamped area in areas with good stamp contact (Fig. 4d, see ESI† for example AFM images used for segmentation). The average band through which molecules diffuse decreases in width from ~600 nm for HCDA to ~50 nm for NCDA. For all four carboxylic acids, the number density of domains was 10–20 per μm^2 within the contact

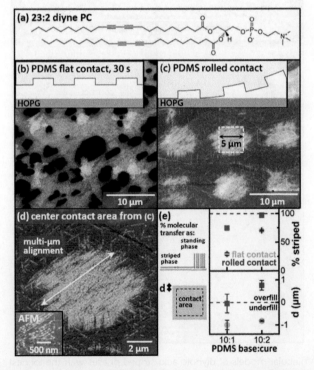

Fig. 5 (a) Structure of diyne PC. (b–d) SEM images of 0.5 mM diyne PC in EtOH transferred to HOPG using (b) 30 s flat contact and (c, d) rolled contact (stamp prepared at 10 : 2 base : crosslinker ratio). (e) Comparison of % striped phase (*vs.* standing phase) molecular transfer with flat and rolled stamp contact, and fill of contact area, for PDMS stamps prepared with 10 : 1 and 10 : 2 base : crosslinker ratios.

area, which is reasonable given that the monomer concentration in the ink solution was the same for each molecule.

Transfer of dual-chain amphiphiles

Commercially available diyne phospholipids have two alkyl chains and a zwitterionic headgroup, which would be expected to modulate molecular transfer and spreading on the substrate in comparison with the single-chain carboxylic acids transferred above. Here, we test the transfer behavior of two diyne phospholipids, 23:2 diyne phosphocholine (diyne PC, Fig. 5) and 23:2 diyne phosphoethanolamine (diyne PE, Fig. 6). The phospholipid structures are identical with the exception that the bulky terminal quaternary ammonium in the PC headgroup (Fig. 5a) limits molecular packing in comparison with PE, which has a smaller terminal primary amine.

Transfer conditions similar to those optimized for single-chain amphiphiles result in a large fraction of standing phase formation (bright areas in square centers) (Fig. 5b, highlighted in yellow as a guide to the eye; also see ESI†). This is reasonable given the large number of alkyl carbons per molecule, promoting interchain interactions leading to standing phase formation. To mechanically

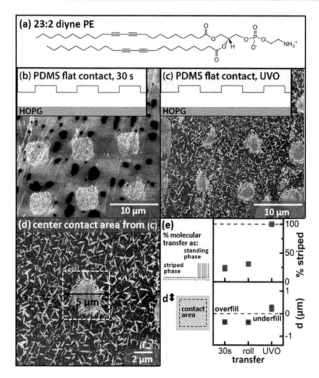

Fig. 6 (a) Structure of diyne PE. (b–d) SEM images of 0.5 mM diyne PE in EtOH transferred to HOPG using (b) 30 s flat contact and (c, d) flat contact with stamp hydrophilicity increased with UV ozone (stamp prepared at 10 : 2 base : crosslinker ratio). (e) Comparison of % striped phase (vs. standing phase) molecular transfer with flat contact, rolled contact, and flat contact with UV ozone, and fill of contact area, for PDMS stamps prepared with 10 : 2 base : crosslinker ratios.

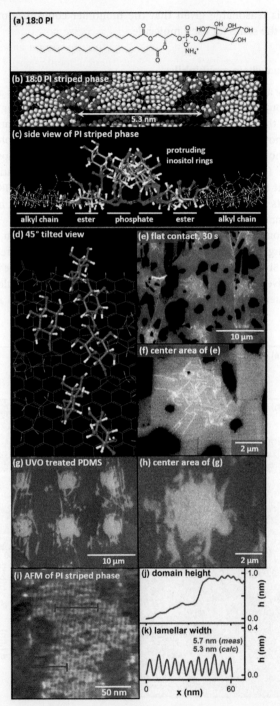

Fig. 7   (a) Structure of 18:0 phosphoinositol (18:0 PI). (b–d) Minimized molecular models of striped phase of 18:0 PI on HOPG, illustrating: (b) lamellar width, (c) projection of inositol rings, in side view, (d) spacing of inositol rings (45° tilted view). (e–h) SEM images of PI striped phases formed using (e, f) rolling contact and (g, h) UV ozone-treated stamps for microcontact printing. (i) AFM image of PI striped phase, and line scans illustrating (j) domain height and (k) lamellar width.

destabilize interchain interactions (*e.g.* standing phases) on the stamp, and to initiate domain growth from a limited area (to increase post-transfer molecular alignment), we tested molecular delivery by rolling the stamp along the HOPG surface (Fig. 5c and d, see ESI† for more experimental detail regarding the rolling procedure). Testing transfer from stamps prepared with both 10 : 1 and 10 : 2 PDMS elastomer base : crosslinker ratios, we found that rolled contact increased the percentage of molecular transfer that produced striped phases (to near 100% for 10 : 2 stamps with rolled contact, Fig. 5e). Flat contact typically resulted in underfilling of the stamp contact area, while rolled contact resulted in average coverage zones extending nearly 1 μm outside the stamp contact area (as visible in Fig. 5d). In some cases (again, see Fig. 5d), rolled contact produced molecular alignment across the stamp contact areas (*i.e.*, lamellar axes aligned from upper left to lower right in Fig. 5d). Using other contact geometries, we have not observed this behavior, so with further optimization, rolled contact may represent a means of achieving long-range molecular alignment in printed striped phases, for applications in which such alignment is desirable.

Diyne PE (Fig. 6a) has a smaller terminal amine group that enables stronger lateral interactions between headgroups in standing phases, in comparison with the PC headgroup (which is bulky enough to limit packing). Importantly, the primary amine can also act as a functional handle for further coupling reactions, of potential utility in elaborating headgroups for glycobiological applications. Microcontact transfers of diyne PE in the conventional flat contact geometry also produced large areas of molecules assembled in standing phases (Fig. 6b). For transfer of diyne PE, the highest percentages of striped phase were observed for transfers in which the stamp surface hydrophilicity was increased by treatment with UV ozone plasma (a process which has been used previously to transfer hydrophilic molecules to create standing phase self-assembled monolayers (SAMs)). While multiple factors may contribute to the observed improvement in striped phase assembly during transfer, one possibility is that the hydrophilic stamp enables PE to assemble with polar headgroups oriented toward the stamp surface, with tails oriented favorably to mediate the initial stages of adsorption to HOPG for striped phase assembly. The differences in transfer behavior observed for molecules as structurally similar as diyne PE and diyne PC suggests a need to carefully balance molecule–stamp, molecule–molecule, and molecule–substrate interaction strengths for transfer of complex amphiphiles such as those relevant to glycobiology.

Striped phases from carbohydrate-conjugated lipids

The procedures developed above are also useful for microcontact printing of phospholipids incorporating carbohydrates in the headgroups. Here, we demonstrate that 1,2-distearoyl-*sn-glycero*-3-phosphoinositol (18:0 PI, Fig. 7a), a phospholipid with an O-linked monosaccharide appended to the phosphate, can assemble into striped phases through microcontact printing (Fig. 7b–d, models; Fig. 7e–h, SEM). As with other phospholipids, bringing the stamp into flat contact with the HOPG substrate resulted in assembly of standing phases (see ESI†), while rolling contact or stamps treated with UV ozone produced striped phases with domain lengths in some cases >2 μm (Fig. 7f). Characterization of domain structure based on SEM images is more challenging for these amphiphiles, since they lack the polymerizable diyne group, and thus do not exhibit cracking defects under the electron beam. However, AFM images (Fig. 7i) reveal a lamellar structure consistent with that predicted by molecular models, with average peak domain heights of ~0.8 nm (Fig. 7j, corresponding to inositol headgroup ridges), and measured lamellar widths of 5.7 nm (Fig. 7j), similar to the modeled values of 5.3 nm.

Conclusions

Here, we have demonstrated that it is feasible to use microcontact printing to create microscale striped patterns of amphiphiles. Stripes were printed using diynoic acids with chain lengths from 21–29 carbons, diyne phospholipids with phosphocholine and phosphoethanolamine headgroups, and phosphoinositol with 18-carbon saturated chains. The lamellar structures assembled in this way present 1 nm-wide stripes of functional headgroups with pitches from 5–10 nm determined by alkyl chain length. In the cell membrane, amphiphiles with diverse headgroup chemistry, including pendant carbohydrates, are used to mediate interactions with other cells and the extracellular matrix. Our findings point to the possibility that similarly diverse headgroup chemistries could be installed in striped phases, either directly through Langmuir–Schaefer conversion, or through post-assembly modification using common coupling chemistries. Overall, this illustrates a new route for controlled molecular-scale clustering of complex ligands such as carbohydrates at interfaces.

Conflicts of interest

There are no conflicts to declare.

Acknowledgements

S. A. C. acknowledges support through a DARPA Young Faculty Award, N66001-17-1-4046.

Notes and references

1 S. A. Claridge, Standing, lying, and sitting: Translating building principles of the cell membrane to synthetic 2D material interfaces, *Chem. Commun.*, 2018, **54**, 6681–6691.
2 J. C. Love, L. A. Estroff, J. K. Kriebel, R. G. Nuzzo and G. M. Whitesides, Self-assembled monolayers of thiolates on metals as a form of nanotechnology, *Chem. Rev.*, 2005, **105**, 1103–1169.
3 J. Aizenberg, A. J. Black and G. M. Whitesides, Control of crystal nucleation by patterned self-assembled monolayers, *Nature*, 1999, **398**, 495–498.
4 Y. J. Han and J. Aizenberg, face-selective nucleation of calcite on self-assembled monolayers of alkanethiols: Effect of the parity of the alkyl chain, *Angew. Chem., Int. Ed.*, 2003, **42**, 3668–3670.
5 B. Pokroy and J. Aizenberg, Calcite shape modulation through the lattice mismatch between the self-assembled monolayer template and the nucleated crystal face, *CrystEngComm*, 2007, **9**, 1219–1225.

6 R. Liang, L. Yan, J. Loebach, M. Ge, Y. Uozumi, K. Sekanina, N. Horan, J. Gildersleeve, C. Thompson, A. Smith, K. Biswas, W. C. Still and D. Kahne, Parallel synthesis and screening of a solid phase carbohydrate library, *Science*, 1996, **274**, 1520–1522.

7 O. Oyelaran and J. C. Gildersleeve, Glycan arrays: Recent advances and future challenges, *Curr. Opin. Chem. Biol.*, 2009, **13**, 406–413.

8 L. L. Kiessling and N. L. Pohl, Strength in numbers: Non-natural polyvalent carbohydrate derivatives, *Chem. Biol.*, 1996, **3**, 71–77.

9 J. E. Gestwicki, C. W. Cairo, L. E. Strong, K. A. Oetjen and L. L. Kiessling, Influencing receptor–ligand binding mechanisms with multivalent ligand architecture, *J. Am. Chem. Soc.*, 2002, **124**, 14922–14933.

10 L. L. Kiessling, J. E. Gestwicki and L. E. Strong, Synthetic multivalent ligands as probes of signal transduction, *Angew. Chem., Int. Ed.*, 2006, **45**, 2348–2368.

11 A. Kumar and G. M. Whitesides, Features of gold having micrometer to centimeter dimensions can be formed through a combination of stamping with an elastomeric stamp and an alkanethiol ink followed by chemical etching, *Appl. Phys. Lett.*, 1993, **63**, 2002–2004.

12 A. Kumar, H. A. Biebuyck and G. M. Whitesides, Patterning self-assembled monolayers: applications in materials science, *Langmuir*, 1994, **10**, 1498–1511.

13 A. Perl, D. N. Reinhoudt and J. Huskens, Microcontact printing: Limitations and achievements, *Adv. Mater.*, 2009, **21**, 2257–2268.

14 Y. N. Xia and G. M. Whitesides, Soft lithography, *Annu. Rev. Mater. Sci.*, 1998, **28**, 153–184.

15 S. Park, J. C. Gildersleeve, O. Blixt and I. Shin, Carbohydrate microarrays, *Chem. Soc. Rev.*, 2013, **42**, 4310–4326.

16 L. L. Kiessling, Chemistry-driven glycoscience, *Bioorg. Med. Chem.*, 2018, **26**, 5229–5238.

17 X. Han, Y. Zheng, C. J. Munro, Y. Ji and A. B. Braunschweig, Carbohydrate nanotechnology: Hierarchical Assembly using nature's other information carrying biopolymers, *Curr. Opin. Biotechnol.*, 2015, **34**, 41–47.

18 A. J. Groszek, Preferential adsorption of normal hydrocarbons on cast iron, *Nature*, 1962, **196**, 531–533.

19 A. J. Groszek, Preferential adsorption of long-chain normal paraffins on MoS_2 WS_2 and graphite from N-heptane, *Nature*, 1964, **204**, 680.

20 J. P. Rabe and S. Buchholz, Commensurability and mobility in 2-dimensional molecular patterns on graphite, *Science*, 1991, **253**, 424–427.

21 P. C. M. Grim, S. De Feyter, A. Gesquiere, P. Vanoppen, M. Rucker, S. Valiyaveettil, G. Moessner, K. Mullen and F. C. De Schryver, Submolecularly resolved polymerization of diacetylene molecules on the graphite surface observed with scanning tunneling microscopy, *Angew. Chem., Int. Ed. Engl.*, 1997, **36**, 2601–2603.

22 D. M. Cyr, B. Venkataraman and G. W. Flynn, STM investigations of organic molecules physisorbed at the liquid–solid interface, *Chem. Mater.*, 1996, **8**, 1600–1615.

23 Y. Okawa and M. Aono, Linear chain polymerization initiated by a scanning tunneling microscope tip at designated positions, *J. Chem. Phys.*, 2001, **115**, 2317–2322.

24 Y. Okawa, M. Akai-Kasaya, Y. Kuwahara, S. K. Mandal and M. Aono, Controlled chain polymerisation and chemical soldering for single-molecule electronics, *Nanoscale*, 2012, **4**, 3013–3028.

25 B. Li, K. Tahara, J. Adisoejoso, W. Vanderlinden, K. S. Mali, S. De Gendt, Y. Tobe and S. De Feyter, Self-assembled air-stable supramolecular porous networks on graphene, *ACS Nano*, 2013, **7**, 10764–10772.

26 J. J. Bang, K. K. Rupp, S. R. Russell, S. W. Choong and S. A. Claridge, Sitting phases of polymerizable amphiphiles for controlled functionalization of layered materials, *J. Am. Chem. Soc.*, 2016, **138**, 4448–4457.

27 T. R. Hayes, J. J. Bang, T. C. Davis, C. F. Peterson, D. G. McMillan and S. A. Claridge, Multimicrometer noncovalent monolayer domains on layered materials through thermally controlled Langmuir–Schaefer conversion for noncovalent 2D functionalization, *ACS Appl. Mater. Interfaces*, 2017, **9**, 36409–36416.

28 T. A. Villarreal, S. R. Russell, J. J. Bang, J. K. Patterson and S. A. Claridge, Modulating wettability of layered materials by controlling ligand polar headgroup dynamics, *J. Am. Chem. Soc.*, 2017, **139**, 11973–11979.

29 T. C. Davis, J. J. Bang, J. T. Brooks, D. G. McMillan and S. A. Claridge, Hierarchically patterned noncovalent functionalization of 2D materials by controlled Langmuir–Schaefer conversion, *Langmuir*, 2018, **34**, 1353–1362.

30 J. J. Bang, A. G. Porter, T. C. Davis, T. R. Hayes and S. A. Claridge, Spatially controlled noncovalent functionalization of 2D materials based on molecular architecture, *Langmuir*, 2018, **34**, 5454–5463.

31 S. W. Choong, S. R. Russell, J. J. Bang, J. K. Patterson and S. A. Claridge, Sitting phase monolayers of polymerizable phospholipids create dimensional, molecular-scale wetting control for scalable solution based patterning of layered materials, *ACS Appl. Mater. Interfaces*, 2017, **9**, 19326–19334.

32 A. G. Porter, T. Ouyang, T. R. Hayes, J. Biechele-Speziale, S. R. Russell and S. A. Claridge, 1 nm-wide hydrated dipole arrays regulate AuNW assembly on striped monolayers in nonpolar solvent, *Chem*, 2019, DOI: 10.1016/j.chempr.2019.07.002.

33 R. K. Workman, A. M. Schmidt and S. Manne, Detection of a diffusive two-dimensional gas of amphiphiles by lateral force microscopy, *Langmuir*, 2003, **19**, 3248–3253.

34 R. K. Workman and S. Manne, Molecular transfer and transport in noncovalent microcontact printing, *Langmuir*, 2004, **20**, 805–815.

35 J. S. Hovis and S. G. Boxer, Patterning and composition arrays of supported lipid bilayers by microcontact printing, *Langmuir*, 2001, **17**, 3400–3405.

36 S. R. Russell, T. C. Davis, J. J. Bang and S. A. Claridge, Spectroscopic metrics for alkyl chain ordering in lying-down noncovalent monolayers of diynoic acids on graphene, *Chem. Mater.*, 2018, **30**, 2506–2514.

37 A. Miura, S. De Feyter, M. M. S. Abdel-Mottaleb, A. Gesquiere, P. C. M. Grim, G. Moessner, M. Sieffert, M. Klapper, K. Mullen and F. C. De Schryver, Light- and STM-tip-induced formation of one-dimensional and two-dimensionalorganic nanostructures, *Langmuir*, 2003, **19**, 6474–6482.

38 R. Giridharagopal and K. F. Kelly, Substrate-dependent properties of polydiacetylene nanowires on graphite and MoS₂, *ACS Nano*, 2008, **2**, 1571–1580.

39 Z. Wang, A. A. Volinsky and N. D. Gallant, Crosslinking effect on polydimethylsiloxane elastic modulus measured by custom-built compression instrument, *J. Appl. Polym. Sci.*, 2014, **131**, 41050.

40 Y. Xia and G. M. Whitesides, Use of controlled reactive spreading of liquid alkanethiol on the surface of gold to modify the size of features produced by microcontact printing, *J. Am. Chem. Soc.*, 1995, **117**, 3274–3275.

41 R. B. A. Sharpe, D. Burdinski, J. Huskens, H. J. W. Zandvliet, D. N. Reinhoudt and B. Poelsema, Spreading of 16-mercaptohexadecanoic acid in microcontact printing, *Langmuir*, 2004, **20**, 8646–8651.

42 L. Libioulle, A. Bietsch, H. Schmid, B. Michel and E. Delamarche, Contact-inking stamps for microcontact printing of alkanethiols on gold, *Langmuir*, 1999, **15**, 300–304.

43 D. J. Graham, D. D. Price and B. D. Ratner, Solution assembled and microcontact printed monolayers of dodecanethiol on gold: A multivariate exploration of chemistry and contamination, *Langmuir*, 2002, **18**, 1518–1527.

44 I. Doudevski, W. A. Hayes and D. K. Schwartz, Submonolayer island nucleation and growth kinetics during self-assembled monolayer formation, *Phys. Rev. Lett.*, 1998, **81**, 4927–4930.

45 R. B. A. Sharpe, D. Burdinski, C. van der Marel, J. A. J. Jansen, J. Huskens, H. J. W. Zandvliet, D. N. Reinhoudt and B. Poelsema, Ink dependence of poly(dimethylsiloxane) contamination in microcontact printing, *Langmuir*, 2006, **22**, 5945–5951.

46 S. A. Claridge, W.-S. Liao, J. C. Thomas, Y. Zhao, H. H. Cao, S. Cheunkar, A. C. Serino, A. M. Andrews and P. S. Weiss, From the bottom up: Dimensional control and characterization in molecular monolayers, *Chem. Soc. Rev.*, 2013, **42**, 2725–2745.

47 C. Srinivasan, T. J. Mullen, J. N. Hohman, M. E. Anderson, A. A. Dameron, A. M. Andrews, E. C. Dickey, M. W. Horn and P. S. Weiss, Scanning electron microscopy of nanoscale chemical patterns, *ACS Nano*, 2007, **1**, 191–201.

48 E. Delamarche, H. Schmid, A. Bietsch, N. B. Larsen, H. Rothuizen, B. Michel and H. Biebuyck, Transport mechanisms of alkanethiols during microcontact printing on gold, *J. Phys. Chem. B*, 1998, **102**, 3324–3334.

PAPER

Charge-tuning of glycosaminoglycan-based hydrogels to program cytokine sequestration

Uwe Freudenberg,†[a] Passant Atallah,†[a] Yanuar Dwi Putra Limasale[a] and Carsten Werner [ID] *[abc]

Received 11th February 2019, Accepted 22nd February 2019

DOI: 10.1039/c9fd00016j

Glycosaminoglycan (GAG)-based biohybrid hydrogels of varied GAG content and GAG sulfation pattern were prepared and applied to sequester cytokines. The binding of strongly acidic and basic cytokines correlated with the integral space charge density of the hydrogel, while the binding of weakly charged cytokines was governed by the GAG sulfation pattern.

Sulfated glycosaminoglycans (GAGs) play a key role in the presentation of soluble signalling molecules in living tissues and recent research has established powerful methodologies for analysing and modulating their binding characteristics.[1-6] In parallel, biohybrid hydrogels containing sulfated GAGs were introduced and successfully applied to direct cell-fate decisions by mimicking the capacity of extracellular matrices (ECM) to reversibly bind, stabilize and sustainably deliver a plethora of cytokines, chemokines and growth factors.[7,8] The approach allowed for the investigation of new therapeutic concepts[9,10] as well as for the development of advanced tissue and disease *in vitro* models.[11-14]

Soluble signalling molecules are known to interact with negatively charged GAGs through their positively charged surface domains, but structural features, in particular the spatial distribution of ionized groups and intermolecular forces beyond electrostatics, were shown to influence the binding as well.[15,16] While the majority of the available binding data originate from analyses of individual molecules, GAG–cytokine interactions in tissues as well as in engineered biomaterials typically occur within three-dimensional hydrated polymer networks where the interplay of electrostatics and steric constraints is governed by ensembles of multiple (bio)polymeric components.

[a]*Leibniz Institute of Polymer Research Dresden, Germany. E-mail: werner@ipfdd.de*
[b]*Cluster of Excellence Center for Regenerative Therapies, TU Dresden, Germany*
[c]*Cluster of Excellence Physics of Life, TU Dresden, Germany*
† These authors contributed equally to this work.

In an attempt to unravel this condition, we herein applied a recently introduced platform of GAG-based biohybrid hydrogels[17,18] to systematically explore the relevance of GAG content and GAG sulfation pattern for the capacity of the polymer matrices to sequester selected cytokines with very different characteristics.

A theory-driven materials design concept provided us with sets of thoroughly defined biohybrid hydrogels that contain GAG units of varied sulfation pattern (*i.e.* heparin or different selectively desulfated heparin derivatives) in binary covalent networks with multi-armed poly(ethylene glycol) (starPEG) units.[19,20] The charge carrier distribution within the swollen hydrogels can be described by two parameters: (P1) the integral space charge density (the number of ionizable sulfate groups per hydrogel volume in [mmol ml^{-1}]) and (P2) the charge density on the GAG component (the number of sulfate moieties per GAG repeating unit divided by the molecular weight of the repeating unit [mol^2 g^{-1}]). To freely vary both parameters, a previously established hydrogel synthesis protocol[18] was extended to combine thiol-terminated starPEG units, maleimide-functionalized GAG units of different sulfation pattern and maleimide-terminated starPEG units in different ratios by a Michael type addition crosslinking scheme (see Table 1) that is nearly quantitative and cytocompatible, *i.e.* allows for cell encapsulation or gel formation in living tissues. Specifically, we used the following gel building blocks: (BB1) maleimide terminated starPEG (4arm, MW 10 000 g mol^{-1} JenKem), (BB2) thiol-terminated starPEG (4arm, MW 10 000 g mol^{-1}, JenKem) and (BB3) heparin, *N*-desulfated heparin or 6*O,N*-desulfated heparin functionalized with six maleimide groups. Heparin derivatives were synthesized from a 15 000 g mol^{-1} heparin (product: 375095, Merck) as recently reported.[20] Hydrogels were formed within 20–60 seconds after mixing the building blocks in the concentrations given in Table 1 and subsequently swollen in PBS. Swelling degree and the storage modulus of the obtained gels were determined as previously described.[20] This extended hydrogel formation scheme allowed us to independently tune the integral space charge density of the gels (as parameterized by P1) and the local charge density of the contained GAG units (as parameterized by P2) at otherwise invariant, precisely defined conditions.

The experimentally determined swelling degree and the concentrations of the building blocks were used to calculate the integral gel space charge densities of the gels (P1, [mmol ml^{-1}], see Table 1). The local charge distribution at the GAG component was parameterized as the number of sulfate groups per molecular weight of the GAG unit as experimentally determined before[19,20] (P2, [mol^2 g^{-1}], see Table 1). The hydrogels showed a gradually varying integral space charge density ranging from 0.001 to 0.12 mmol ml^{-1} while the storage moduli were kept rather similar (2–4.5 kPa). The hydrogel mesh size was estimated from storage moduli data by the affine network model[21] to 10–13 nm, which clearly exceeds the radius of all cytokines used in the subsequent sequestration studies (molecular weight range 6–38 kDa, see Table 2). Thus, steric constraints hardly restrict the transport of cytokines in either hydrogel type and their sequestration accordingly reflects the relevance of the above parameters P1 and P2 only.

A defined mixture of biomedically important cytokines with distinct charge characteristics was used in the related binding studies (Table 2). The included cytokines were grouped according to their charge characteristics into (A) strongly basic proteins of an IEP $\geq$ 9.0 (VEGF-A, bNGF, IL-4, IFN-gamma, Gro-alpha, IL-8,

Table 1   Formation and properties of biohybrid hydrogels[a]

Gel type	PEG	PEG-HEP	PEG-HEP	PEG-HEP	PEG-6ONDHEP	PEG-HEP	PEG-HEP	PEG-NDHEP	PEG-HEP
P1 $mmol\,ml^{-1}$	0	0.001	0.01	0.02	0.06	0.09	0.11	0.12	0.12
P2 $mol^2\,g^{-1}$	0	0.005	0.005	0.005	0.002	0.005	0.005	0.0035	0.005
G' (kPa)	4.5 ±0.11	2.3 ±0.26	4.2 ±0.28	4.5 ±0.1	2.3 ±0.0	2.1 ±0.1	4.1 ±0.6	4.0 ±1.1	2.0 ±1.9
Q_v	1.15 ±0.11	1.26 ±0.18	1.19 ±0.03	0.95 ±0.01	1.71 ±0.04	1.17 ±0.09	1.8 ±0.19	1.62 ±0.03	1.77 ±1.70
BB1 (µM)	1680	1780	1620	1200	0	0	0	0	0
BB2 (µM)	1500	1200	1500	1500	4500	1300	3000	3900	3000
BB3 (µM)	0	15	150	500	7300	1500	3000	7900	9100

[a] Gels are prepared from three building blocks: BB1 (maleimide-terminated starPEG), BB2 (thiol-terminated starPEG) and BB3 (maleimide-functionalized heparin, N- or 6O,N-desulfated heparin) were crosslinked in different ratios by a Michael type addition reaction between maleimide and thiol groups to vary P1 (the integral space charge density of the swollen gels) and P2 (the number of sulfate groups per molecular weight of the GAG unit). (G': storage modulus of the swollen gel; Q_v: volume swelling degree of swollen gel per volume of gel reaction mixture; BB1, BB2, BB3: concentrations of building blocks within the hydrogel reaction mixture).

MCP-1, and SDF-1-alpha), (B) acidic proteins containing a positively charged heparin binding domain (MIP-1-alpha and MIP-1-beta, IEP 4,8), (C) weakly charged or neutral proteins of an IEP range 5.9 to 7.6 (IL-6, TNF-alpha, IL-1b, and IL-10) and (D) acidic proteins of IEP < 5.5 (EGF and GM-CSF).

The recombinant proteins (reconstituted using several ProcartaPlex Simplex protein standards, Cat. No. EPX01A-xxxx-901 ThermoFisher) were dissolved in PBS with 1% BSA and 0.1% proclin (Sigma) at different biologically relevant concentrations in the range of 4.5–54.3 ng ml^{-1} 10 µl of each gel type ($n = 3$) were incubated with 250 µl of the protein mixture for 24 hours and the cytokine concentration in the supernatant was subsequently analyzed with Procartaplex

Table 2 Properties of cytokines used in sequestration experiments[a]

Group	Protein	M_{wt}/kDa	IEP (pH units)	Deployed amount and conc./pmol and ng ml^{-1}	
A	VEGF-A	38.2	9.2	0.097	14.75
	bNGF	13.5	9	0.302	16.31
	IL-4	15	9.3	0.388	23.25
	IFN-gamma	16.2	9.5	0.372	24.1
	GRO-alpha	7.9	9.5	0.243	7.68
	IL-8	8.9	9.2	0.233	8.31
	MCP-1	8.7	9.4	0.391	13.62
	SDF-1 alpha	8.5	10.1	1.597	54.31
B	MIP-1 alpha	7.7	4.8	0.244	7.52
	MIP-1 beta	7.8	4.8	0.634	19.79
C	IL-6	20.8	6.2	0.230	19.15
	TNF-alpha	17.4	7	0.246	17.1
	IL-1 beta	17.4	5.9	0.065	4.5
	IL-10	18.6	7.6	0.068	5.05
D	EGF	6.2	4.8	0.275	6.81
	GM-C SF	14.5	5.2	0.478	27.75

[a] Group: protein classification according to charge characteristics – see text; M_{wt}: molecular weight in kDa; IEP: isoelectric point in pH units; deployed amount and conc.: molar and weight concentration of cytokine solutions used in sequestration experiments with a volume of 10 µl hydrogel.

Simplex multiplex kits (ThermoFischer) using a Bioplex 200 (Biorad). The detected amounts of the proteins after gel exposure of the solution were compared to the detected protein amounts in control experiments without hydrogels to determine the percentage of the deployed protein sequestered by the gels (Fig. 1). Since the number of GAG units in the gels exceeded the deployed protein amount by several orders of magnitude, gel saturation effects could be excluded for all experiments.

The obtained results corroborate the potential of GAG-containing (but not GAG-free) polymer matrices to effectively sequester many biologically important signalling proteins with up to quantitative binding of the deployed cytokines for some of the investigated systems. However, the binding was found to be clearly graduated according to the protein and gel types:

Acidic cytokines of group (D) hardly bound to any hydrogel type, whereas the strongly basic cytokines of group (A) bound in direct correlation to the integral space charge (sulfate) density of the hydrogels (P1) (see Fig. 1), no matter whether the GAG units were fully sulfated heparin units or selectively N- and $6O,N$-desulfated heparin derivatives (P2). For group (A) proteins, we observed that the integral space charge density of the hydrogels but not the local sulfation pattern of the GAG unit determined the binding to the compared hydrogel types. This is in accordance with our earlier VEGF binding data to hydrogels made from different heparin derivatives.[22] Accordingly, the bound concentration of group (A) and (D) cytokines can be easily predicted and effectively adjusted by tuning P1, the integral space charge density of the gels.

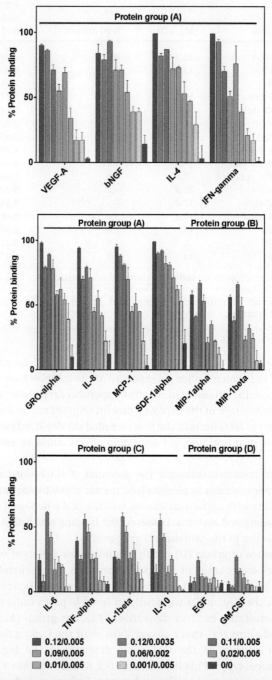

Fig. 1 Sequestered percentage of deployed cytokines to hydrogels of different GAG content and GAG sulfation (parameterized as P1/P2). The compared proteins are grouped according to their charging (groups A–D, see text).

For group (C) cytokines, the protein binding correlated with the integral space charge density (P1) of the hydrogels made from fully sulfated heparin (green bars, P2 0.005) but exhibited a lower binding compared to group (A) proteins. Group (C) proteins further showed a rather similar but lower binding to hydrogels made of the two compared desulfated heparin derivatives (orange bars, P2 0.0035 and 0.002) when compared to gels made of fully sulfated heparin, confirming that GAG specific sulfation can influence the binding to these weakly charged proteins.[23,24] The comparable binding of the weakly charged group (C) proteins suggests that non-electrostatic specific interactions govern their binding to less strongly charged GAGs and, thus, to our compared hydrogels containing selectively desulfated heparin units. The involved intermolecular interactions may include geometrically matching dipole–dipole interactions, van der Waals forces and hydrophobic interactions.[15]

For the binding of group (B) proteins to the hydrogels, the integral space charge density determines the binding characteristics of gels containing the fully sulfated heparin (green bars, P2 0.005) whereas for gels containing selectively desulfated heparin derivatives a less pronounced but similar trend (orange bars, P2 0.0035 and 0.002) was observed. MIP1-alpha/beta, acidic proteins of IEP 4.8, are chemokines of the CCL3/4-structure that carry a positively charged GAG binding pocket on their otherwise negatively charged surface resulting in a binding selectivity to GAGs according to the pattern and spacing of their sulfated saccharides.[25] MIP1-alpha/beta was found to bind to the compared hydrogels with a dependence on their integral space charge density (P1) and the degree of sulfation of the contained GAG units (P2) – the sequestration of group (B) proteins is determined by both P1 and P2.

In sum, our systematic analysis shows that the binding of strongly basic proteins (group A) or strongly acidic proteins (group D) to GAG-containing bio-hybrid hydrogels is quantitatively controlled by the integral space charge density of the gels (P1), while the binding of acidic proteins with a positively charged binding domain (group B) is determined by both the integral space charge density of the gels (P1) and the local charge density on the GAG unit (P2), and the binding of weakly charged or neutral proteins (group C) is controlled by the local charge density on the GAG unit (P2) or non-electrostatic interactions.

Of note, while the current study did not investigate any protein concentration dependence of the sequestration performance of the compared gel systems, the obtained data can be considered representative as they cover biologically relevant cytokine concentration ranges, and as for the chosen conditions only a minor (<1%) fraction of the protein binding capacity of the hydrogel materials is utilized.

Conclusions

Our reported results demonstrate the importance of the integral space charge density of GAG-containing cell-instructive matrices for the binding of strongly charged proteins. Furthermore, the local accumulation of charge carriers on individual GAG components was found to be decisive for the binding of several weakly charged proteins. These results not only enable the design of hydrogel systems with tailored protein sequestration but may also shed new light on the binding principles of cytokines to ECM in living tissues.

The introduced approach provides a powerful means to explore persisting questions on the matrix-mediated control of signalling molecules in space and time. Ongoing studies aim at applying the established methodology to record the interactions of a broader range of proteins and (synthetic as well as tissue-derived) matrix systems. Data science tools are being used to fully exploit the resulting options in the design of customized biofunctional materials.

Conflicts of interest

There are no conflicts to declare.

Acknowledgements

This work was supported by the Deutsche Forschungsgemeinschaft through CRC TR 67 (C. W., P. A. and U. F.), FR 3367/2-1 (U. F.) and ZI 1238/4-1 (Y. L.).

References

1 R. P. Richter, N. S. Baranova, A. J. Day and J. C. Kwok, *Curr. Opin. Struct. Biol.*, 2018, **50**, 65.

2 R. Sasisekharan, Z. Shriver, G. Venkataraman and U. Narayanasami, *Nat. Rev. Cancer*, 2002, **2**, 521.

3 U. Häcker, K. Nybakken and N. Perrimon, *Nat. Rev. Mol. Cell Biol.*, 2005, **6**, 530.

4 M. C. Z. Meneghetti, A. J. Hughes, T. R. Rudd, B. Nader, A. K. Powell, E. A. Yates, M. A. Lima, E. A. Yates, H. B. Nader, A. K. Powell, E. A. Yates and M. A. Lima, *J. R. Soc. Interface*, 2015, **12**, 0589.

5 C. Noti and P. H. Seeberger, *Chem. Biol.*, 2005, **12**, 731.

6 J. E. Turnbull, *Nat. Methods*, 2018, **15**, 867.

7 U. Freudenberg, Y. Liang, K. L. Kiick and C. Werner, *Adv. Mater.*, 2016, **28**, 8861.

8 G. A. Hudalla and W. L. Murphy, *Adv. Funct. Mater.*, 2011, **21**, 1754.

9 N. Lohmann, L. Schirmer, P. Atallah, E. Wandel, R. A. Ferrer, C. Werner, J. C. Simon, S. Franz and U. Freudenberg, *Sci. Transl. Med.*, 2017, **9**, 1–12.

10 M. F. Maitz, U. Freudenberg, M. V. Tsurkan, M. Fischer, T. Beyrich and C. Werner, *Nat. Commun.*, 2013, **4**, 2168.

11 K. Chwalek, M. V. Tsurkan, U. Freudenberg and C. Werner, *Sci. Rep.*, 2014, **4**, 4414.

12 C. Papadimitriou, H. Celikkaya, M. I. Cosacak, V. Mashkaryan, L. Bray, P. Bhattarai, K. Brandt, H. Hollak, X. Chen, S. He, C. L. Antos, W. Lin, A. K. Thomas, A. Dahl, T. Kurth, J. Friedrichs, Y. Zhang, U. Freudenberg, C. Werner and C. Kizil, *Dev. Cell*, 2018, **46**, 85.

13 H. M. Weber, M. V. Tsurkan, V. Magno, U. Freudenberg and C. Werner, *Acta Biomater.*, 2017, **57**, 59–69.

14 L. J. Bray, M. Binner, A. Holzheu, J. Friedrichs, U. Freudenberg, D. W. Hutmacher and C. Werner, *Biomaterials*, 2015, **53**, 609.

15 I. Capila and R. J. Linhardt, *Angew. Chem., Int. Ed.*, 2002, **41**, 390.

16 H. G. Garg, R. J. Linhardt and C. A. Hales, *Chemistry and Biology of Heparin and Heparan Sulfate*, Elsevier, 2005.

17 U. Freudenberg, J.-U. Sommer, K. R. Levental, P. B. Welzel, A. Zieris, K. Chwalek, K. Schneider, S. Prokoph, M. Prewitz, R. Dockhorn and C. Werner, *Adv. Funct. Mater.*, 2012, **22**, 1391.

18 M. V. Tsurkan, K. Chwalek, S. Prokoph, A. Zieris, K. R. Levental, U. Freudenberg and C. Werner, *Adv. Mater.*, 2013, **25**, 2606.

19 A. Zieris, R. Dockhorn, A. Röhrich, R. Zimmermann, M. Müller, P. B. Welzel, M. V Tsurkan, J. Sommer, U. Freudenberg and C. Werner, *Biomacromolecules*, 2014, **15**, 4439.

20 P. Atallah, L. Schirmer, M. Tsurkan, Y. D. Putra Limasale, R. Zimmermann, C. Werner and U. Freudenberg, *Biomaterials*, 2018, **181**, 227.

21 M. Rubinstein and R. H. Colby, *Polymer Physics*, Oxford University Press, Pennsylvania, USA, 2003.

22 U. Freudenberg, A. Zieris, K. Chwalek, M. V. Tsurkan, M. F. Maitz, P. Atallah, K. R. Levental, S. A. Eming and C. Werner, *J. Controlled Release*, 2015, 220.

23 S. Salek-Ardakani, J. R. Arrand, D. Shaw and M. Mackett, *Blood*, 2000, **96**, 1879.

24 L. Ramsden and C. C. Rider, *Eur. J. Immunol.*, 1992, **22**, 3027.

25 T. M. Handel, Z. Johnson, S. E. Crown, E. K. Lau, M. Sweeney and A. E. Proudfoot, *Annu. Rev. Biochem.*, 2005, **74**, 385.

Faraday Discussions

DISCUSSIONS

New directions in surface functionalization and characterization: general discussion

Helena S. Azevedo, Adam B. Braunschweig, Ryan C. Chiechi, Shelley A. Claridge, Leroy Cronin, Yuri Diaz Fernandez, Ten Feizi, Laura Hartmann, Mia Huang, Yoshiko Miura, Matteo Palma, Xinkai Qiu, Bart Jan Ravoo, Stephan Schmidt, W. Bruce Turnbull, Carsten Werner, Zijian Zheng and Dejian Zhou

DOI: 10.1039/C9FD90064K

Yuri Diaz Fernandez opened discussion of the paper by Zijian Zheng: In your schematic diagram of the initiator transfer process, you located the initiator on top of the first polymer layer (DOI: 10.1039/c9fd00013e). Considering that your initiator molecule contains thiols, have you observed any exchange at the gold surface, between the initiator and the polymer? A molecular exchange mechanism could also explain the time dependence observed in the thickness of the final polymer brush, controlled by the initiator-transfer step.

Zijian Zheng answered: We did not observe an exchange between the thiol molecules and the polymer. The polymer should be more thermally stable than a thiol molecule.

Adam Braunschweig opened discussion of the paper by Matteo Palma: One of the major challenges in using nanopatterns to study biological properties is that often hundreds of different parameters must be explored, and to achieve statistical significance, this will involve repeats. So, making any significant statements should involve thousands of different data points. Does your system have the versatility and the throughput to acquire these thousands of data points?

Matteo Palma responded: One of the major challenges in using nanopatterns to study biological properties is the ability to systematically change different parameters with single-molecule control, nanoscale spatial resolution, and multivalent ability; this is what the DNA nano arrays presented here permit. Additionally, the system has the versatility to acquire multiple and numerous data points when merged with high throughput fabrication methods. Future research will be focused on merging this bottom-up ability with top-down lithography, allowing insight into many different chemical and physical properties, on the same chip.

Laura Hartmann continued the discussion of the paper by Zijian Zheng: Did you test the adsorption of the initiator molecules and the potential to graft on biopolymers or hydrogels?

Zijian Zheng noted: We tried hydrogels but the thiol molecules can easily absorb into the porous network of the hydrogel, meaning that no polymer growth was observed.

Laura Hartmann returned to the discussion of Matteo Palma's paper: You very nicely demonstrated control in spacing in the x-y-direction; is it also possible to control positioning in the z-direction?

Matteo Palma commented: As of today, we cannot fabricate 3D nano arrays with the same control demonstrated here in x and y. Moving forward, we are thinking about potential strategies to achieve this

Dejian Zhou asked: The use of DNA origami displaying a well-controlled number of specific ligands to study interactions with cells appears to be very attractive. However, given that the origami are DNA-based, they can be degraded by various nuclease enzymes, especially under intra-cellular or *in vivo* conditions; have you tested their stability under such conditions? Are these DNA origami readily taken up by cells and endocytosed into cells during incubation?

Matteo Palma responded: The DNA origami we emptied were immobilised (covalently) on surfaces. The origami were exposing peptides that were interacting with trans-membrane cell proteins. We do not have evidence of degradation of the origami. When trying to work with the same DNA origami in solution, we did not see evidence of cell uptake – for that, you would rather employ 3D DNA origami, such as tetrahedrons.

Dejian Zhou followed up by asking: You have produced three labels on each DNA origami, with relatively large inter-label spacing. Can you also make denser tether sites, so as to create higher label valencies on each origami scaffold? I think they could also be used as potential scaffolds to controllably display varying numbers of our recently developed polyvalent glycan nanoparticles[1] and study how these stimulate dendritic cell immune responses. We will be happy to collaborate on this if you are interested.

1 Y. Guo, I. Nehlmeier, E. Poole, C. Sakonsinsiri, N. Hondow, A. Brown, Q. Li, S. Li, J. Whitworth, Z. Li, A. Yu, R. Brydson, W. B. Turnbull, S. Pöhlmann and D. Zhou, *J. Am. Chem. Soc.*, 2017, **139**, 11833–11844.

Matteo Palma answered: We labelled multiple sites on individual origami with single-molecule control – here, we did so with up to 18 molecules per origami. In particular, we picked 60, 30, and 20 nm spatial separation for investigating integrin activation (avβ3 integrins have, in the past, been shown to exhibit a preferential spacing of 60 nm; this was done by employing nano-patterned metal dots functionalised with RGD peptides). I think there is potential to explore what you suggest: combing your glycan nanoparticles and the DNA origami. I am definitely interested in exploring this possibility together with you.

Adam Braunschweig continued the general discussion: I am very sceptical about the rush to adopt machine learning methods as a substitute for more reliable experimental data. Garbage in, garbage out. What is needed are thousands of data points and more reliable data. One of the first things that you learn in a statistics course is that significance increases with n, the number of data points. So, a priority with all these novel surface modification strategies must be to develop approaches where thousands of different data points can be acquired, ideally, on the same surface in order to minimize batch-to-batch variation.

Zijian Zheng agreed: Without reliable data and accurate data analysis, it would be very difficult to build up a database for machine learning. This requires a tremendouss amount of work.

Matteo Palma added: I also agree. I would add that the level of control is important (single-molecule, nanoscale, and multivalent), and this is what the nanoarray platform I have presented permits .

Ryan Chiechi remarked: The problem with machine learning vis-à-vis classifying images of cell growth and adhesion is indeed a small, heterogeneous set of training data from which probabilities are extracted. The comparison to machine vision systems in, for example, self-driving cars is not really apt because a car only needs to differentiate between a person and a hedgehog with a reasonable degree of certainty over many frames of video. In five seconds of video, it may classify the person as a penguin or an elephant in a few frames, before concluding that, on average, it is most likely a person. That sort of classification does not seems useful for studying the morphology of cells.

Zijian Zheng commented: It is difficult to comment or to give a clear answer to that. But indeed, it is way more complicated to understand biological structure and behaviour.

Adam Braunschweig returned to the discussion of Zijian Zheng's paper: Achieving feature diameters of 25 nm is extremely impressive, especially in additive scanning probe methods like yours, where you put a material onto a surface. The first question is, how did you achieve such small feature diameters? Also, given the importance of multiplexing and testing various different conditions to resolve biological problems, how do you envision creating patterns with multiple different inks while maintaining this ~25 nm resolution? Could you, as well, get different materials with controlled spacing between different materials?

Zijian Zheng responded: The sharp tips of scanning probes provide us with the ability to pattern very small features. In principle (as well as in several nice examples), one can pattern the surface with multiplexed materials at sub-100 nm resolution.[1,2] The difficulty is how to resolve the registration between the different features/materials in a easy way. Additive printing provides a promising solution to this, and it would be important to control the ink flow/transport onto and out from the tips.

1 L. Chen, Z. Xie, T. Gan, Y. Wang, G. Zhang, C. A. Mirkin and Z. Zheng, *Small*, 2016, **12**, 3400–3406.
2 Z. Zheng, W. L. Daniel, L. R. Giam, F. Huo, A. J. Senesi, G. Zheng and C. A. Mirkin, *Angew. Chem., Int. Ed.*, 2009, **48**, 7626–7629.

Mia Huang continued the general discussion: The cell and its components are already nanoscale and multivalent. What are the possibilities for developing tools to study the precise nature of their nanoscale and multivalent organization?

Matteo Palma answered: The platform we published in ACS Nano[1] and in this Faraday Discussions paper (DOI: 10.1039/c9fd00023b) is exactly the sort of tool needed to study biomolecular interactions with nanoscale spatial resolution, single-molecule control, and multivalent capability. In particular, we focused on cell matrix interactions dictated by trans-membrane protein integrins and their synergistic work with other membrane receptors to modulate normal cell and cancer cell behavior, as well as cardiomyocyte adhesion. It has been already demonstrated, employing metal nano dots functionalised with peptides (RGD), that there is nanoscale preferential spacing between integrins that favour cell adhesion (60 nm: see work by Joachim Spatz)[2,3] and that there is a minimal adhesion unit of at least 4 integrins per cluster in fibroblast cell adhesion, *i.e.* clustering is important.[4] What was missing was the ability to perform these kinds of experiments with multivalent capability and real single-molecule control, maintaining nanoscale spatial resolution: our DNA origami nanoarray platform is a tool to do exactly this. (By the way, the cell is microscale; nanoscale would include, for example, the integrin head (10 nm) and the spacing between integrins that favour cell adhesion; and nanoscale and multivalent are the interactions at the cell-ECM interface, as well as the interactions present in many other biologically relevant systems).

1 D. Huang, K. Patel, S. Perez-Garrido, J. F. Marshall and M. Palma, *ACS Nano*, 2019, **13**, 728–736.
2 M. Arnold, E. A. Cavalcanti-Adam, R. Glass, J. Blummel, W. Eck, M. Kantlehner, H. Kessler and J. P. Spatz, Activation of Integrin Function by Nanopatterned Adhesive Interfaces, *ChemPhysChem*, 2004, **5**, 383.
3 E. A. Cavalcanti-Adam, *et al.* Cell spreading and focal adhesion dynamics are regulated by spacing of integrin ligands, *Biophys. J.*, 2007, **92**, 2964–2974.
4 M. Schvartzman, M. Palma, J. Sable, J. Abramson, X. Hu, M. P. Sheetz and S. J. Wind, *Nano Lett.*, 2011, **11**, 1306–1312.

Zijian Zheng added: For surface analysis, a tool should be flexible enough to create structures with multiple functionalities at high resolution and registration. This is the critical challenge in nanoscale patterning of functional materials.

Adam Braunschweig opened discussion of the paper by Bart Jan Ravoo: I really appreciate the use of supramolecular methods to address biological questions, specifically because they give you the ability to systematically vary certain parameters that you may not be able to do with natural systems. Your cyclodextrin vesicles are a perfect example of this. One phenomenon that occurs is that the lipids move around in response to binding to form rafts. I was wondering if the lipids in your cyclodextrin vesicles are similarly mobile.

Bart Jan Ravoo responded: The cyclodextrin vesicles are highly dynamic structures, much like typical liposomes. There is rapid lateral diffusion in the bilayer as long as the vesicles are kept above their T_g (which is around 0 °C for the vesicles used here). Furthermore, there is flip-flop and even some exchange of molecules between vesicles, although this is slow compared to lipids due to the high molecular weight and low water solubility of these amphiphiles with seven hydrophobic chains. We also know that the cyclodextrins can be mixed with other components in the membrane, such as lipids[1] or other amphiphilic macrocycles,[2] which can give rise to interesting dynamic phenomena, such as ligand-induced receptor clustering to bind multivalent guest molecules. Furthermore, one should keep in mind that the host–guest complexes of cyclodextrins are also highly dynamic, so that guest molecules can rapidly diffuse along the vesicles surface.

1 U. Kauscher, M. C. A. Stuart, P. Drücker, H.-J. Galla and B. J. Ravoo, *Langmuir*, 2013, **29**, 7377–7383.
2 Z. Xu, S. Jia, W. Wang, Z. Yuan, B. J. Ravoo and D.-S. Guo, *Nat. Chem.*, 2019, **11**, 86–93.

Xinkai Qiu remarked: The hydrogel showed quite significant fatigue over the four magneto- and photo-switching cycles, in the sense that the storage modulus of the hydrogel decreases dramatically after each cycle. Does the speaker perhaps have more insight to the origin of the fatigue in his system?

Bart Jan Ravoo replied: The observed fatigue is most likely caused by inhomogeneities in the hydrogels. The magneto-response tends to diminish over time due to the precipitation of a fraction of the magnetic nanoparticles during each magnetization cycle. The photo-response tends to diminish as the cycle time is slightly too short for complete recovery after green light irradiation. Furthermore, it is difficult to optimize the dual magneto- and photo-response, *i.e.* larger changes in moduli and less fatigue are obtained if the hydrogel is optimized towards a single response.

Laura Hartmann asked: In your experiment, testing the effect of the magnetic field on the mechanical properties of the hydrogel, based on the proposed alignment of magnetic nanoparticles along the magnetic field, do you expect there to be a difference if the magnetic field and shear are aligned in parallel or perpendicular?

Bart Jan Ravoo noted: There should be a large difference indeed. In our experimental set-up, the field is perpendicular to the shear direction, so that in principle we would expect a maximum effect because the alignment of nanoparticles would also occur perpendicular to the shear direction. It would be interesting to systematically vary the direction of alignment, but unfortunately our set-up does not allow that.

Yoshiko Miura opened discussion of the paper by Carsten Werner: Glycosaminoglycan-based hybrid gels have lots of potential in nanomedicine and tissue engineering. Considering the role of GAGs in natural systems, it is possible to develop a GAG gel that has specific affinity to cytokine or a growth factor. Have you investigated the binding specificity of cytokines for these GAG gels? Have you investigated growth factor binding to the GAG gel?

Carsten Werner answered: While our reported approach aims at designing selectively cytokine-affine GAG–hydrogel systems the obtained specificity will hardly be as high as, for example, in receptor–ligand or antibody–antigen recognition. For growth factor binding to different biohydride gel types (containing different heparin derivatives) see ref. 1–3.

1 A. Zieris, K. Chwalek, S. Prokoph, K. R. Levental, P. B. Welzel, U. Freudenberg and C. Werner, *Journal of Controlled Release*, 2011, **156**, 28–36.
2 A. Zieris, R. Dockhorn, A. Röhrich, R. Zimmermann, M. Müller, P. B. Welzel, M. V. Tsurkan, J.-U. Sommer, U. Freudenberg and C. Werner, *Biomacromolecules*, 2014, **15**, 4439–4446.
3 U. Freudenberg, Y. Liang, K. L. Klick and C. Werner, *Adv. Mater.*, 2016, **28**, 8861–8891.

Dejian Zhou enquired: Have you investigated how binding affects the cytokine diffusing through the hydrogel? Is there any correlation between the cytokine–hydrogel binding affinity and their diffusion rate in the hydrogel?

Carsten Werner answered: Yes, we have studied cytokine transport. It not only depends on the specific binding to the gel (as described by P1 and P2, DOI: 10.1039/c9fd00016j), but also on the geometric constraints (mesh size) of the polymer network.

Adam Braunschweig opened discussion of the paper by Shelley Claridge: I have several questions related to pattering using your lipids. The first: the polar groups on your lipids are involved in forming the monolayers. In one example, you showed an inositol headgroup, which is a proto-carbohydrate. Presumably, you want to use these in binding. Are your headgroups available for binding and have you tried any such experiments? Second, a really interesting aspect of your monolayers is the ability to precisely control spacing on the nanoscale, while making patterns that extend across the macroscale. Have you developed a method yet to form patterns composed of different materials?

Shelley Claridge replied: I agree, the availability of the headgroup to interact with the environment is critical. The first experiments we did with striped phases related to wetting,[1–3] which is very much about the availability of the headgroup. What we have found is that some surprising aspects of the striped phase structure, like the position of the polymerizable group within the chain, can impact how available the headgroup is to impact wetting. Wetting of carboxylic acid stripes also decreases substantially in buffers that contain small concentrations of divalent cations, which would be consistent with the headgroups complexing the ions, a very simple form of binding. Similarly, the structure of a phospholipid *versus* a simpler single-chain amphiphile is important, since it can help the terminal functional group project from the interface, presumably something it is doing in a related but different way at the membrane periphery. More recently, we are beginning to ask questions about headgroup binding to more complex adsorbates in the environment, which also appears to be possible.

1 J. J. Bang, K. K. Rupp, S. R. Russell, S. W. Choong and S. A. Claridge, Sitting phases of polymerizable amphiphiles for controlled functionalization of layered materials, *J. Am. Chem. Soc.*, 2016, **138**, 4448–4457.

2 T. A. Villarreal, S. R. Russell, J. J. Bang, J. K. Patterson and S. A. Claridge, Modulating wettability of layered materials by controlling ligand polar headgroup dynamics, *J. Am. Chem. Soc.*, 2017, **139**, 11973–11979.

3 S. W. Choong, S. R. Russell, J. J. Bang, J. K. Patterson and S. A. Claridge, Sitting phase monolayers of polymerizable phospholipids create dimensional, molecular scale wetting control for scalable solution based patterning of layered materials, *ACS Appl. Mater. Interfaces*, 2017, **9**, 19326–19334.

In terms of making multi-component patterns, we know it is possible to co-transfer multiple amphiphiles, which is a step in that direction, and certainly it would be possible to lithographically pattern multiple classes of molecules.

Stephan Schmidt reopened the discussion of Carsten Werner's paper: In your experience, what are the advantages and challenges of 3D *vs.* 2D cell assays?

Carsten Werner responded: 3D culture is often a more realistic tissue model, however, the exogenous cues acting on cells are more difficult to control in 3D.

Mia Huang addressed Shelley Claridge, Bart Jan Ravoo, and Carsten Werner: Serendipitously, I think the three of you can make an artificial cell! Bart can provide the encapsulation provided by cell membranes, Shelley can provide long-range order found in actin filaments, and Carsten can impart three-dimensional order. I think there are tremendous applications for your technologies that could be employed to understand outstanding questions in biology. For Carsten, especially, using polysialic acid-based hydrogels would be very enabling in understanding neuronal interactions.

Carsten Werner responded: Exploring biohybrid gels containing polysialic acid is an excellent idea - we will definitely do that. Thanks!

Shelley Claridge replied: Thanks, Mia. We have not explored adhesion to striped phases in the sense of actin, but we have definitely been interested recently in the adhesive properties of the stripes more generally, and we are always open to collaboration.

Xinkai Qiu returned to the discussion of the paper by Shelley Claridge: The speaker introduced a useful method to pattern molecules that show long-range order for potential application. In the field of molecular electronics, self-assembled monolayers of organic molecules on polycrystalline substrates, more often than not, show stable and reproducible charge transport properties in the absence of long-range order of the molecules. In that sense, the long-range order of the molecules does not have any impact on the practicality of SAMs. Could the speaker comment on how we should justify long-range order of molecules when concerning the practicality of SAMs?

Shelley Claridge responded: That is a great point. Absolutely, there are applications like you are describing that do not require long-range ordering. At the same time, for many applications, it would be vital. Take, for instance, a situation where you want to use patterns in the nanometer-scale chemistry of the monolayer in the xy plane to control ordering of other objects at the interface. In some of our work, we are using nanometer-wide stripes of functional groups to orient inorganic nanostructures that are microns in length, as prototype circuit elements.[1] In order to orient each micrometer of the nanostructure, you need an aligned row about 1000 molecules in length, which is already reasonably long-range ordering. For interaction with biological structures using something like a carbohydrate-functionalized striped phase, you might need ordering over distances representing tens to hundreds of thousands of molecules in the monolayer. It all depends on what you want to do with the surface. The nice surprise has been that it is possible to generate such long-range order, which is really opening up the range of problems we can address.

1 A. G. Porter, T. Ouyang, T. R. Hayes, J. Biechele-Speziale, S. R. Russell and S. A. Claridge, *Chem*, 2019, **5**, 2264–2275.

Laura Hartmann asked: After polymerization of the lipids on the surface, is it possible to take them off the surface and isolate them as ribbons or sheets?

Shelley Claridge replied: We have been curious about that as well. It certainly seems possible that this would be a way of creating amphiphilic structures that would be hard to synthesize in other ways. A significant limitation using our current methods would be throughput – a square meter of monolayer of this type represents just a milligram of material. But perhaps scaling can be achieved using graphite flakes or other high surface area supports.

Adam Braunschweig addressed Carsten Werner and Bart Jan Ravoo: These nanoscale systems will be perennial models until we can make them heterogeneous mixtures, to which we can add different components. Carsten, in the case of your system, you have to worry about how additives affect gelation, and Bart, in yours, you have to be concerned about the ability of heterogeneous mixtures to form vesicles. My question is: to what extent have you attempted to make heterogeneous gels and vesicles by incorporating materials that may add new functionality.

Carsten Werner responded: We are working on the formation of multiphasic microgel-in-gel materials with microgels of different characteristics and have incorporated collagen fibrils in the gels. Moreover, we use the rapidly (Michaeal-type addition) crosslinking variant of the gelation to apply the biohybrids as bioinks in additive manufacturing schemes.

Bart Jan Ravoo replied: I would say that a highly attractive feature of our system is that we use a modular strategy in which components can be easily added, substituted, or modified to add or change functionality without a complete redesign and optimization of the entire system. We know that we can change the composition of the vesicles, *e.g.* by adding lipids or by adsorbing magnetic nanoparticles, as shown in the present paper (DOI: 10.1039/c9fd00012g). This also applies to the peptide gelators: we can vary the percentage of guest-tagged peptide to tailor the cross-linking density and we are also exploring the incorporation of two orthogonal guests, one responsive and one inert, so that we can make gels with shape memory.

Adam Braunschweig continued the discussion of the paper by Bart Jan Ravoo: Many vesicles are used to deliver contents that are encapsulated within the interior. Do your vesicles encapsulate different molecules, and have you thought of using your vesicles for a type of stimuli-responsive triggered release?

Bart Jan Ravoo said: The cyclodextrin vesicles are rather leaky but we have explored various ways to make them less permeable, *e.g.* by mixing the cyclo-dextrins with phospholipids[1] or by wrapping the vesicles in a polymer shell.[2] We have also shown that the vesicles can enter cells rather efficiently through endocytosis and we have made redox-responsive polymer shells containing disulfide crosslinks, so that a triggered intracellular release is obtained.[2] In an alternative approach, we have decorated the vesicles with photoswitchable guest molecules that bind, for example, proteins, so that these can be captured and released in response to irradiation.[3]

1 U. Kauscher, M. C. A. Stuart, P. Drücker, H.-J. Galla and B. J. Ravoo, *Langmuir*, 2013, **29**, 7377–7383.
2 W. C. de Vries, D. Grill, M. Tesch, A. Ricker, H. Nüsse, J. Klingauf, A. Studer, V. Gerke and B. J. Ravoo, *Angew. Chem. Int. Ed.*, 2017, **56**, 9603–9607.
3 A. Samanta, M. C. A. Stuart and B. J. Ravoo, *J. Am. Chem. Soc.*, 2012, **134**, 19909–19914.

Ten Feizi addressed Carsten Werner: How would you approach bringing cells in your system?

Carsten Werner answered: Cells are incorporated in our gel system by mixing the cell suspension with maleimide-prefunctionalized GAG component, gel formation results within seconds after combining this mixture with the solution of peptide-functionalized PEG. This procedure was, for example, in the bio-printing of cell-containing gel materials (see, e.g., ref. 1).

1 R. Zimmermann, C. Hentschel, F. Schrön, D. Moedder, T. Büttner, P. Atallah, T. Wegener, T. Gehring, S. Howitz, U. Freudenberg and C. Werner, *Biofabrication*, 2019, **11**, 045008.

Ten Feizi continued by asking: What is the diffusivity of your materials into the hydrogels?

Carsten Werner answered: Transport phenomena of cytokines - and GAG-binding as well as non-binding proteins within our bihybrid GAG-PEG gels are being studied by experiments and by simulations. A systematic study will be published soon.

Ten Feizi asked: Are you looking at any different glycoso-amino-glycans?

Carsten Werner answered: So far we have been using heparin, and different selectively desulfated heparin derivatives as building blocks of our gels. Moreover, we have been using hyaluronic acid, dermatan sulfate and chondroitine sulfate for gel formation with similar crosslinking chemistries. Beyond that, we are planning to use CRISPR-based GAG variants for the preparation of gel matrix libraries.

Helena Azevedo communicated in response to the discussion of Carsten Wern-er's paper: It is known that, in general, positively charged materials tend to be cytotoxic and cell viability assays show cell death after 24 h. However, if cells are maintained in culture for longer periods of time (1 week), cells are able to recover as proteins from the serum and produced by cells adsorb onto the materials shielding the charge. We may need to revise how we assess biocompatibility of materials *in vitro*.

Stephan Schmidt continued the discussion of the paper by Shelley Claridge: What is the advantage of self-assembly pattering methods in comparison to mass-scale top-down lithography methods in the semiconductor industry?

Shelley Claridge answered: So, if Intel is getting to nodes at or below 10 nm,[1] why bother, right? There are some pretty important reasons, actually. The fabrication processes required to get down to feature scales that small are incredibly expensive, and often require several sequential lithographic steps, so it is not like you can just draw any arbitrary pattern you want, and generate it. In contrast, biomolecules routinely generate patterns at sub-10 nm scales. However, I think the other, even more significant reason (at least for me), is the chemistry. The surface of the cell membrane or a protein binding pocket creates exquisitely precise chemical environments that are the basis for molecular recognition and catalysis. These chemical environments come from positioning functional groups not just in big blocks, but in well-defined geometries relative to other complementary functional groups. To me, that is one of the end goals of molecular lithography, combining that sort of precision with scalability.

1 https://www.extremetech.com/computing/291029-intel-will-launch-7nm-chips-in-2021-ice-lake-ships-in-june.

Mia Huang enquired: Have you thought about using your microscale system to study piezo proteins?

Shelley Claridge replied: No, but we are definitely interested in mechanical signal transduction, so that would be a great thing to look into. Thanks!

W. Bruce Turnbull commented: I really like the idea of patterning the glycolipids on a nanometer scale that could match the distance between binding sites for various lectins, as it could provide a different strategy for selective binding to different lectins. Please can you comment on whether the patterning methodology could be extended beyond phospholipids to natural glycolipids with ceramide tails? Is it possible to functionalize the hydrophobic stripes on the surface to reduce any potential non-specific binding to proteins?

Shelley Claridge replied: Much of the ordering in striped phases arises from the interaction between the alkyl chain and the substrate, so from that perspective, I would also expect ceramides to be capable of forming stripes. The relative width of the head and tail segments of the molecule can also be important in long-range ordering, so, for instance, a molecule with a very wide headgroup might need to be co-assembled with another molecule to achieve a stable striped phase. Usually, if we were going to try assembly of a new molecule that was precious, we would model the likely stripe structure first to see if there are obvious steric factors that would preclude a stable assembled structure. In regards to functionalizing the hydrophobic stripes, there are indications that it is possible to do reactions with the polymer backbone, which would be one way of changing the chemistry to reduce non-specific binding.

Conflicts of interest

There are no conflicts to declare.

Faraday Discussions

PAPER

Nanolithography of biointerfaces

Ten Feizi ⓘ

Received 12th August 2019, Accepted 16th August 2019

DOI: 10.1039/c9fd00082h

This article is based on the Concluding remarks made at the Faraday Discussion meeting on Nanolithography of Biointerfaces, held in London, UK, 3–5th July 2019.

Introduction

The Faraday Discussion meeting was held in London in July 2019. I would like to begin by thanking the co-chairs of the organizing committee, Adam Braunschweig and Laura Hartmann, as well as the other members of the Scientific Organizing Committee – Ryan Chiechi, Lee Cronin, Stephan Schmidt, and Sébastien Vidal – and the representatives from the RSC, including Lucy Balshaw, Michael Spencelayh, Heather Montgomery, and Rosalind Searle, who brought together an outstanding collection of investigators performing research at the interface of nanolithography and biology. I feel privileged to provide the Concluding remarks lecture at the first Faraday Discussion meeting I have attended. I have very much appreciated the admirable style of the meeting with succinct talks followed by the generous time allocation for discussions. The principle is a model that is eminently adaptable for future scientific meetings.

My remarks are intended to highlight the presentations at the meeting with something of a personal perspective and a viewpoint from my field of glyco-science. Indeed, subtly pervading the meeting were themes (in 11 of the 16 talks and discussions) relevant to and potentially applicable in the study of glycosylation in cells, tissues, and extracellular matrices.

Peter Seeberger's Introductory lecture (Mende *et al.*, DOI: 10.1039/c9fd00080a) set the scene with coverage of the automated synthesis of glycans, which he has pioneered.[1] A highlight was the ability to synthesize structurally homogeneous polysaccharides for diverse applications. From natural sources, these are notoriously heterogeneous and almost impossible to resolve into homogeneous forms. The other theme of the lecture was the incorporation of glycans into different glycan array platforms. These are admirably elaborated in the proceedings of this discussion meeting.

Faculty of Medicine, Imperial College London, London, UK. E-mail: t.feizi@imperial.ac.uk

Multidimensional micro- and nano-printing technologies

In her presentation on high-throughput enzyme nanopatterning, Elisa Riedo discussed thermolithography for depositing proteins (Liu *et al.*, DOI: 10.1039/c9fd00025a). These she foresees to have potential for the future fabrication of biochips on an industrial scale, and they would be applicable in cell adhesion studies. During the discussion that followed, the possibility was raised that the approach would be applicable in studies of glycoconjugates of cell surfaces.

UV-responsive cyclic peptide progelator bio-inks were the subject of Nathan Gianneschi's talk (Carlini *et al.*, DOI: 10.1039/c9fd00026g). Using this approach, porous gel networks can be generated. This may be an interesting avenue to explore multilayered 4D printing of variable morphologies useful in nano-lithography, with applications in generating microfluidic devices, as well as artificial tissue scaffolds for studying reactions of single cells or proteins.

Alshakim Nelson described cuboidal hydrogel lattices that constitute micro-chambers for whole-cell catalysis, which, for example, can be used in yeast fermentation studies (Johnston *et al.*, DOI: 10.1039/c9fd00019d).[4] The chambers are reusable. There was a discussion of the possible applications of the chambers for cultures of other cell types. However, it was pointed out that the chambers are permeable and allow the diffusion of solvents.

Glycan interactions on glycocalyx mimetic surfaces

The cellular glycocalyx was the subject of Kamil Godula's talk (Honigfort *et al.*, DOI: 10.1039/c9fd00024k). This is a dynamic carbohydrate-rich macromolecular system at the surface of cells, composed of glycolipids, glycoproteins, and proteoglycans. Among these, mucin-type glycoproteins can extend tens to hundreds of nanometers above the plasma membrane, providing a cell with a physical protective barrier, while also facilitating cellular interactions with its environment. Kamil discussed results from a model of the epithelial cell surface onto which an artificial mucin-sized yet inert macromolecule was inserted. This caused crowding in the glycocalyx. The result was differential hindrance of the accessibility of glycan ligands for two lectins analysed, *Sambucus nigra* agglutinin (SNA) and concanavalin A (Con A), depending on whether they were at the peripheral or central region (for SNA) or embedded within the deeper parts of the glycocalyx (Con A).

Porous glycopolymers consisting of polyacrylamide to which α-mannose monomer was coupled were described by Yoshiko Miura (Miura *et al.*, DOI: 10.1039/c9fd00018f). These were referred to as glycomonoliths, and I imagine this is a sophisticated term for a type of well controlled adsorbent column. The porous structures were induced by the solubility differences of the polymer using poro-genic alcohols. The diameter of the pores was dependent on the type of alcohol. The lower porogenic alcohol gave pores with small diameters and a large surface area, whereas the higher porogenic alcohol gave pores with large diameters and a small surface area. The mannose monoliths were highly specifically bound by Con A. Yoshiko concludes that monoliths with small pores and large surface areas and a high density of ligands would be suitable for a variety of bioseparation applications.

Helena Azevedo presented an elegant way to display hyaluronic acid (HA) of different lengths in order to measure their bioactivities (Pang *et al.*, DOI: 10.1039/c9fd00015a). Use was made of a peptide with the sequence GAHWQFNALTVR (named Pep-1) which has been previously identified by phage display as having a strong binding affinity for HA. This was thiolated at the N-terminal to form self-assembled monolayers (SAMs) on gold (Au) substrates. Microcontact printing (mCP) was used to develop patterned surfaces for the controlled spatial presentation of HA. Acetylated Pep-1 and a scrambled sequence of Pep-1 were used as controls. These were used as mimics of the endothelial glycocalyx. Cell culture experiments with human endothelial venule cells (HUVECs) demonstrated that smaller sized HA (5 kDa and 60 kDa) stimulated cell spreading, migration, and viability better than higher molecular weight HA. Helena and her colleagues predict that the knowledge gained from these studies will take us a step closer to developing new HA-based biomaterials as potential therapeutic solutions for vascular diseases.

New directions in surface functionalization and characterization

Another application of polymers was presented by Zijian Zheng (Chen *et al.*, DOI: 10.1039/c9fd00013e). He and his colleagues have developed a method (which they refer to as an initiator stickiness method) for fabricating brushes composed of cell-adhesive and cell-inert polymers (in 3-dimensions) to create micro- and nano-scale multicomponent surface patterns. They are used for a range of purposes, such as cell arrays, cell biological biosensors, and for nano-chemical studies. Zijian said that the approach has great potential. The time-consuming aspects were discussed, as well as the need for more data from different laboratories to address issues of reproducibility.

Matteo Palma presented an impressive method that addresses well the theme of this discussion meeting, namely, the fabrication of nanochips that enables studies of the interactions of cells with single molecules, such as ligands for cell surface receptors (Hawkes *et al.*, DOI: 10.1039/c9fd00023b). In this platform, the ligands are displayed on a triangular arrangement of DNA referred to as a biomimetic origami nanoarray. Matteo and colleagues have previously applied this platform to study cancer cell spreading with spatial resolution of multiple different ligands (integrins and epidermal growth factor) presented as single molecules. They are currently applying the system to nanoscale studies of ligand clustering dynamics in cardiomyocytes. This was presented as the ideal platform for studying the spreading and adhesion of different cell types to different receptor ligands in a highly controlled manner – alone or in combination with other cell surface receptors.

Bart Jan Ravoo's presentation was on soft nanomaterials that respond to external non-invasive stimuli, such as light and a magnetic field (Nowak *et al.*, DOI: 10.1039/c9fd00012g). In this paragraph, I am quoting words in Bart Jan's manuscript where he notes that these are highly desirable properties with potential applications in soft robotics, tissue engineering, and life-like materials:

"Hydrogels are an emerging subclass of soft nanomaterials driven by their unique mechanical properties, as well as their inherent biocompatibility and therefore find widespread applications in tissue engineering. In general, hydrogels are either polymeric gels or supramolecular gels based on the self-assembly of low molecular weight gelators (LMWGs). Among the latter, short peptides with aromatic protecting groups are especially attractive due to their ease of modification and preparation *via* solid phase peptide synthesis. Glycans (cyclodextrin vesicles) can be embedded in the gel or displayed at the surface of the gel using host–guest chemistry so that both the material stiffness and its biorecognition properties can be tuned using a combination of co-assembly components and orthogonal physical stimuli." Bart Jan and colleagues envisage that, in this way, biointerfaces can be assembled with tunable and responsive stiffness and ligand display.

An approach for presenting carbohydrates on glycolipids in the form of patterned striped phases of polymerized lipids was described by Shelley Claridge (Davis *et al.*, DOI: 10.1039/c9fd00022d). This was developed in order to have controlled carbohydrate presentation at interfaces. To cite her manuscript, "striped phase monolayers, in which alkyl chains extend across the substrate, have larger, more complex lattices: nanometer-wide stripes of headgroups with 0.5 or 1 nm lateral periodicity along the row, separated by wider (5 nm) stripes of exposed alkyl chains. The monolayers are not covalently bound to the substrate. Such monolayers can be photopolymerized, increasing robustness. With appropriate modifications, microcontact printing can be used to generate well-defined microscopic areas of striped phases of both single-chain and dual-chain amphiphiles (phospholipids), including one (phosphoinositol) with a carbohydrate in the headgroup, prototyping a strategy of potential relevance for glycobiology." I think that the effects of different ceramides could well be examined in this platform.

The interactions of glycosaminoglycan (GAG) chains with growth factors, morphogens, chemokines, and cytokines are crucial in embryonic development, cell growth, differentiation, and in inflammatory cascades. Carsten Werner and colleagues have a programme to unravel such interactions and they have applied a recently introduced platform of GAG-based biohybrid hydrogels to systematically explore the relevance of GAG content and GAG sulfation patterns for the capacity of the polymer matrices to sequester selected cytokines that have very different charge characteristics (Freudenberg *et al.*, DOI: 10.1039/c9fd00016j). Acidic cytokines are hardly bound to any hydrogel type, whereas strongly basic cytokines are bound in direct correlation to the integral space charge (sulfate) density of the hydrogels, no matter whether the GAG units were fully sulfated heparin units or selectively 6*O*- and *N*-desulphated heparin derivatives. For charged proteins, they observed that the integral space charge density of the hydrogels, and not the local sulfation pattern of the GAG unit, determined the binding to the compared hydrogel types. Accordingly, the bound concentration of a group of acidic and strongly basic cytokines could be easily predicted and effectively adjusted by tuning the integral space charge density of the gels. These results demonstrate the importance of the integral space charge density of GAG-containing cell-instructive matrices for the binding of strongly charged proteins. Furthermore, the local accumulation of charge carriers on individual GAG components was found to be decisive for the binding of several weakly charged proteins. These results not only enable the design of hydrogel systems with tailored protein sequestration, but may also shed new light on the binding principles of cytokines to the extracellular matrix in living tissues.

Preparation of multivalent glycan micro- and nano-arrays

Daniel Valles described careful studies of the binding avidities of the plant lectin concanavalin A (ConA) toward mannosides presented at defined densities (Valles *et al.*, DOI: 10.1039/c9fd00028c). These are very good models to understand how the degree of clustering of the ligands influences the strength of the binding signals elicited. This presentation led to a discussion of the variability of the display of glycans in different tissues, and even different cells within a given tissue and also in extracellular matrices, where precise measurements would be challenging to make.

Continuing the glycan binding theme, Clare Mahon described an ingenious means of isolating cholera toxin from lysates of the bacterium *Vibrio cholerae* (Mahon *et al.*, DOI: 10.1039/c9fd00017h). Multiple copies of the glycan ligand for the cholera toxin, GM1, were coupled to a temperature responsive polymer. This served as an absorbent of the toxin. Below the lower critical solution temperature (LCST), the polymer forms a complex with the toxin with nanomolar affinity. When heated above the LCST, the polymer undergoes a reversible coil to globule transition that renders a proportion of the tethered glycan ligand inaccessible to the toxin. There is thus a thermally modulated decrease in the toxin-glycan avidity on the adsorbent. This has been used to reversibly capture the toxin from solution, thereby enabling its convenient isolation from a complex mixture.

A talk on a topic very close to my own interests was by Jeff Gildersleeve. This was a well-considered treatise on glycan array data from analyses in which the same glycan sequences are presented differently on surfaces (Temme *et al.*, DOI: 10.1039/c9fd00021f). The first is presentation in multivalent display on the carrier protein bovine serum albumin (BSA), as in the Gildersleeve platform. The second is by chemical (covalent) bonding of discrete glycans onto glass slides, as in the platform of the Consortium of Functional Glycomics (CFG). Overall, the two array platforms had a very high degree of internal consistency. Jeff focused on differences in binding signals that were large, more than 50-fold from one array to the other. Seventy one of 101 instances of large differences discussed involved comparisons with an identical glycan but a different linker. The majority of the large discrepancies between the two arrays were attributed to (a) the use of different linkers of glycans in the covalent CFG arrays, and (b) the difference in densities (clustering) of the glycans in the arrayed BSA spots. Jeff elaborated on the major effect of clustered glycan display in the relative intensities of the binding signals elicited.

This was a cue to my reviewing of the background to the establishment of glycan arrays and their applications for glycan ligand discoveries.

Inception and evolution of glycan microarrays

Glycan arrays have their origin in 1985, when it became clear that there was a need for a micro method for binding analyses with glycans released from glycoproteins.

Until then, the analyses were by inhibition, using purified glycans as inhibitors of the binding of antibodies and glycan-binding receptors. The amount of glycans required to make assignments of the determinants recognized by the proteins was very high, of the order of milligrams (micromoles) of purified glycans per tube.[1,2] The method that my colleagues and I introduced involves the conjugation of glycans in the purified state or as mixtures to a lipid molecule to have neo-glycolipids, NGLs,[3,4] as depicted in Fig. 1.

A special feature of NGLs is that the glycans in the conjugates remain as discrete entities and, unlike the free glycans, they can be immobilized in a clustered state on solid matrices (plastic microwells and silica gel chromatograms) and can be resolved by TLC for direct binding analyses in conjunction with mass spectrometric analysis for sequence determination. The majority of glycan-

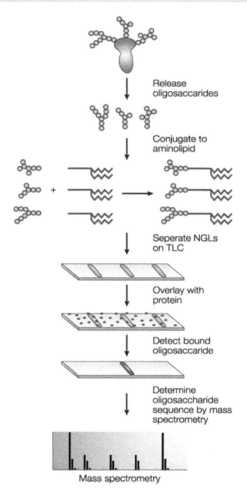

Fig. 1 Schematic presentation of the NGL technology whereby glycans, singly or as mixtures, are conjugated to an aminolipid by reductive amination, resolved as necessary by high performance TLC, probed for binding by carbohydrate recognizing proteins, and analyzed by mass spectrometry; further details are depicted in Scheme 1. Taken from ref. 5.

binding proteins have low binding affinities to glycans; the clustered presentation of the glycans in NGLs was considered of paramount importance in the design of the technology thereby increasing avidities and detectabilities of binding.

NGL technology is:

- applicable to microgram (nanomole) amounts of starting oligosaccharides;
- is applicable to glycolipids and their glycans, which can be analyzed in parallel with NGLs;
- affords the flexibility to clinch the role of glycans and ceramides of glycolipids in binding signals;
- has provision for generating 'designer' microarrays from natural sources, glycomes; and
- in liposomal formulations, the lipid-linked probes behave as planar membranes that mimic cell surfaces, and they can be incorporated into live cells to evaluate the biological significance of binding data.

Glycans in the reducing state are conjugated to an amino lipid by reductive amination,[3,4] or alternatively, as established by Yan Liu, by oxime ligation,[6] as depicted in Scheme 1.

A considerable number of assignments were enabled using NGL technology prior to the miniaturization of glycan arrays (Fig. 2). These included differentiation antigens recognized by monoclonal antibodies and endogenous carbohydrate-recognizing receptors, as well as the demonstration of the applicability to assign glycans bound by bacteria, reviewed in ref. 7 and 8. The discovery of yeast type O-mannosyl glycans (mannose-linked O-glycans) was made while assigning the sulphoglucuronyl antigen (HNK 1) on brain glycoproteins.[9]

The technology has undergone a number of advances (Fig. 3), and in 2002, with Wengang Chai, Shigeyuki Fukui, and colleagues, we introduced glycan arrays based on the NGL technology.[11] Since 2006, the NGL probes derived from natural sources or synthesized chemically are arrayed in a liposomal formulation on nitocellulose-coated glass slides, side by side with glycolipids, at low fmol levels per spot (Fig. 4). The sensitivity is 10^9 times greater than in classical inhibition studies.

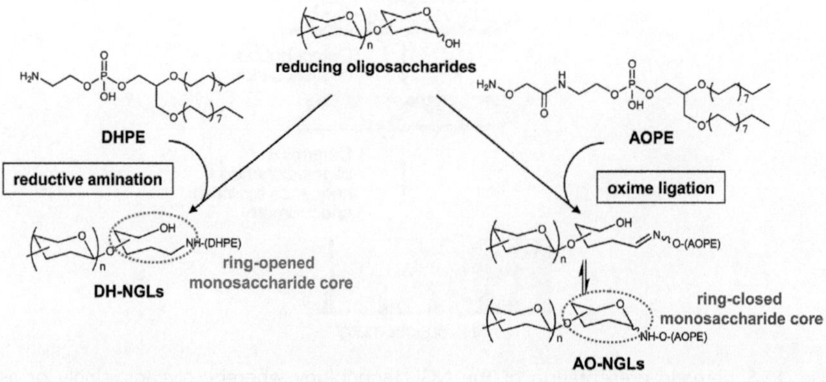

Scheme 1 Principles of generating NGLs by conjugating reducing oligosaccharides by reductive amination to an aminolipid 1,2-dihexadecyl-*sn-glycero*-3-phosphoethanolamine (DHPE) (ref. 4) or by oxime ligation to an aminooxy-functionalized DHPE (AOPE) (ref. 6).

1985 – Tang *et al.* NGLs to study antigenicities and receptor functions of carbohydrate chains	Assignments of ligands for endogenous lectins
Elucidation of carbohydrate differentiation antigens e.g. 'neural induction' antigen L5 Streit et al (1996); and scrapie lesion antigen 10E4 Leteux et al (2001)	• Conglutinin Loveless *et al* (1989); Mizuochi *et al* (1989); Solis *et al* (1994) • Mannose binding proteins Childs *et al* (1989); Childs *et al* (1990) • Collectin-43 Loveless *et al* (1995) • Pulmonary surfactant protein Childs *et al* (1992)
Demonstration of carbohydrate receptors for bacteria e.g. uropathogenic *E. coli* and *P. aeruginosa* from cystic fibrosis Rosenstein et al 1988 &1992	• Galectins 1 and 3 Solomon *et al* (1989); Feizi *et al* (1994) • Amyloid P protein Loveless *et al* (1992) • Cysteine-rich domain of the macrophage mannose receptor Leteux *et al* (2000)
Contributions of NGL technology • *O*-GalNAc *O*-glycan sequencing • discovery of *O*-Mannose chains Yuen et al (1997)	• E-, L-, P-selectins Yuen *et al* (1992); Larkin *et al* (1992); Green *et al* (1992) (1995); Osanai *et al* (1996); Galustian *et al* (1999); Alon *et al* (1995)

Fig. 2   NGL technology: contributions (1985–2002) prior to the introduction of micro-arrays – a validated system resulting in more than 30 publications.[10]

Beyond data storage and display as charts and tables, it is desirable to be able to sort and filter glycans bound or not bound by particular recognition systems according to their structural features. Unique software tools with such functions were developed by Mark Stoll[14] and have been the mainstay of microarray data storage, presentation, and reporting at the Wellcome Trust-supported

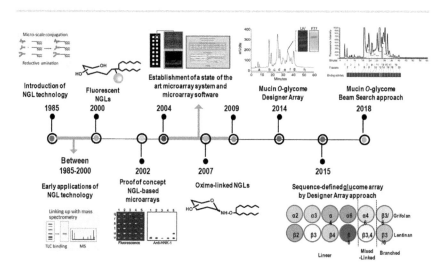

Fig. 3   The history of NGL technology, which led to the development of microarrays of sequence defined glycans and to Designer Array and Beam Search approaches in order to pinpoint, isolate, and assign glycan ligands in biologically relevant microenvironments. Taken from ref. 12.

Carbohydrate Microarray Facility in the Glycosciences Laboratory at Imperial College (160 published and over 8000 internal data sets which can be retrieved at will, http://www.imperial.ac.uk/glycosciences/) (Fig. 5).

Libraries of sequence-defined glycans are expanding. The glycan sequences thus far assembled (in the order of 900) are indeed a minuscule proportion of those in mammalian glycomes and those of microbiota and plants (Fig. 6).

Glycan array technologies have gained momentum and have become essential tools revolutionizing the unravelling of protein–carbohydrate interactions in health and disease processes, both infective and non-infective[15] (Fig. 7).

Advances in chemical and chemoenzymatic syntheses are contributing enormously toward populating glycan libraries with bespoke interrelated glycan sequences for close comparisons of their displays of recognition motifs. Concurrently with these developments are Designer and Beam Search array strategies[16–18] (Fig. 4 and 8), which capitalize upon a robotic array of glycan fractions of NGLs for glycan ligand discovery, and also the Shotgun array approach,[19] revealing hitherto unsuspected or unknown ligands in natural microenvironments. Hand in hand with these strategies is an advanced and distributable software we are developing, CarbArrayART (Carbohydrate microArray Analysis and Reporting Tool), for glycan array data storage, interpretation, and management. We are also working toward the establishment of a public repository under the auspices of GlyGen, the glycobioinformatics initiative under the auspices of the NIH-supported Commons Fund.

The Beam Search approach (Fig. 8) is of orders of magnitude more sensitive than traditional methods and is capable of identifying extremely minor components previously undetected within the O-glycomes[18] of mucins that are ligands

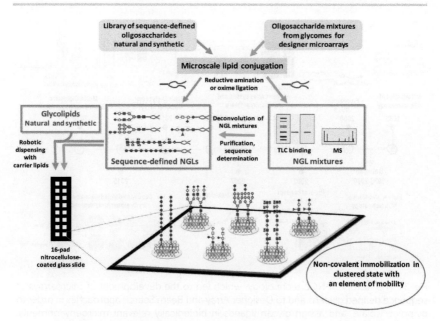

Fig. 4 Schematic representation of the NGL-based microarray system. Taken from ref. 13.

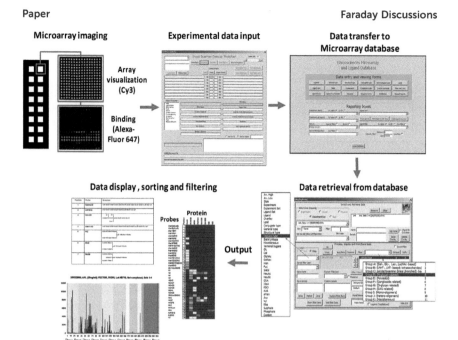

Fig. 5 Software for data input, storage, and retrieval for presentation as charts, tables, and heatmaps.

for the adhesive proteins of infective agents. There is important complementarity between the increasing repertoire of sequence-defined glycan arrays and the arrays generated *de novo* from particular natural sources. Analyses using

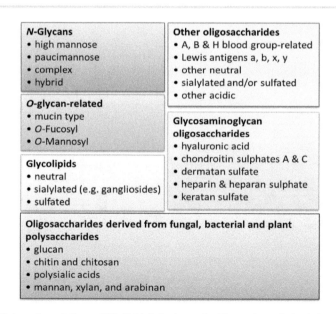

Fig. 6 Categories of the ~900 lipid-linked saccharide probes derived from natural sources or chemically synthesized in the Glycosciences Laboratory's library (https://glycosciences.med.ic.ac.uk/glycanLibraryIndex.html).

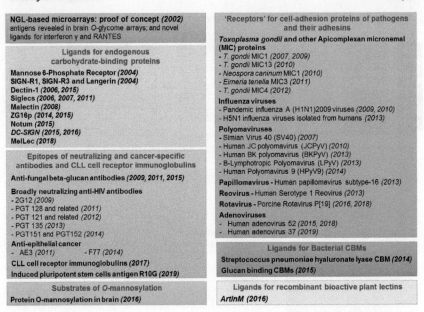

NGL-based microarrays: proof of concept *(2002)*
antigens revealed in brain *O*-glycome arrays; and novel
ligands for interferon γ and RANTES

**Ligands for endogenous
carbohydrate-binding proteins**

Mannose 6-Phosphate Receptor *(2004)*
SIGN-R1, SIGN-R3 and Langerin *(2004)*
Dectin-1 *(2006, 2015)*
Siglecs *(2006, 2007, 2011)*
Malectin *(2008)*
ZG16p *(2014, 2015)*
Notum *(2015)*
DC-SIGN *(2015, 2016)*
MelLec *(2018)*

**Epitopes of neutralizing and cancer-specific
antibodies and CLL cell receptor immunoglobulins**

Anti-fungal beta-glucan antibodies *(2009, 2011, 2015)*

Broadly neutralizing anti-HIV antibodies
- 2G12 *(2009)*
- PGT 128 and related *(2011)*
- PGT 121 and related *(2012)*
- PGT 135 *(2013)*
- PGT151 and PGT152 *(2014)*

Anti-epithelial cancer
- AE3 *(2011)*   - F77 *(2014)*

CLL cell receptor immunoglobulins *(2017)*
Induced pluripotent stem cells antigen R10G *(2019)*

Substrates of *O*-mannosylation
Protein O-mannosylation in brain *(2016)*

**'Receptors' for cell-adhesion proteins of pathogens
and their adhesins**

Toxoplasma gondii and other Apicomplexan micronemal
(MIC) proteins
- *T. gondii* MIC1 *(2007, 2009)*
- *T. gondii* MIC13 *(2010)*
- *Neospora caninum* MIC1 *(2010)*
- *Eimeria tenella* MIC3 *(2011)*
- *T. gondii* MIC4 *(2012)*

Influenza viruses
- Pandemic influenza A (H1N1)2009 viruses *(2009, 2010)*
- H5N1 influenza viruses isolated from humans *(2013)*

Polyomaviruses
- Simian Virus 40 (SV40) *(2007)*
- Human JC polyomavirus (JCPyV) *(2010)*
- Human BK polyomavirus (BKPyV) *(2013)*
- B-Lymphotropic Polyomavirus (LPyV) *(2013)*
- Human Polyomavirus 9 (HPyV9) *(2014)*

Papillomavirus - Human papillomavirus subtype-16 *(2013)*

Reovirus - Human Serotype 1 Reovirus *(2013)*

Rotavirus - Porcine Rotavirus P[19] *(2016, 2018)*

Adenoviruses
- Human adenovirus 52 *(2015, 2018)*
- Human adenovirus 37 *(2019)*

Ligands for Bacterial CBMs
Streptococcus pneumoniae hyaluronate lyase CBM *(2014)*
Glucan binding CBMs *(2015)*

Ligands for recombinant bioactive plant lectins
ArtinM (2016)

Fig. 7 Assignments made with the NGL-based microarray system published since 2002.

sequence-defined glycan arrays provide reference compounds not only to determine specificities of binding, but also to reveal unsuspected recognition structures that are represented in different physiological contexts. The Beam Search approach paves the way towards *O*-glycome recognition studies in a wide range of basic and medical settings to give new insight into glycan recognition structures in natural microenvironments.

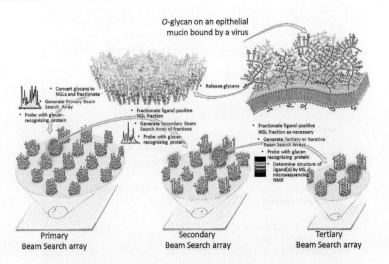

Fig. 8 Schematic depiction of the Beam Search strategy as applied to pinpointing, isolating, and structurally characterizing the glycan ligand(s) among the innumerable *O*-glycans on an epithelial mucin (picture designed by Jin Yu, the Glycosciences Laboratory, Imperial College London).

Conclusion: new opportunities for interdisciplinary interactions

The special feature of the Faraday Discussion meeting was that it brought together scientists from diverse disciplines, biophysicists, biochemists, and biologists, who do not necessarily frequent the same conferences. Thus, we often heard unfamiliar terminologies and principles for the first time. A criticism is that speakers did not always start with definitions of the specialized terms in the systems they are studying. Their relevance to one another's field of research may not have always been apparent immediately. However, there were ample opportunities to address these during the discussion periods. Surely, follow-on interactions and collaborations will be taking place after further reflection, and I know that some are already taking place.

Conflicts of interest

There are no conflicts to declare.

Acknowledgements

I am indebted to the late Elvin A. Kabat, who introduced me to glycoscience in the course of my studies of the I antigen, the target of autoantibodies in cold agglutinin disease and, as we determined with him, a backbone of the major blood group antigens. I am also indebted to Yuan Chuan Lee, with whom we discussed during several conference breaks possible strategies for immobilizing glycans for microscale binding analyses, following which we settled on their conjugation to lipids to produce neoglycolipids. I acknowledge the support I had for our glycan array work from successive programme grants from the Medical Research Council, the joint Research Councils' Basic Technology initiative and the Wellcome Trust. For the establishment of the neoglycolipid-based microarray system, special thanks are due to former and current colleagues in the Glycoscience Laboratory: Ping W. Tang, Robert A. Childs, Wengang Chai, Shigeyuki Fukui, Maria Campanero, Yan Liu, Mark Stoll, Angelina Palma, Lisete Silva, and many other colleagues and collaborators.

References

1 E. A. Kabat, Contributions of quantitative Immunochemistry to knowledge of blood group A, B, H, Le, I and i antigens, *Am. J. Clin. Pathol.*, 1982, **78**, 281–292.
2 T. Feizi, E. A. Kabat, G. Vicari, B. Anderson and W. L. Marsh, Immunochemical studies on blood groups XLIX. The I antigen complex: Specificity differences among anti-I sera revealed by quantitative precipitin studies; partial structure of the I determinant specific for one anti-I serum, *J. Immunol.*, 1971, **106**, 1578–1592.
3 P. W. Tang, H. C. Gooi, M. Hardy, Y. C. Lee and T. Feizi, Novel approach to the study of the antigenicities and receptor functions of carbohydrate chains of glycoproteins, *Biochem. Biophys. Res. Commun.*, 1985, **132**, 474–480.

4 M. S. Stoll, T. Mizuochi, R. A. Childs and T. Feizi, Improved procedure for the construction of neoglycolipids having antigenic and lectin-binding activities from reducing oligosaccharides, *Biochem. J.*, 1988, **256**, 661–664.

5 T. Feizi and W. Chai, Oligosaccharide microarrays to decipher the glyco code, *Nat. Rev. Mol. Cell Biol.*, 2004, **5**(7), 582–588.

6 Y. Liu, W. Chai, P. R. Crocker, H. M. I. Osborn and T. Feizi, Neoglycolipids prepared *via* oxime-ligation for microarray analysis of carbohydrate–protein interactions, *Glycobiology*, 2006, **16**(11), 1140.

7 T. Feizi, M. S. Stoll, C.-T. Yuen, W. Chai and A. M. Lawson, Neoglycolipids: probes of oligosaccharide structure, antigenicity and function, *Methods Enzymol.*, 1994, **230**, 484–519.

8 R. A. Childs, K. Drickamer, T. Kawasaki, S. Thiel, T. Mizuochi and T. Feizi, Neoglycolipids as probes of oligosaccharide recognition by recombinant and natural mannose-binding proteins of the rat and man, *Biochem. J.*, 1989, **262**, 131–138.

9 C.-T. Yuen, W. Chai, R. W. Loveless, A. M. Lawson, R. U. Margolis and T. Feizi, Brain contains HNK-1 immunoreactive O-glycans of the sulfoglucuronyl lactosamine series that terminate in 2-linked or 2,6-linked hexose (mannose), *J. Biol. Chem.*, 1997, **272**, 8924–8931.

10 T. Feizi, Progress in deciphering the information content of the 'glycome'– a crescendo in the closing years of the millennium, *Glycoconjugate J.*, 2000, **17**(7–9), 553–565.

11 S. Fukui, T. Feizi, C. Galustian, A. M. Lawson and W. Chai, Oligosaccharide microarrays for high-throughput detection and specificity assignments of carbohydrate–protein interactions, *Nat. Biotechnol.*, 2002, **20**(10), 1011–1017.

12 Z. Li and T. Feizi, The neoglycolipid (NGL) technology-based microarrays and future prospects, *FEBS Lett.*, 2018, **17**(1), 121–133.

13 A. S. Palma, T. Feizi, R. A. Childs, W. Chai and Y. Liu, The neoglycolipid (NGL)-based oligosaccharide microarray system poised to decipher the *meta*-glycome, *Curr. Opin. Chem. Biol.*, 2014, **18**, 87–94.

14 M. S. Stoll and T. Feizi, Software tools for storing, processing and displaying carbohydrate microarray data, *Proceeding of the Beilstein Symposium on Glyco-Bioinformatics, 4-8 October, 2009, Potsdam, Germany*, ed. C. Kettner, Beilstein Institute for the Advancement of Chemical Sciences, Frankfurt, Germany, 2009, pp. 123–140.

15 C. D. Rillahan and J. C. Paulson, Glycan microarrays for decoding the glycome, *Annu. Rev. Biochem.*, 2011, **80**, 797–823.

16 A. S. Palma, T. Feizi, Y. Zhang, M. S. Stoll, A. M. Lawson, E. Diaz-Rodríguez, A. S. Campanero-Rhodes, J. Costa, G. D. Brown and W. Chai, Ligands for the beta-glucan receptor, Dectin-1, assigned using 'designer' microarrays of oligosaccharide probes (neoglycolipids) generated from glucan polysaccharides, *J. Biol. Chem.*, 2006, **281**(9), 5771–5779.

17 C. Gao, Y. Liu, H. Zhang, Y. Zhang, M. N. Fukuda, A. S. Palma, R. P. Kozak, R. A. Childs, M. Nonaka, Z. Li, D. L. Siegel, P. Hanfland, D. M. Peehl, W. Chai, M. I. Greene and T. Feizi, Carbohydrate Sequence of the Prostate Cancer-associated Antigen F77 Assigned by a Mucin O-Glycome Designer Array, *J. Biol. Chem.*, 2014, **289**(23), 16462–16477.

18 Z. Li, C. Gao, Y. Zhang, A. S. Palma, R. A. Childs, L. M. Silva, Y. Liu, X. Jiang, Y. Liu, W. Chai and T. Feizi, O-Glycome Beam Search Arrays for Carbohydrate Ligand Discovery, *Mol. Cell. Proteomics*, 2018, **17**(1), 121–133.

19 X. Song, Y. Lasanajak, B. Xia, J. Heimburg-Molinaro, J. M. Rhea, H. Ju, C. Zhao, R. J. Molinaro, R. D. Cummings and D. F. Smith, Shotgun glycomics: a microarray strategy for functional glycomics, *Nat. Methods*, 2011, **8**(1), 85–90.

Poster titles

Self-generating soft biophotovoltaic devices, **X. Qiu, O. Castaneda Ocampo, A. Herrmann and R. C. Chiechi**, *University of Groningen, Netherlands*

Smart glyco-nanomaterials as diagnostic tools for cancer, **E. Hunt, D. Benito-Alifonso and M. Carmen Galan**, *University of Bristol, UK*

Bio-inspired glycan-nanoprobes as antimicrobial labels, **J. Samphire, Y. Takebayashi, J. Spencer and M. Carmen Galan**, *University of Bristol, UK*

Probing multivalent lectin–glycan interactions using multifunctional glycan-nanoparticles, **U. S. Akshath, D. Budhadev, E. Kalverda, N. Hondow, W. B. Turnbull, Y. Guo and D. Zhou**, *University of Leeds, UK*

Effect of morphology and surface chemistry on interactions of silicon microparticles with macrophages and lung epithelium, **G. Bruce, S. Stolnik-Trenkic, M. Arjona, M. Duch, J. A. Plaza and L. Perez-Garcia**, *University of Nottingham, UK*

Probing structural basis of multivalent DC-SICN–glycan interactions using polyvalent glycan-quantum dots, **J. Hooper, D. Budhadev, Y. Liu, W. B. Turnbull, D. Zhou and Y. Guo**, *University of Leeds, UK*

Precision synthesis and nanofabrication using nanoscale electrochemistry, **P. Wilson**, *University of Warwick, UK*

Specific adhesion of pathogens on glycocalyx mimetic hydrogels, **T. Paul, F. Jacobi, H. Wang, C. Spormann, T. K. Lindhorst and S. Schmidt**, *Heinrich-Heine-Universität, Germany*

High resolution patterned thin films based on plant-derived building blocks, **M. Gestranius, I. Otsuka, S. Halila, D. Hermida Merino, E. Solano, R. Borsali and T. Tammelin**, *VTT Technical Research Centre of Finland Ltd, Finland*

Transparent nanoneedles enable live-cell studies of the dynamic response of epithelial cells and fibroblasts to nanostructured surfaces, **D. Boland, S. G. Higgins, J. E. Sero, H. Seong, M. Becce and M. M. Stevens**, *Imperial College London, UK*

Glycomimetic polymers with brush-like structures, **F. Shamout, M. Giesler and L. Hartmann**, *Heinrich-Heine-Universität, Germany*

Unravelling the structure of glycosyl cations *via* cold-ion infrared spectroscopy, **M. Marianski, E. Mucha, F. Xu, D. A. Thomas, G. Meijer, G. von Helden, P. Seeberger and K. Pagel**, *City University of New York, USA*

 This journal is © The Royal Society of Chemistry 2019

Developing potentially tumor targeting carbohydrate based antitumor agents, **P. Gonzalez and D. R. Mootoo**, *Hunter College, City University of New York, USA*

Chemical editing of nanoscale proteoglycan architecture and organization to control cellular differentiation, **M. L. Huang, T. N. Stephenson and T. O'Leary**, *Scripps Research, USA*

The affiliation of the presenting author is listed above.

The *Faraday Discussions* Poster Prize for the best poster was jointly awarded to Miss Jennifer Samphire of Bristol University, UK, for her poster on bio-inspired glycan-nanoprobes as antimicrobial labels and Mr Fadi Shamout of Heinrich-Heine-Universität, Germany, for his poster on glycomimetic polymers with brush-like structures.

List of participants

Mr Samir Aoudjane, *University College London, United Kingdom*
Dr Helena Azevedo, *Queen Mary University of London, United Kingdom*
Miss Lucy Balshaw, *Royal Society of Chemistry, United Kingdom*
Mr Daniel Boland, *Imperial College London, United Kingdom*
Professor Adam Braunschweig, *City University of New York, USA*
Mr Gordon Bruce, *University of Nottingham, United Kingdom*
Dr Joseph Byrne, *National University of Ireland, Galway, Ireland*
Professor Ryan Chiechi, *University of Groningen, Netherlands*
Professor Shelley Claridge, *Purdue University, USA*
Professor Leroy Cronin, *University of Glasgow, United Kingdom*
Dr Yuri Diaz Fernandez, *University of Liverpool, United Kingdom*
Professor Ten Feizi, *Imperial College :London, United Kingdom*
Dr Antonio Fernandez Mato, *Loughborough University, United Kingdom*
Dr Wayne George, *Illumina, United Kingdom*
Professor Nathan Gianneschi, *Northwestern University, USA*
Dr Jeff Gildersleeve, *National Cancer Institute, USA*
Dr Kamil Godula, *University of California San Diego, USA*
Dr Nils Goedecke, *SwissLitho AG, Switzerland*
Ms Patricia Gonzalez, *Hunter College, City University of New York, USA*
Professor Laura Hartmann, *Heinrich-Heine-Universität, Germany*
Mr James Hooper, *University of Leeds, United Kingdom*
Professor Mia Huang, *Scripps Research, USA*
Ms Eliza Hunt, *University of Bristol, United Kingdom*
Professor Eero Kontturi, *Aalto University, Finland*
Dr Vajinder Kumar, *University of Leeds, United Kingdom*
Dr Clare Mahon, *University of York, United Kingdom*
Professor Mateusz Marianski, *Hunter College, City University of New York, USA*
Professor Yoshiko Miura, *Kyushu University, Japan*
Dr Heather Montgomery, *Royal Society of Chemistry, United Kingdom*
Dr Alshakim Nelson, *University of Washington, USA*
Dr Matteo Palma, *Queen Mary University of London, United Kingdom*
Miss Xinqing Pang, *Queen Mary University of London, United Kingdom*
Mrs Tanja Paul, *Heinrich-Heine-Universität Düsseldorf, Germany*
Ms Hollie Pietsch, *US Army CCDC Atlantic, United Kingdom*
Mr Xinkai Qiu, *University of Groningen, Netherlands*
Professor Bart Jan Ravoo, *Westfälische Wilhelms-Universität Münster, Germany*
Professor Elisa Riedo, *New York University, USA*
Mr Kenaniah Rona, *Kingston University, United Kingdom*
Miss Ainur Sabirova, *King Abdullah University of Science and Technology, Saudi Arabia*
Miss Jennifer Samphire, *Bristol University, United Kingdom*
Professor Stephan Schmidt, *Heinrich-Heine-Universität, Germany*
Mrs Rosalind Searle, *Royal Society of Chemistry, United Kingdom*

 This journal is © The Royal Society of Chemistry 2019

Professor John Seddon, *Imperial College London, United Kingdom*
Professor Peter H. Seeberger, *Max Planck Institute of Colloids and Interfaces, Germany*
Mr Fadi Shamout, *Heinrich-Heine-Universität Düsseldorf, Germany*
Dr Michael Spencelayh, *Royal Society of Chemistry, United Kingdom*
Mr Alexander Strzelczyk, *Heinrich-Heine-Universität Düsseldorf, Germany*
Professor Tekla Tammelin, *VTT Technical Research Centre of Finland Ltd, Finland*
Professor Bruce Turnbull, *University of Leeds, United Kingdom*
Dr Akshath Uchangi Satyaprasad, *University of Leeds, United Kingdom*
Mrs Serap Ueclue, *Heinrich-Heine-Universität Düsseldorf, Germany*
Mr Daniel Valles, *City University of New York, USA*
Dr Paul Wilson, *University of Warwick, United Kingdom*
Dr Jin Yu, *Imperial College :London, United Kingdom*
Professor Zijian Zheng, *The Hong Kong Polytechnic University, Hong Kong*
Professor Dejian Zhou, *University of Leeds, United Kingdom*